AF581729

Monitoring, Modelling and Management of Water Quality

Monitoring, Modelling and Management of Water Quality

Editor

Matthias Zessner

MDPI • Basel • Beijing • Wuhan • Barcelona • Belgrade • Manchester • Tokyo • Cluj • Tianjin

Editor
Matthias Zessner
Institute for Water Quality and
Resource Management
TU Wien
Vienna
Austria

Editorial Office
MDPI
St. Alban-Anlage 66
4052 Basel, Switzerland

This is a reprint of articles from the Special Issue published online in the open access journal *Water* (ISSN 2073-4441) (available at: www.mdpi.com/journal/water/special_issues/Water_Quality_Management).

For citation purposes, cite each article independently as indicated on the article page online and as indicated below:

LastName, A.A.; LastName, B.B.; LastName, C.C. Article Title. *Journal Name* **Year**, *Volume Number*, Page Range.

ISBN 978-3-0365-1548-9 (Hbk)
ISBN 978-3-0365-1547-2 (PDF)

Contents

Editorial

Monitoring, Modeling and Management of Water Quality

Matthias Zessner 

Institute for Water Quality and Resources Management, TU-Wien, 1040 Vienna, Austria;
mzessner@iwag.tuwien.ac.at

Abstract: In this special issue, we are able to present a selection of high-level contributions showing the manifold aspects of the monitoring, modeling, and management of water quality. Monitoring aspects range from cyanobacteria in water using spectrophotometry via wide-area water quality monitoring and exploiting unmanned surface vehicles, to using sentinel-2 satellites for the near-real-time evaluation of catastrophic floods. Modeling ranges from small scale approaches by deriving a Bayesian network for assessing the retention efficacy of riparian buffer zones, to national scales with a modification of the MONERIS (Modeling Nutrient Emissions in River Systems) nutrient emission model for a lowland country. Management is specifically addressed by lessons learned from the long-term management of a large (re)constructed wetland and the support of river basin management planning in the Danube River Basin.

Keywords: effectiveness of measures; scenarios and forecasts; socioeconomic context; sources and pathways of water pollution; system understanding; water governance; water quality statuses and trends; water pollution control

Citation: Zessner, M. Monitoring, Modeling and Management of Water Quality. *Water* **2021**, *13*, 1523. https://doi.org/10.3390/w13111523

Received: 13 October 2020
Accepted: 27 May 2021
Published: 28 May 2021

1. Introduction

Different types of pressures, such as nutrients [1], micropollutants [2], microbes [3], nanoparticles [4], microplastics [5], and antibiotic-resistant genes [6], endanger the quality of water bodies. Evidence-based pollution control needs to be built on the three basic elements of water governance: Monitoring, modeling, and management [7]. Monitoring sets the empirical basis by providing space- and time-dependent information on substance concentrations and loads, as well as driving boundary conditions for assessing water quality trends, water quality statuses, and providing necessary information for the calibration and validation of models [2,7]. Modeling needs proper system understanding and helps to derive information for times and locations where no monitoring is done or possible. Possible applications are risk assessment for the exceedance of quality standards, assessment of the regionalized relevance of sources and pathways of pollution, effectiveness of measures, bundles of measures or policies, and assessment of future developments as scenarios or forecasts [8]. Management relies on this information and translates it in a socioeconomic context into specific plans for implementation [9]. The evaluation of the success of management plans again includes well-defined monitoring strategies [7]. This special issue provides an important overview of a hot topic in this context as it is summarized in the following.

2. Issue Contents

2.1. Monitoring

In order to measure the chromaticity of water and the content of dissolved matter more accurately, effectively, and cheaply, a chromaticity measurement system based on the image method was proposed and applied by Cao et al. [10]. The measurement system used a designed acquisition device and image processing software to obtain the red-green-blue (RGB) values of the image and converted the color image from RGB color space to hue-saturation-intensity (HSI) space to separate the chromaticity and brightness. According to the definition of chromaticity, the hue (H), saturation (S) values, and chromaticity of standard chromaticity solution images were fitted by a non-linear surface, and a three-dimensional chromaticity measurement model was established based on the H and S values of water images. For the measurement of a standard chromaticity solution, the proposed method has a higher accuracy than spectrophotometry. For actual water sample measurements, there is no significant difference between the results of the tested method and the common spectrophotometer method. This verified the validity of the chromaticity method. In addition, the system was tested for measuring the concentration of ammonia nitrogen, phosphate, and chloride in water, with satisfactory results [10].

Management of cyanobacteria blooms and their negative impact on human and ecosystem health requires effective tools for monitoring their concentration in water bodies. Agberien et al. [11] investigated the potential of derivative spectrophotometry for the detection and monitoring of cyanobacteria using toxigenic and non-toxigenic strains of Microcystis aeruginosa. Microcystis aeruginosa was quantified in deionized water and surface water using traditional spectrophotometry and the first derivative of absorbance. The first derivative of absorbance was effective in improving the signal of traditional spectrophotometry; however, it was not adequate for differentiating between signal and noise at low concentrations. Savitzky–Golay coefficients for the first derivative were used to smooth the derivative spectra and improve the correlation between concentration and noise at low concentrations. Derivative spectrophotometry improved the detection limit by as much as eight times in deionized water and as much as four times in surface water. The lowest detection limit measured in surface water with traditional spectrophotometry was 392,982 cells/mL, while the Savitzky–Golay first derivative of absorbance was 90,231 cells/mL. The method provided herein provides a promising tool for the real-time monitoring of cyanobacteria concentration [11].

Water environment pollution is an acute problem, especially in developing countries, so water quality monitoring is crucial for water protection. Cao et al. [12] developed an intelligent three-dimensional wide-area water quality monitoring and online analysis system. The proposed system was composed of an automatic cruise intelligent unmanned surface vehicle (USV), a water quality monitoring system (WQMS), and a water quality analysis algorithm. An automatic positioning cruising system was constructed for the USV. The WQMS consists of a series of low-power water quality detecting sensors and a lifting device that can collect the water quality monitoring data at different water depths. These data are analyzed by the proposed water quality analysis algorithm based on the ensemble learning method to estimate the water quality level. Then, a real experiment was conducted in a lake to verify the feasibility of the proposed design. The experimental results obtained in a real application demonstrated the good performance and feasibility of the proposed monitoring system [12].

Flooding is among the most common natural disasters in our planet and one of the main causes of economic and human life loss worldwide. Evidence suggests an increase in floods at a European scale, with the Mediterranean coast being critically vulnerable to this risk. The devastating event in the West Mediterranean during the second week of September 2019 is a clear case of this risk, when a record-breaking flood (locally called the "Cold Drop" (Gota Fría)) was swollen into a catastrophe in the southeast of Spain and surpassing previous all-time records [13]. By using a straightforward approach with the Sentinel-2 twin satellites from the Copernicus Programme and the ACOLITE atmospheric correction processor, Caballero et al. [13] accomplished an initial approximation of the delineated flooded zones, including agricultural and urban areas, in quasi-real-time. This robust and flexible approach requires no ancillary data for rapid implementation. A composite of pre- and post-flood images was obtained to identify changes and mask water pixels. Sentinel-2 identifies not only impacts on land but also on water ecosystems and their services, providing information on water quality deterioration and the concentration of suspended matter in highly sensitive environments. Subsequent water quality deterioration occurred in large portions of Mar Menor, the largest coastal lagoon in the Mediterranean

This study demonstrated the potentials brought by the free and open-data policy of Sentinel-2, a valuable source of rapid synoptic spatio-temporal information at a local or regional scale for supporting scientists, managers, stakeholders, and society in general during and after an emergency [13].

2.2. Monitoring and Modeling

The increasing deterioration of aquatic environments has attracted more attention to water quality monitoring techniques, with most researchers focusing on the acquisition and assessment of water quality data, but seldom on the discovery and tracing of pollution sources. In the study of Wang et al. [14], a semantic-enhanced modeling method for ontology modeling and rules building is proposed, which can be used for river water quality monitoring and relevant data observation processing. The observational process ontology (OPO) method can describe the semantic properties of water resources and observation data. In addition, it can provide the semantic relevance among the different concepts involved in the observational process of water quality monitoring. A pollution alert can be achieved using the reasoning rules of the water quality monitoring stations. In this study, a case is made for the usability testing of the OPO models and reasoning rules by using a water quality monitoring system. The system contributes to the water quality observational monitoring process and traces the source of pollutants using sensors, observation data, process models, and observation products that users can access in a timely manner [14].

Urban river catchments face multiple water quality challenges that threaten the biodiversity of riverine habitats and the flow of ecosystem services. Medupin et al. [15] examined two water quality challenges: runoff from increasingly impervious land covers, and effluent from combined sewer overflows, within a temperate zone river catchment in Greater Manchester, North-West UK. Sub-catchment areas of the River Medlock were delineated from digital elevation models using a Geographical Information System. By combining flow accumulation and high-resolution land cover data within each sub-catchment and water quality measurements at five sampling points along the river, they identified which land cover(s) are key drivers of water quality. Impervious land covers increased downstream and were associated with higher runoff and poorer water quality. Of the impervious covers, transportation networks had the highest runoff ratios and therefore the greatest potential to convey contaminants to the river. We suggest more integrated management of imperviousness to address water quality and flood risk, while urban well-being could be achieved working with greater catchment partnerships [15].

Hepp and Zessner [16] present a simple mapping key suitable for quick and systematic assessments of the type of agricultural and civil engineering structures present in a certain agricultural catchment, as well as the impact they may have on the spatial distribution of critical source areas. An application of this mapping key to three small sub-catchments of a case study catchment, with an area of several hundred square kilometers (one-stage cluster sampling), in Austria clearly revealed that road embankments with subsurface drainage can exert a major influence on the emission and transport pathways of sediment-bound pollutants such as particulate phosphorus (PP). Due to this, the semi-empirical, spatially distributed PhosFate model is extended to separately model PP emissions into surface waters via storm drains along road embankments. Furthermore, the overall share of road embankments with subsurface drainage on all road embankments in the case study catchment was inferred with the help of a Bayesian hierarchical model. The combination of the results of these two models showed that the share of storm drains at road embankments on total PP emissions ranges from about one fifth to one third in the investigated area [16].

Water quality in urban streams is highly influenced by emissions from waste water treatment plants (WWTP) and from sewer systems, particularly by overflows from combined systems. During storm events, this causes random fluctuations in discharge and pollutant concentrations over a wide area. The study by Dittmer et al. [17] focuses on the environmental impact of micropollutant loads emitted from combined sewer systems. For this purpose, high-resolution time series of river concentrations were generated by combining a detailed calibrated model of a sewer system with the measured discharge of a small natural river to a virtual urban catchment. This river base flow represents the remains of the natural hydrological system in the urban catchment. River concentrations downstream of the outlets were simulated based on mixing ratios of base flow, WWTP effluent, and CSO discharge. The results showed that the standard method of time proportional sampling of rivers does not capture the risk of critical stress on aquatic organisms. The ratio between average and peak concentrations and the duration of elevated concentrations strongly depends on the source and the properties of the particular substance. The design of sampling surveys and evaluation of data should consider these characteristics and account for their effects [17].

Bayesian networks (BN) have increasingly been applied in water management but not to estimate the efficacy of riparian buffer zones (RBZ). The methodical study of Gericke et al. [18] aims at evaluating the first BN for predicting RBZ efficacy in retaining sediment and nutrients (dissolved, total, and particulate nitrogen and phosphorus) from widely available variables (width, vegetation, slope, soil texture, flow pathway, nutrient form). To evaluate the influence of the parent nodes and how the number of states affected the prediction errors, they used a predefined general BN structure, collected 580 published datasets from North America and Europe, and performed classification tree analyses and multiple 10-fold cross-validations of different BNs. These errors ranged from 0.31 (two output states) to 0.66 (five states). The outcome remained unchanged without the least influential nodes (flow pathway, vegetation). Lower errors were achieved when the parent nodes had more than two states. The number of efficacy states influenced most strongly by the prediction error as its lowest and highest states were better predicted than the intermediate states. While the derived BNs could support or replace simple design guidelines, they are limited for more detailed predictions. More representative data on vegetation or additional nodes, such as preferential flow, would probably improve the predictive power [18].

2.3. Monitroing, Modeling, and Management

The contamination of water with nutrients, especially nitrogen and phosphorus originating from various diffuse and point sources, has become a worldwide issue in recent decades. Due to the complexity of the processes involved, watershed models are gaining an increasing role in their analysis. The goal set by the EU Water Framework Directive to reach "good status" for all water bodies requires spatially detailed information on the fate of contaminants. In a study by Jolánkai et al. [19], the watershed nutrient model MONERIS was applied to the Hungarian part of the Danube River Basin. The spatial resolution was 1078 water bodies (mean area of 86 km^2), and two subsequent 4 year periods (2009–2012 and 2013–2016) were modeled. Various elements/parameters of the model were adjusted and tested against surface and subsurface water quality measurements taken from all over the country, namely (i) the water balance equations (surface and subsurface runoff), (ii) the nitrogen retention parameters of the subsurface pathways (excluding tile drainage), (iii) the shallow groundwater phosphorus concentrations, and (iv) the surface water retention parameters. The study revealed that (i) digital-filter-based separation of surface and subsurface runoff yielded different values for these components, but this change did not influence nutrient loads significantly; (ii) shallow groundwater phosphorus concentrations in the sandy soils of Hungary differ from those of the MONERIS default values; (iii) a significant change of the phosphorus in-stream retention parameters was needed to approach measured in-stream phosphorus load values. Local emissions and pathways were analyzed and compared with previous model results [19].

Environmental management decisions should be made based on solid scientific evidence, and which relies on monitoring and modeling. In practice, changing economic, societal, and political boundary conditions often interfere with management during large, long, and complex projects. The result may be a sub-optimal development path that may finally diverge from the original intentions and be economically or technically ineffective

Nevertheless, unforeseen benefits may be created in the end [20]. The Kis-Balaton wetland system is a typical illustration of such a case and has been extensively studied by Honti et al. [20]. Despite tremendous investments and huge efforts put in monitoring and modeling, the sequence of decisions during implementation can hardly be considered optimal. A catchment model and a basic water quality model have been used to coherently review the impacts of management decisions during the 30-year history. Due to the complexity of the system, science mostly excelled in finding explanations for observed changes after the event, instead of predicting the impacts of management measures a priori. In parallel, the political setting and sectoral authorities experienced rearrangements during the system implementation. Despite being expensive as a water quality management investment, originally targeting nutrient removal, the Kis-Balaton wetland system created a huge ecological asset, and thereby became worth the price [20].

3. Conclusions

In this special issue, we are able to present a selection of high-level contributions showing the manifold aspects of monitoring, modeling, and management of water quality. If we look at the chosen subjects we see that four out of the eleven contributions are specifically addressing monitoring aspects and five contributions focus on the interface of modeling and associated monitoring, delivering the scientific basis for water quality management. Only two contributions directly address management aspects in their research focus, indicating that this element of water governance is somehow underrepresented in this special issue. In spite of the small size on the sample, it still points out that the gap between science in its conventional sense and science in an inter- and transdisciplinary understanding is not yet completely closed.

Scientists publishing in a scientific journal still tend to focus on "pure" scientific questions, and use management and policy aspects more for arguing the motivation of their research or as an appendix on what should be considered further, rather than directly including them in their research focus. Therefore additional efforts are needed to bridge the gap between science and policy.

Nevertheless, directly addressing management in the title of a special issue of a scientific journal clearly gives the right sign, and this special issue provides an important overview on a hot topic in water related research. Finally, I would like to thank all the authors for their great contributions and remind you that "he (or she) not busy being born is busy dying" [21].

Informed Consent Statement: Not applicable.

Data Availability Statement: Not applicable.

Conflicts of Interest: The author declares no conflict of interest.

References

1. Steffen, W.; Richardson, K.; Rockstrom, J.; Cornell, S.E.; Fetzer, I.; Bennett, E.M.; Sörlin, S. Planetary boundaries: Guiding human development on a changing planet. *Science* **2015**, *347*, 1259855. [CrossRef] [PubMed]
2. Zoboli, O.; Clara, M.; Gabriel, O.; Scheffknecht, C.H.; Humer, M.; Brielmann, H.; Kulcsar, S.; Trautvetter, H.; Kittlaus, K.; Amann, A.; et al. Occurrence and levels of micropollutants across environmental and engineered compartments in Austria. *J. Environ. Manag.* **2019**, *232*, 636–653. [CrossRef] [PubMed]
3. Mayer, R.E.; Reischer, G.H.; Ixenmaier, S.K.; Derx, J.; Blaschke, A.P.; Ebdon, J.E.; Linke, R.; Egle, L.; Ahmed, W.; Blanch, A.R.; et al. Global Distribution of Human-Associated Fecal Genetic Markers in Reference Samples from Six Continents. *Environ. Sci. Technol.* **2018**, *52*, 5076–5084. [CrossRef] [PubMed]
4. Limbach, L.K.; Bereiter, R.; Müller, E.; Krebs, R.; Gälli, R.; Stark, W.J. Removal of oxide nanoparticles in a model wastewater treatment plant: Influence of agglomeration and surfactants on clearing efficiency. *Environ. Sci. Technol.* **2008**, *42*, 5828–5833. [CrossRef] [PubMed]
5. Rios Mendoza, L.M.; Balcer, M. Microplastics in freshwater environments: A review of quantification and assessment. *Trends Anal. Chem.* **2019**, *113*, 402–408. [CrossRef]

6. Berendonk, T.U.; Manaia, C.M.; Merlin, C.; Fatta-Kassinos, D.; Cytryn, E.; Walsh, F.; Bürgmann, H.; Sørum, H.; Norström, M.; Pons, M.-N.; et al. Tackling antibiotic resistance: The environmental framework. *Nat. Rev. Microbiol.* **2015**, *13*, 310–317. [CrossRef] [PubMed]
7. Zessner, M.; Zoboli, O.; Hepp, G.; Kuderna, M.; Weinberger, C.; Gabriel, O. Shedding Light on Increasing Trends of Phosphorus Concentration in Upper Austrian Rivers. *Water* **2016**, *8*, 404. [CrossRef]
8. Zessner, M.; Schönhart, M.; Parajka, J.; Trautvetter, H.; Mitter, H.; Kirchner, M.; Hepp, G.; Blaschke, A.P.; Strenn, B.; Schmid, E. A novel integrated modelling framework to assess the impacts of climate and socio-economic drivers on land use and water quality. *Sci. Total Environ.* **2017**, *579*, 1137–1151. [CrossRef]
9. Schönhart, M.; Trautvetter, H.; Parajka, J.; Blaschke, A.P.; Hepp, G.; Kirchner, M.; Mitter, H.; Schmid, E.; Strenn, B.; Zessner, M. Modelled impacts of policies and climate change on land use and water quality in Austria. *Land Use Policy* **2018**, *76*, 500–514. [CrossRef]
10. Cao, P.; Zhu, Y.; Zhao, W.; Liu, S.; Gao, H. Chromaticity Measurement Based on the Image Method and Its Application in Water Quality Detection. *Water* **2019**, *11*, 2339. [CrossRef]
11. Agberien, A.V.; Örmeci, B. Monitoring of Cyanobacteria in Water Using Spectrophotometry and First Derivative of Absorbance. *Water* **2020**, *12*, 124. [CrossRef]
12. Cao, H.; Guo, Z.; Wang, S.; Cheng, H.; Zhan, C. Intelligent Wide-Area Water Quality Monitoring and Analysis System Exploiting Unmanned Surface Vehicles and Ensemble Learning. *Water* **2020**, *12*, 681. [CrossRef]
13. Caballero, I.; Ruiz, J.; Navarro, G. Sentinel-2 Satellites Provide Near-Real Time Evaluation of Catastrophic Floods in the West Mediterranean. *Water* **2019**, *11*, 2499. [CrossRef]
14. Wang, X.; Wei, H.; Chen, N.; He, X.; Tian, Z. An Observational Process Ontology-Based Modeling Approach for Water Quality Monitoring. *Water* **2020**, *12*, 715. [CrossRef]
15. Medupin, C.; Bark, R.; Owusu, K. Land Cover and Water Quality Patterns in an Urban River: A Case Study of River Medlock, Greater Manchester, UK. *Water* **2020**, *12*, 848. [CrossRef]
16. Hepp, G.; Zessner, M. Assessing the Impact of Storm Drains at Road Embankments on Diffuse Particulate Phosphorus Emissions in Agricultural Catchments. *Water* **2019**, *11*, 2161. [CrossRef]
17. Dittmer, U.; Bachmann-Machnik, A.; Launay, M.A. Impact of Combined Sewer Systems on the Quality of Urban Streams: Frequency and Duration of Elevated Micropollutant Concentrations. *Water* **2020**, *12*, 850. [CrossRef]
18. Gericke, A.; Nguyen, H.H.; Fischer, P.; Kail, J.; Venohr, M. Deriving a Bayesian Network to Assess the Retention Efficacy of Riparian Buffer Zones. *Water* **2020**, *12*, 617. [CrossRef]
19. Jolánkai, Z.; Kardos, M.K.; Clement, A. Modification of the MONERIS Nutrient Emission Model for a Lowland Country (Hungary) to Support River Basin Management Planning in the Danube River Basin. *Water* **2020**, *12*, 859. [CrossRef]
20. Honti, M.; Gao, C.; Istvánovics, V.; Clement, A. Lessons Learnt from the Long-Term Management of a Large (Re)constructed Wetland, the Kis-Balaton Protection System (Hungary). *Water* **2020**, *12*, 659. [CrossRef]
21. Dylan, B. It's Alright, Ma (I'm Only Bleeding). In *The Album Bringing It All Back Home*; Columbia Recording: New York, NY, USA, 1965

Article

Chromaticity Measurement Based on the Image Method and Its Application in Water Quality Detection

Pingping Cao [1], Yuanyang Zhu [1], Wenzhu Zhao [1], Sheng Liu [1,*] and Hongwen Gao [2]

1 College of Computer Science and Technology, Huaibei Normal University, Anhui 235000, China; 15077983852@139.com (P.C.); 15556116568@139.com (Y.Z.); 15212626136@139.com (W.Z.)

2 College of Environmental Science and Engineering, Tongji University, Shanghai 200092, China; hwgao@tongji.edu.cn

* Correspondence: liurise@139.com; Tel.: +86-183-6523-9378

Received: 4 October 2019; Accepted: 5 November 2019; Published: 8 November 2019

Abstract: In order to measure the chromaticity of water and the content of dissolved matter more accurately, effectively, and cheaply, a chromaticity measurement system based on the image method was proposed and applied. The measurement system used the designed acquisition device and image processing software to obtain the Red-Green-Blue (RGB) values of the image and converted the color image from RGB color space to Hue-Saturation-Intensity (HSI) space to separate the chromaticity and brightness. According to the definition of chromaticity, the hue (H), saturation (S) values, and chromaticity of standard chromaticity solution images were fitted by a non-linear surface, and a three-dimensional chromaticity measurement model was established based on the H and S values of water images. For the measurement of a standard chromaticity solution, the proposed method has higher accuracy than spectrophotometry. For actual water sample measurements, there is no significant difference between the results of this method and the spectrophotometer method, which verified the validity of the method. In addition, the system was tried to measure the concentration of ammonia nitrogen, phosphate, and chloride in water with satisfactory results.

Keywords: water quality; analysis method; chromaticity measurement; surface fitting; concentration of dissolved matter

1. Introduction

Pollution can lead to changes in water color, especially in textile, leather, paper, pharmaceutical, printing, and dyeing industries [1,2]. Therefore, the pollution degree of water can be monitored by measuring the chromaticity of water [3]. Chromaticity is an index for the quantitative determination of the color of natural water or treated water, usually in the unit of degree (°). Natural water often shows light-yellow, light-brown, yellow-green, and other different colors. The cause of color is due to humus, organic or inorganic substances dissolved in water [4].

China's current national standard method for water chromaticity measurement, "Water quality-determination of colority" (GB/T 11903-1989), is the platinum-cobalt colorimetric method and dilution multiple method [5]. Among them, platinum-cobalt colorimetric refers to the international standard "Clear liquids—Estimation of color by the platinum-cobalt scale—Part 1: Visual method" (ISO 6271-1-2004) [6]. At present, the international methods of measuring chromaticity include a spectrophotometer [7], a three-wavelength transmittance method [8], and a support vector machine regression prediction method [9]. In practical application, the platinum-cobalt colorimetric method has no obvious difference between 5 and 25, is not easy to judge, and its error is large. In the dilution multiple method, the response of different color tones to optic nerve stimulation is different because of the great difference of water sample tones, which creates personal subjectivity in discriminating. Finally, it is difficult for the dilution end point (colorless) to have a unified standard, and the final result will have a great error. Spectrophotometry is a method for the qualitative or quantitative analysis of the absorbance of light of a specific wavelength or a certain wavelength range. Because the substances in the solution have selectivity to absorb light, the absorbance and absorption spectra of different substances are different at different wavelengths, which make the solution present different colors [10]. Therefore, spectrophotometry can identify substances or measure their content according to the absorption spectra of different substances. A spectrophotometer is used to determine the chromaticity value of water samples by establishing the correlation between the absorbance value of a standard chromaticity solution at a characteristic wavelength or the peak area of the absorption spectrum and the chromaticity value. This method improves the accuracy of the water chromaticity measurement, but the price of a professional spectrophotometer is higher, and the characteristic wavelength of a chromaticity solution is no uniform standard, which inevitably leads to disagreements in measurement. The method of water chromaticity measurement based on three-wavelength transmittance is also a spectrophotometric method, which measures the true chromaticity of water quality by measuring the transmittance at three wavelengths. However, the selection of measuring points (three wavelengths) can be different because of different measurers, and the selection of measuring points will also be different, so the measurement results will also be different. The application of support vector machine regression to predict water chromaticity can obtain a high accuracy prediction model, but a large number of learning samples are needed. A large number of training samples will increase the workload of the measurement, leading to low measurement efficiency.

Excessive ammonia nitrogen in the water environment will cause much harm to water bodies, such as reducing the dissolved oxygen concentration in water and accelerating water eutrophication [11]. Phosphate can disrupt the ecological balance of the water environment by promoting the proliferation of algae (called eutrophication) and the consequent consumption of dissolved oxygen (when algae decay) [12]. Chloride dissolved in water is toxic and can cause dizziness, nausea, dyspnea, and even death [13]. Therefore, the measurement of ammonia nitrogen, phosphate, and chloride content in water is very important. At present, the methods for the determination of the properties and contents of organic matter in water are ultraviolet visible spectroscopy [14], mobile mass spectrometry [15], linear regression and artificial neural network [16], the sensor method [17], the trace element tracing method [18,19] and the potentiometric titration method [20]. Among them, ultraviolet visible spectroscopy, mobile mass spectrometry, and the sensor method need professional analytical instruments that come with high measurement cost; the linear regression and artificial neural network method needs a large number of training data to ensure the accuracy of the measurement, and the measurement efficiency is low; the trace element tracing method and potentiometric titration method have low operability and need professional technicians to operate in water quality detection.

Cameras have been widely used in various fields. In physics, a digital camera is used to study the trajectory of the water jet, the outline of the suspension chain, and the defocus figure reflected on the mirror of different shapes [21]. Digital cameras are also used to describe the color and physical properties of soil samples [22]. In chemistry, Red-Green-Blue (RGB) values can be converted into other color spaces, and then, the conversion functions of soil organic carbon and iron can be derived by using these color spaces to measure their concentrations quickly and accurately [23]. Moreover, a digital camera was used to determine iron and residual chlorine in water using N,N-diethylphenylenediamine [24]. In Maputo Bay, Mozambique, digital cameras are also used to measure the suspended sediment concentration, which prompted discussion on the possibility of using digital photography to measure the suspended sediment concentration in other coastal waters [25]. The development of using digital cameras within the field of water quality detection can provide a new measurement method for water quality detection. In order to measure water chromaticity more accurately, quickly, and cheaply, a water chromaticity measurement system based on the image method is proposed. The proposed chromaticity measurement system based on a digital camera can be used to measure the content of ammonia nitrogen, phosphate, and chloride in water. And it can simplify the design of measuring instruments and improve the accuracy of measurement, which is of great significance in water quality detection.

The designed chromaticity measurement system uses an image acquisition device to collect the image of a standard chromaticity solution. The solution image captured by the camera is the color image of the standard chromaticity solution. The color can be separated into brightness and chromaticity [26]. Chromaticity is the property of color that does not include brightness. It reflects the hue and saturation of color. The image processing software designed in this paper actually converts the color solution image from RGB color space to Hue-Saturation-Intensity (his) space. Then, the brightness can be separated. Finally, a three-dimensional relationship model between hue (H), saturation (S), and chromaticity and a standard curve between H, S-standard-deviation (△HS), and solute concentration are established for the measurement of water chromaticity and solute content.

2. Materials and Methods

2.1. Design of a Measurement System Structure

The water chromaticity measurement system was designed to include a water image acquisition device and image processing software, as shown in Figure 1.

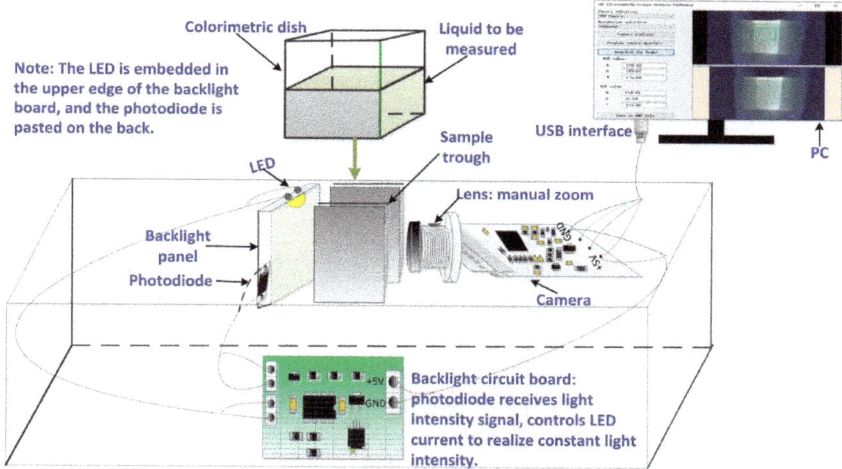

Figure 1. Structure of a water chromaticity measurement system.

The acquisition device is used to collect a liquid image, which is processed by the image processing software to get RGB and HSI values and to achieve chromaticity separation.

2.1.1. Image Acquisition Device

The water image acquisition device consists of a high-color-rendering LED, a backlight panel, a constant light source circuit, a digital camera, and a sealed box. The light color of the LED is warm white, the color temperature is in the range of 2600–4500 K, the corresponding luminous intensity is 9.5 cd(lm), and the forward voltage is 2.9–3.5 V.

Adding a backlight panel between the light source and the solution can bring uniformity to the light and improve the image quality, avoid the uneven brightness of the collected image, and affect the color value of the image. The constant light source circuit makes the light intensity of the light source constant, which is a closed-loop control circuit, as shown in Figure 2. Within the circuit, D2 detects the light intensity of D1 and generates a voltage signal, which is buffered by U1B, then controls the working current of D1 to stabilize the light intensity. Although the design is relatively simple, the effect of constant light intensity is very good [27].

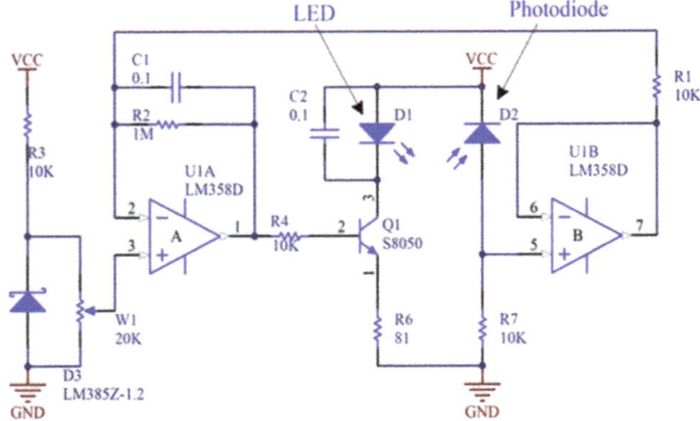

Figure 2. Constant light intensity circuit diagram.

The digital camera used is a JD-300 digital camera produced by Shenzhou Jiuding Technology Co., Ltd. in Beijing, China, and its structure is shown in Figure 3.

Figure 3. Camera printed circuit board details.

The camera can adjust brightness, contrast, hue, saturation, and white balance, and use manual adjustments for exposure and zoom. When using this camera to acquire chromaticity images, it is necessary to ensure that the various parameters are constant and adjust the exposure and focus to the appropriate size.

2.1.2. Software Design

The image processing software designed in the measurement system is developed on the platform of Visual Studio 2012 based on the open source camera development toolkit and C# language. The software digitalizes the collected water image, that is, to get the pixel value of the image.

The edge of the liquid image captured by the camera will be geometrically distorted, and the RGB value of the image will be affected when the image is digitized. In order to avoid the effect of edge distortion on the accuracy of data, the RGB values of 400 pixels in the central region of the image are selected, and then the average RGB values of these pixels are obtained. The image was then transformed from RGB color space to HSI color space to separate the chromaticity and brightness by the image processing software. The conversion algorithm is as follows.

Given the RGB value of the acquired water image, the H component can be obtained by Formula (1):

$$H = \begin{cases} \theta, (B \le G) \\ 360 - \theta, (B > G) \end{cases}, \tag{1}$$

in which,

$$\theta = \arccos\left\{ \frac{\frac{1}{2}[(R-G)+(R-B)]}{\left[(R-G)^2+(R-G)(G-B)\right]^{\frac{1}{3}}} \right\}. \tag{2}$$

The saturation component S is derived from Formula (3):

$$S = 1 - \frac{3}{(R+G+B)}[\min(R,G,B)], \tag{3}$$

and the final intensity component Intensity (I) is derived from Formula (4):

$$I = \frac{1}{3}(R+G+B). \tag{4}$$

The color space conversion is shown in Figure 4.

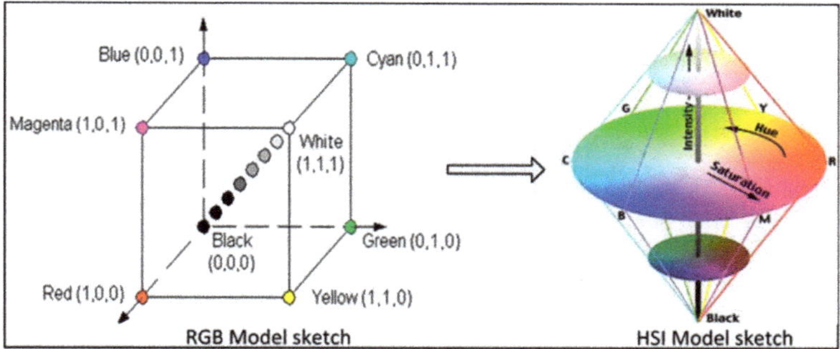

Figure 4. Diagram of the color space model from Red-Green-Blue (RGB) to Hue-Saturation-Intensity (HSI).

The software designed can control the exposure, contrast, hue, saturation, and other properties of the camera. It is necessary to adjust and fix the camera attributes when measuring a water image. The designed software interface and camera property control are shown in Figure 5.

Figure 5b shows the main interface of the image processing software, which contains buttons such as camera settings and image capture. The camera setting button controls the properties of the camera. Figure 5a shows the camera parameters of this experiment. The function of the image capture button is to get the RGB and HSI values of the current image. Combining the image processing software and the image acquisition device, it can be used to measure the chromaticity of water.

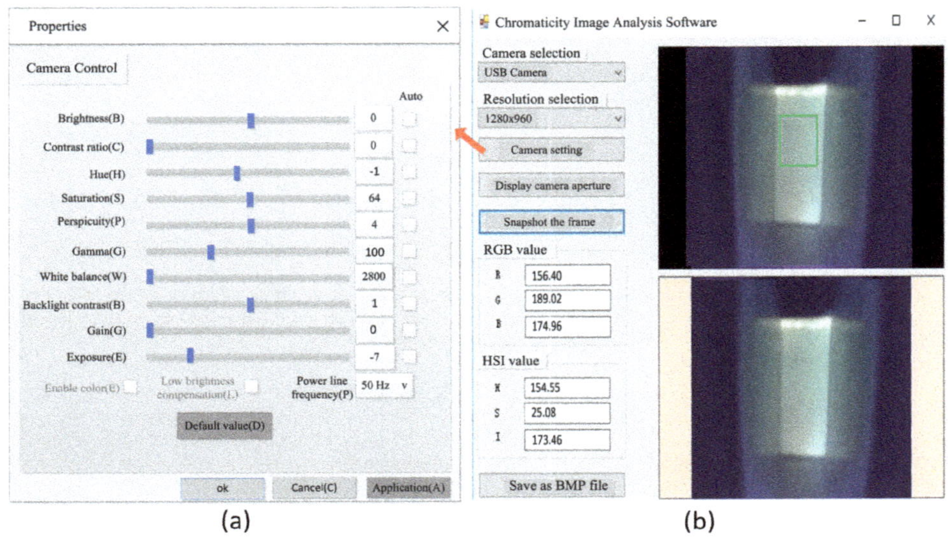

Figure 5. Software interface and camera property settings. (**a**) is the camera parameters of this experiment. (**b**) is the main interface of the image processing software.

2.2. Preparation and Information Acquisition of Chromaticity Standard Solution

In this design, 1.245 g of potassium chloroplatinate (MESCO Chemical Co., Ltd, Tianjin, China) and 1.000 g of cobalt chloride (Jingshiji mall, Changsha, China) were dissolved in 200 mL hydrochloric acid (Mingcheng Chemical Co., Ltd, Qidong, China) with a concentration of 6 mol/L, and then diluted to 1000 mL with deionized water. The standard platinum-cobalt chromaticity solution with a chromaticity of 500 was prepared. Thirty standard solutions of different chromaticity in the range of 0–500 were obtained by diluting the 500-degree chromaticity solution according to a gradient. The standard chromaticity solution was poured into the colorimetric dish (size is 34 × 15 × 43 mm^3, and the optical path is 30 mm) and then put into the sample trough of the designed chromaticity measurement system to measure. The RGB and HSI values of the chromaticity image are obtained by the designed chromaticity measurement system. Some image data are shown in Figure 6.

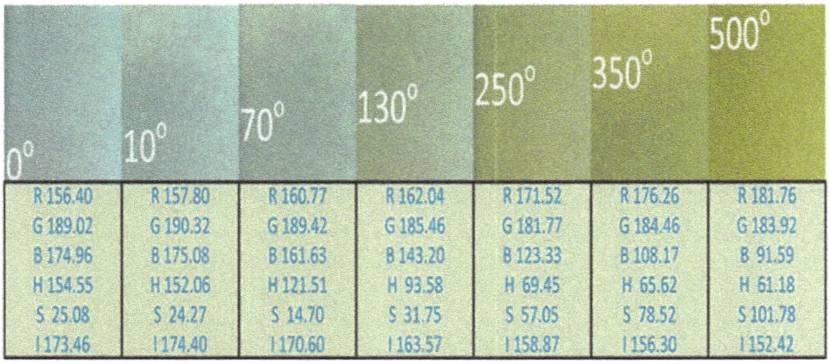

0^0	10^0	70^0	130^0	250^0	350^0	500^0
R 156.40	R 157.80	R 160.77	R 162.04	R 171.52	R 176.26	R 181.76
G 189.02	G 190.32	G 189.42	G 185.46	G 181.77	G 184.46	G 183.92
B 174.96	B 175.08	B 161.63	B 143.20	B 123.33	B 108.17	B 91.59
H 154.55	H 152.06	H 121.51	H 93.58	H 69.45	H 65.62	H 61.18
S 25.08	S 24.27	S 14.70	S 31.75	S 57.05	S 78.52	S 101.78
I 173.46	I 174.40	I 170.60	I 163.57	I 158.87	I 156.30	I 152.42

Figure 6. Partial image data and the corresponding color components.

With the increase of solution chromaticity, the RGB and HSI values of corresponding images also change, reflecting the corresponding relationship between RGB and HSI values and chromaticity. The obtained HS value is fitted with the standard chromaticity, and the no-linear relationship between the chromaticity and the HS value is established. The chromaticity value of the solution to be measured is calculated according to the fitting expression.

2.3. Preparation and Data Acquisition of Standard Solutions of Ammonia Nitrogen, Phosphate, and Chloride

The 1000 mg/L ammonia-nitrogen standard solution: ammonium chloride 3.819 g (NH_4Cl, superior purity, China Jiehui Chemical Reagent, Wuhan, China) dried at 100 °C was dissolved in a small amount of deionized water. It was then transferred into a 1000 mL capacity bottle and diluted to 1000 mL by adding deionized water. The mixture was uniform. The standard ammonia-nitrogen solutions with concentrations of 0 mg/L, 0.2 mg/L, 0.4 mg/L, 0.7 mg/L, 1.0 mg/L, and 1.5 mg/L were prepared by further dilution. Ammonia nitrogen was detected by Nessler's reagent color reaction, such as Formulas (5) and (6):

$$HgI_2 + 2KI \rightarrow K_2HgI_4, \tag{5}$$

$$2K_2HgI_4 + 3KOH + NH_3 \rightarrow NH_2Hg_2OI + 7KI + 2H_2O(\text{yellow-brown complex}), \tag{6}$$

the designed chromaticity measurement system is used to obtain image information of the complex solution, as shown in Figure 7a.

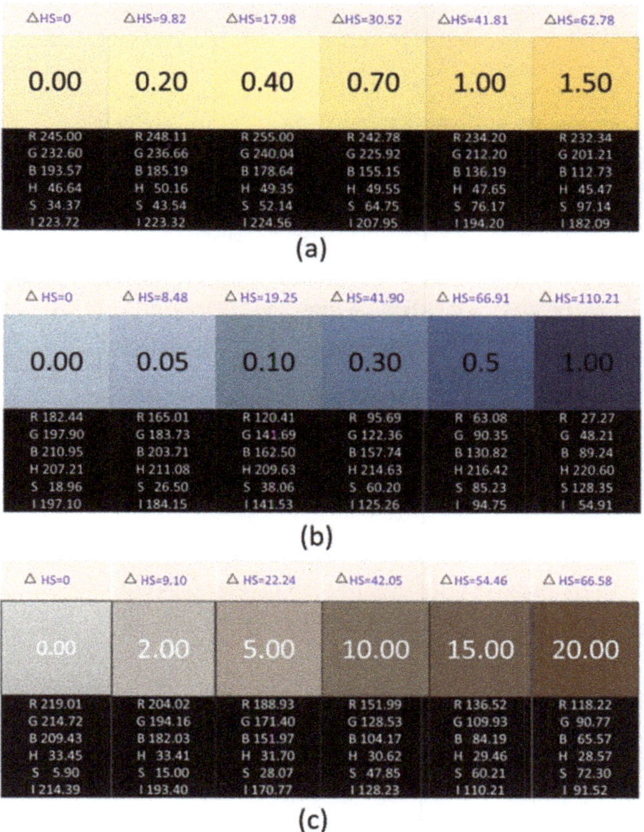

Figure 7. Image information of solutions with different concentrations. (**a**) is the image information of the yellow-brown complex solution. (**b**) is the image information of the blue phosphomolybdenum blue solution. (**c**) is the image information of the brick red complex solution.

The 500 mg/L phosphorus standard solution: potassium dihydrogen phosphate was dried at 110 °C for four hours (KH_2PO_4, superior purity, Shanghai Reagent Plant, Shanghai, China). Then, 5.444 g was accurately weighed. It was dissolved in a small amount of deionized water, transferred to a 500 mL capacity bottle, and diluted to 500 mL calibration with deionized water. The standard phosphorus solutions with concentration of 0 mg/L, 0.05 mg/L, 0.1 mg/L, 0.3 mg/L, 0.5 mg/L, and 1.0 mg/L were prepared by further dilution. Under acidic conditions, the active phosphate reacts with ammonium molybdate to form yellowish phosphomolybdenum yellow, which is reduced to blue phosphomolybdenum blue by stannous chloride, such as Formulas (7) and (8):

$$H_3PO_4 + 12(NH_4)_2MoO_4 + 21HNO_3 = (NH_4)_3H_4[P(Mo_2O_7)_6] + 21NH_4NO_3 + 10H_2O, \quad (7)$$

$$(NH_4)_3H_4[P(Mo_2O_7)_6] + 2SnCl_2 + 4HCL = (NH_4)_2H_4[(Mo_2O_7)_5] + 2SnCl_4 + 2H_2O, \quad (8)$$

The image information of the generated blue solution is obtained by the chromaticity measurement system proposed in this design, as shown in Figure 7b.

A chloride standard solution (1000 mg/L) was purchased from Zhongke Beijing Instrument Consumables Standard Materials Store of China. The chloride standard solution was diluted to 0 mg/L, 2 mg/L, 5 mg/L, 10 mg/L, 15 mg/L, and 20 mg/L. In neutral or weak alkaline solution, when potassium chromate is used as the indicator and silver nitrate is used to titrate chloride, because the solubility of silver chloride is less than that of chromate, after the chloride ion is completely precipitated, the chromium ion is precipitated in the form of silver chromate, and a brick red substance is produced. The color of the red precipitate is proportional to the chloride content. The precipitation titration reactions are Formulas (9) and (10):

$$Ag^+ + Cl \rightarrow AgCl \downarrow \text{(white)}, \quad (9)$$

$$2Ag^+ + CrO_4{}^{2-} \rightarrow Ag_2CrO_4 \downarrow \text{(brick red)}. \quad (10)$$

The brick red precipitate solution is placed in the designed chromaticity measurement system to obtain the image information of the solution, as shown in Figure 7c.

Figure 7a shows the image information and HS standard deviation of different concentrations of ammonia nitrogen, and Figure 7b,c show the image information and HS standard deviation of different concentrations of phosphate and chloride, respectively. Among them, ΔHS is the standard deviation between the HS value of each dissolved matter image and the HS value of the blank sample image, which is calculated by formula (11):

$$\Delta HS = \sqrt{(H - H_0)^2 + (S - S_0)^2}. \quad (11)$$

3. Results

The HS value of standard chromaticity solution was used to fit the non-linear relationship with the chromaticity, and a three-dimensional chromaticity measurement model based on HS was established. The model was applied to the measurement of the standard chromaticity solution to verify the accuracy of the measurement method. When comparing the proposed method with spectrophotometry, the accuracy of this method is higher than that of spectrophotometry for the standard chromaticity solution; for the actual water sample measurement, there is no significant difference between the measurement result of this method and that of spectrophotometry.

3.1. Fitting Results

The color includes chromaticity and brightness, and the chromaticity of the color is obtained by separating the brightness of the color. After separating the brightness, the color image only contains H and S values, which correspond to the chromaticity, so the relationship model between HS and chromaticity can be established. The data fitting software Origin 8.0 (OriginLab, Northampton, MA, USA) is used to establish surface fitting with H and S as X and Y axes and chromaticity as Z axes, respectively. The fitting results are shown in Figure 8.

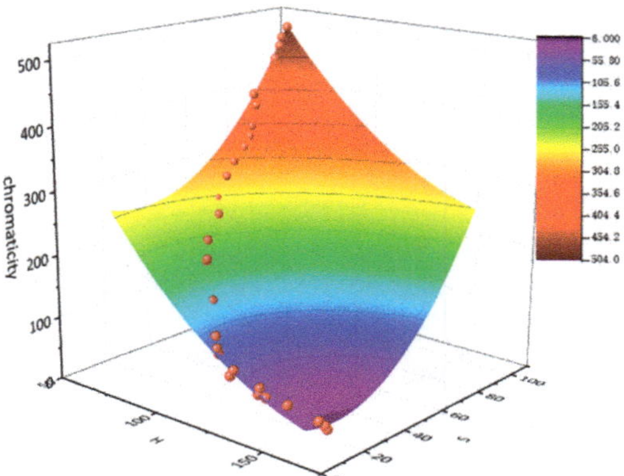

Figure 8. Fitting results of hue (H) and saturation (S) values and chromaticity.

The chromaticity data in the graph are all distributed on the surface, which is the fitting surface of HS and chromaticity. The expression is as shown in Formula (12), and the fitting degree is as high as 0.99845:

$$Z = 634.4507 - 6.73548x - 2.61133y + 0.01882x^2 + 0.04596y^2. \quad (12)$$

In Formula (12), x represents the input value H, y is the input value S, and Z is the chromaticity of the corresponding solution. By substituting the HS value of the solution to be measured into the expression, the chromaticity of the solution to be measured can be calculated.

3.2. Measurement Results

In order to verify the accuracy of the designed chromaticity measurement system, the model was used to measure the standard chromaticity solution with known chromaticity. The results of measurement and comparison are shown in Table 1.

Table 1. Measurements of the standard chromaticity solutions.

	Chromaticity (°)				
Standard solution	20	55	150	265	430
HS surface fitting	18.07	50.00	151.63	264.54	456.72
Error	1.93	5.00	1.63	0.46	26.72

From Table 1, the results of the proposed method are close to the standard chromaticity solution, and the error is small. Therefore, the designed chromaticity measurement system can effectively measure the chromaticity of water.

3.3. Contrast Experiment of Standard Chromaticity Solution

In order to verify the accuracy of the proposed method, the chromaticity measurement model is compared with the latest spectrophotometric method.

In this comparative experiment, a 721 G visible spectrophotometer produced by Shanghai Instrument and Electrical Analysis Instrument Co., Ltd., was used as the comparative object.

When using a spectrophotometer to measure the chromaticity of the solution, it is necessary to determine the characteristic wavelength of the solution to be measured at first. Because of the absorption wave of natural water reaches its peak at 380 nm, and the wavelength of the spectrophotometer is set to 380 nm. A 3 cm colorimetric dish was selected to hold the solution to be measured, and a higher accuracy could be obtained. The wavelength of the spectrophotometer was adjusted to 380 nm. The standard chromaticity solution was put into the spectrophotometer to measure the absorbance. Finally, the standard curve of absorbance (A) and chromaticity was established, as shown in Figure 9.

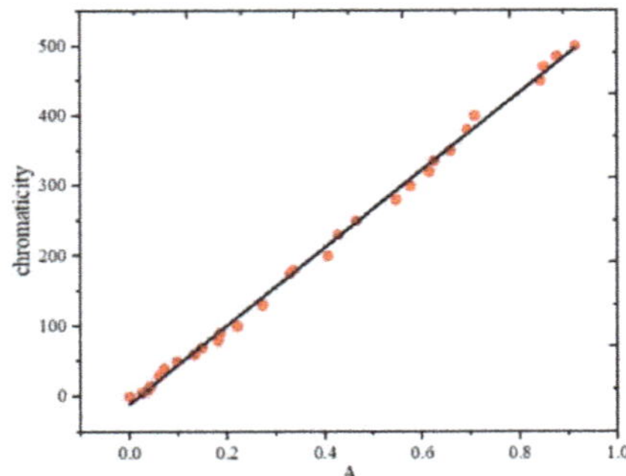

Figure 9. Fitting curves of absorption and chromaticity.

The fitting expression is Formula (13), and the fitting degree is as high as 0.99741:

$$y = 555.2641x - 10.33369. \quad (13)$$

According to the fitting curve of absorbance and chromaticity, the measurement results of spectrophotometry for the standard chromaticity solution can be obtained. The results are compared with the proposed chromaticity measurement system. Some of the comparison results are as shown in Table 2.

Table 2. Comparison of measurement results by two methods.

Methods	Standard Solution (°)					Mean Error (°)	Std Dev (°)
Standard solution	20	55	150	265	430		
The proposed method	18.07	50.00	151.63	264.54	456.72	7.15	5.46
Spectrophotometry	10.77	60.74	156.25	270.63	453.87	10.14	5.51

Table 2 shows that the mean deviation and standard deviation of the standard chromaticity solution measured by spectrophotometry are larger than those proposed in this paper, and the proposed method has higher accuracy. The comparison results of the two methods are shown in Figure 10.

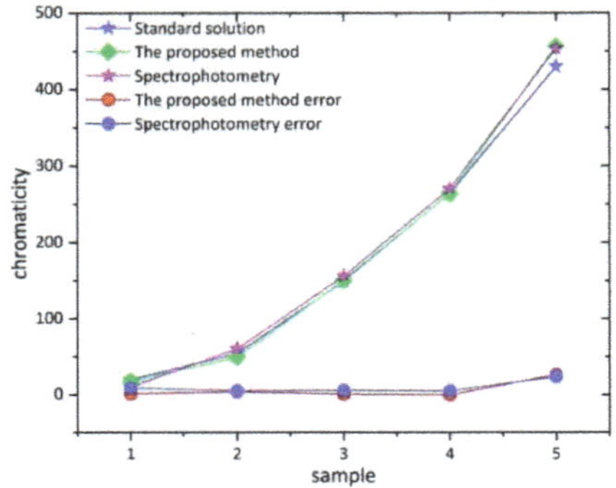

Figure 10. Comparisons of two measurement methods.

In Figure 10, samples one to five represent 20°, 55°, 150°, 265°, and 430° standard chromaticity solutions. The measurement curve of the proposed method almost coincides with the standard chromaticity curve, and the measurement error fluctuates around 0. The deviation between the spectrophotometric measurement curve and the standard chromaticity curve is relatively large, and the fluctuation range of the measurement error curve is larger than that of the method proposed in this design. Therefore, the accuracy of the proposed method is higher than that of the spectrophotometric method.

3.4. Contrast Experiment of Actual Water Sample

The local tap water (Sample 1), lake water (Sample 2 to 4), and river water (Sample 5) were taken as the actual samples of the validation experiment, and the proposed method and spectrophotometric measurement were used, respectively. First, the HS value and absorbance of actual water samples were measured by the designed chromaticity measuring device and 721 G visible spectrophotometer. Some image data acquisition results are shown in Figure 11.

Sample 1	Sample 2	Sample 3	Sample 4	Sample 5
H 158.42 S 25.66 A 0.024	H 153.42 S 23.92 A 0.035	H 143.51 S 20.36 A 0.057	H 152.17 S 24.27 A 0.037	H 154.18 S 24.52 A 0.031

Figure 11. Partial actual water sample data.

The HS and A (absorbance) values of the actual water samples were substituted into the chromaticity measurement model and the fitting expression of the spectrophotometry, and the results of the two methods for the actual water samples were obtained as shown in Table 3.

Table 3. Comparison of measurement results by two methods.

Methods	Sample 1	Sample 2	Sample 3	Sample 4	Sample 5
The Proposed Method (°)	3.08	7.99	21.40	9.08	7.04
Spectrophotometry (°)	3.00	9.10	21.32	10.21	6.88
Error (°)	0.08	1.11	0.08	1.13	0.16

A comparison of the two measurement methods for an actual water sample is shown in Figure 12, and the results are tested by an independent sample t-test. The test results are shown in Table 4.

In Figure 12, the measurement results of the proposed method have the same trend as those of spectrophotometry, and the error is small. In Table 4, the independent sample t-test of the two methods shows that the p value is 0.93 (>0.05), which shows that there is no significant difference between the two methods. The validity of the design in the actual water sample measurement is verified.

Table 4. Comparison of measurement results by two methods.

Methods	Chromaticity (°)	t	p
The Proposed Method	9.72 ± 3.09	−0.09	0.93
Spectrophotometry	10.10 ± 3.06		

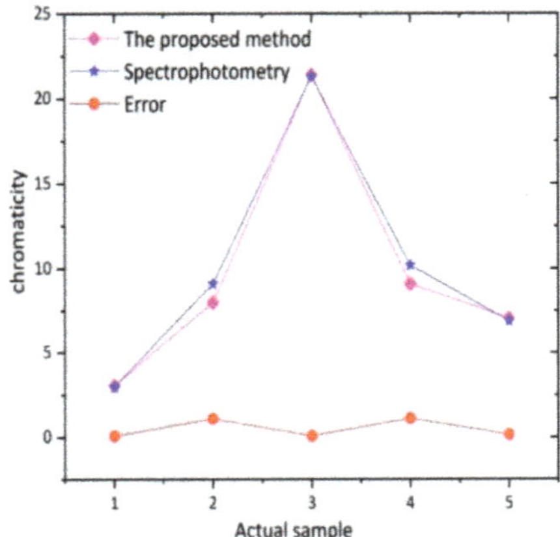

Figure 12. Comparison of measuring results for actual water samples.

4. Discussion and Further Application

4.1. Camera Combined with Image Processing Technology instead of Photoelectric Instrument to Detect Colorable Substances in Water

The main purpose of water quality detection is to detect and analyze the composition, nature, and content of pollutants in water to ensure the safety of water quality [28]. In order to detect water quality more effectively and conveniently, digital camera and image processing technology can be used instead of traditional photoelectric instruments to measure the content of colored organic matter in water. Most of the digital cameras use Complementary Metal-Oxide-Semiconductor (CMOS) sensors. As a semiconductor element, the CMOS photosensitive element is used to record the light changes, which is mainly composed of silicon and germanium. Their advantages are the high integration, low power consumption, and low cost. Each pixel of the camera is equivalent to a photoelectric detection element. The signal processing unit integrated by the camera can replace the digital to analog conversion circuit and the signal processing circuit of the photoelectric detection instrument. Using a camera and the HS color component of a picture to measure water quality can avoid the design of the photoelectric detection circuit, signal processing circuit, digital to analog conversion circuit, and display circuit. In theory, the detection of trace components based on color reaction can be realized by using digital camera and image processing technology.

4.2. Application in the Detection of Ammonia Nitrogen, Phosphate, and Chloride in Water

Generally, the polluted water contains solutes such as ammonia nitrogen, chloride, phosphate, and sulfide [29]. The proposed chromaticity measurement system was applied to the measurement of dissolved substances such as ammonia nitrogen, phosphate, and chloride in water for water quality detection, and the standard curve of HS standard deviation and dissolved substance content in water was established.

The RGB and HSI values of the pictures of ammonia nitrogen, phosphate, and chloride solutions with different concentrations were obtained by using the designed chromaticity measurement system. The standard deviation of HS for each concentration of solute and blank solution was obtained, as shown in Figure 7. The standard deviation of HS was fitted with the concentration of solute in water, and the standard curve was established. The standard curve expression and the fitted results are shown in Figure 13.

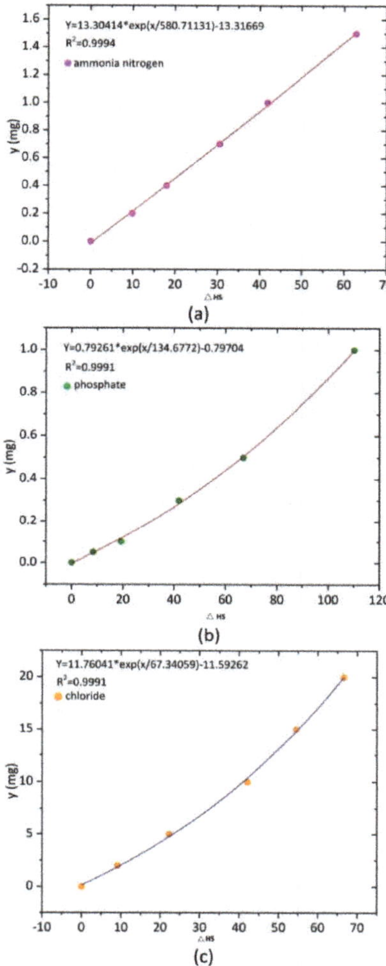

Figure 13. Fitting curve and result of soluble content in water. (**a**) is the fitting curves of ammonia nitrogen. (**b**) is the fitting curves of phosphate. (**c**) is the fitting curves of chloride.

Figure 13a–c shows the fitting curves of ammonia nitrogen, phosphate, and chloride concentration, respectively. The fitting curve is used to measure the standard organic compound solution and test the accuracy of the fitting curve. The different concentrations of ammonia nitrogen, phosphate, and chloride standard solutions were taken, respectively, and used in the chromaticity measurement system to obtain image information of the standard solution, as shown in Figure 14.

Figure 14a–c shows the image information of different concentrations of ammonia nitrogen, phosphate, and chloride solution. The established concentration measurement curve was used to measure the standard organic compound solution, and the measurement results are shown in Figure 15.

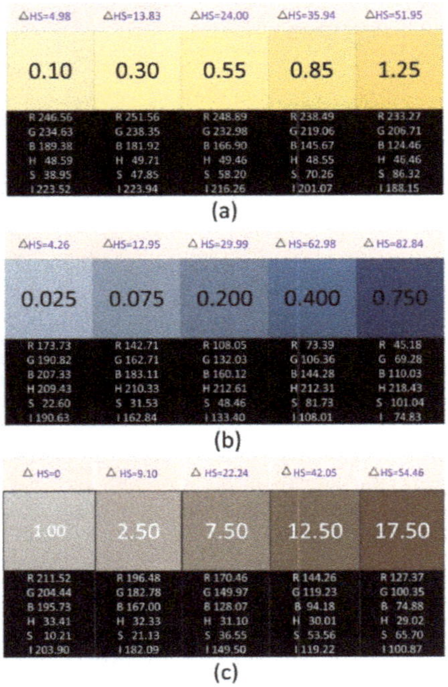

ΔHS=4.98	ΔHS=13.83	ΔHS=24.00	ΔHS=35.94	ΔHS=51.95
0.10	0.30	0.55	0.85	1.25
R 246.56	R 251.56	R 248.89	R 238.49	R 233.27
G 234.63	G 238.35	G 232.98	G 219.06	G 206.71
B 189.38	B 181.92	B 166.90	B 145.67	B 124.46
H 48.59	H 49.71	H 49.46	H 48.55	H 46.46
S 38.95	S 47.85	S 58.20	S 70.26	S 86.32
I 223.52	I 223.94	I 216.26	I 201.07	I 188.15

(a)

ΔHS=4.26	ΔHS=12.95	Δ HS=29.99	Δ HS=62.98	Δ HS=82.84
0.025	0.075	0.200	0.400	0.750
R 173.73	R 142.71	R 108.05	R 73.39	R 45.18
G 190.82	G 162.71	G 132.03	G 106.36	G 69.28
B 207.33	B 183.11	B 160.12	B 144.28	B 110.03
H 209.43	H 210.33	H 212.61	H 212.31	H 218.43
S 22.60	S 31.53	S 48.46	S 81.73	S 101.04
I 190.63	I 162.84	I 133.40	I 108.01	I 74.83

(b)

Δ HS=0	Δ HS=9.10	Δ HS=22.24	Δ HS=42.05	ΔHS=54.46
1.00	2.50	7.50	12.50	17.50
R 211.52	R 196.48	R 170.46	R 144.26	R 127.37
G 204.44	G 182.78	G 149.97	G 119.23	G 100.35
B 195.73	B 167.00	B 128.07	B 94.18	B 74.88
H 33.41	H 32.33	H 31.10	H 30.01	H 29.02
S 10.21	S 21.13	S 36.55	S 53.56	S 65.70
I 203.90	I 182.09	I 149.50	I 119.22	I 100.87

(c)

Figure 14. Image information of standard organic compound solution. (**a**) is the image information of different concentrations of ammonia nitrogen solution. (**b**) is the image information of different concentrations of phosphate solution. (**c**) is the image information of different concentrations of chloride solution.

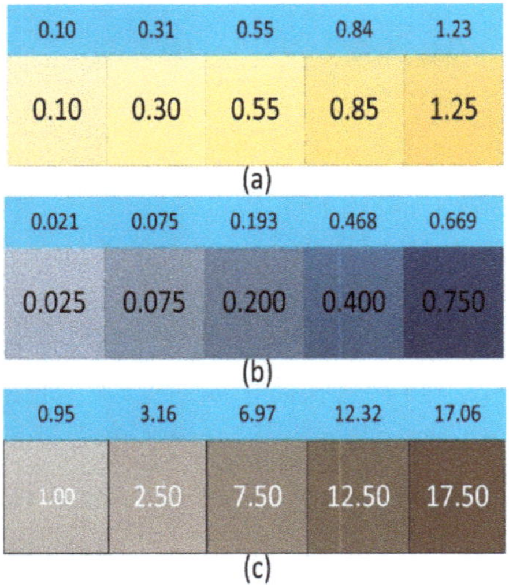

0.10	0.31	0.55	0.84	1.23
0.10	0.30	0.55	0.85	1.25

(a)

0.021	0.075	0.193	0.468	0.669
0.025	0.075	0.200	0.400	0.750

(b)

0.95	3.16	6.97	12.32	17.06
1.00	2.50	7.50	12.50	17.50

(c)

Figure 15. Measurement results of standard solution for ammonia nitrogen, phosphate, and chloride. (**a**) is the measurement results of standard solution for ammonia nitrogen. (**b**) is the measurement results of standard solution for phosphate. (**c**) is the measurement results of standard solution for chloride.

The errors between the measurement results and the standard concentration of solute are shown in Table 5.

Table 5. Comparison of fitted concentration and standard concentration.

Solutes	Maximum Error (mg/L)	Minimum Error (mg/L)	Mean Deviation (mg/L)
Ammonia nitrogen	0.02	0.00	0.008
Phosphate	0.081	0.000	0.031
Chloride	0.66	0.05	0.372

Table 5 shows that the average difference between ammonia nitrogen concentration measured by the fitting curve and standard concentration is 0.008, phosphate concentration is 0.031, and chloride concentration is 0.372. The error between the fitted concentration of solute and the corresponding standard concentration is small, which indicates that the designed color measurement system can be applied to the measurement of solute content in water and has higher accuracy.

4.3. Comparison of Water Quality Detection Methods based on Camera Image

In recent years, a water quality detection method based on the image method has been proposed, mainly based on the ratio and reflection of three primary colors.

M. white et al. [30] found that linear relations between both the ratio of red/green digital output (O/P) values (at a particular camera exposure) and the difference in green–red digital camera O/P were found with the mineral suspended solids (MSS) concentration. A good comparison was also found between the ratios of red/green upwelling light measured with the camera and a conventional irradiance sensor. However, semi-empirical or analytical relationships between camera O/P and the inherent optical properties of the water could not be established. Then, Goddijn [31] proposed a method to measure the content of yellow substance and chlorophyll in water by using a digital camera. The RGB value of an underwater picture was obtained by using a digital camera and the linear relationship between colored dissolved organic matter and the R/B camera output. The log–log relationship between chlorophyll and the G/B camera output was established. These two methods are based on the RGB ratio of the water image.

Based on the water reflection model, a linear relationship between water column radiation and water concentration is established [32] for water quality detection. Based on a similar principle, Thomas Leeuw et al. [33] put forward an App called Hydro Color. The program first obtains three images with cameras, which are used to calculate the red, green, and blue wide band remote sensing reflection. Finally, it estimates the water reflection coefficient (the reflection coefficient can be reversed to estimate the concentration of absorbed and scattered substances in water) and turbidity value according to the remote sensing reflection. These two methods are based on the RGB reflection of water images.

From RGB to HSI color space, a water quality detection method based on the HSI component of a water image was proposed, and a sealed image acquisition device with a constant light source was designed to acquire water image, which avoids the interference of external environment on image information and ensures the accuracy of the measurement. Compared with the traditional optical instrument measurement method, the proposed method has higher accuracy and is effectively applied to the measurement of actual water samples. Compared with the above methods, the proposed method is based on the chromaticity method, which eliminates the influence of brightness, and can be used for the measurement of almost all colored substances in water. Due to a constant light source and closed measurement environment, the proposed method has high stability and practicability and is not affected by the external environment, which can be used as a perfect water quality detection instrument. In the future, we need to study various color components of water, obtain more measurement values, establish an effective and perfect water quality measurement model and apply the method to medicine, food, and other fields.

5. Conclusions

Using a digital camera to acquire a solution image to measure the chromaticity and solute content in water has the advantages of high accuracy, low cost, and strong practicability. It provides a new method for water quality detection. The RGB and HSI values of solution images can be obtained by the designed image acquisition system and processing software. The chromaticity measurement model based on H and S values was established to measure the chromaticity of water and compare it with spectrophotometry. For the measurement of a standard chromaticity solution, the proposed method has higher accuracy than spectrophotometry. For the actual water sample measurement, there is no significant difference between the results of this method and the spectrophotometer method. A curve based on the H and S standard deviation was established. The fitting results accorded with the exponential law, and the fitting degree was more than 0.999. It can be used to measure the solute content in water. This method can be widely used in almost all the determination of trace components based on color reaction, which can replace spectrophotometry without the wavelength setting and is easy to operate. The method is not only suitable for direct color analysis of natural water, but also suitable for wastewater detection. However, wastewater samples must be pretreating, such as digestion, distillation, precipitation, etc., and is preferred to separate and remove interferences.

Author Contributions: Conceptualization, P.C. and S.L.; data curation, S.L.; formal analysis, Y.Z.; funding acquisition, H.G.; investigation, W.Z.; methodology, Y.Z.; project administration, S.L.; resources, P.C. and Y.Z.; software, P.C.; supervision, H.G.; validation, W.Z. and H.G.; visualization, P.C.; writing—original draft, P.C.; Writing—review and editing, S.L.

Funding: This research was funded by the National Natural Science Foundation of China (No. 61671434) and the Natural Science Fund for Colleges and Universities of Anhui Province (No. KJ2017ZD32).

Acknowledgments: The authors would like to thank Hongwen Gao for their assistance on this project.

Conflicts of Interest: The authors declare no conflict of interest.

References

1. Xiao, Y.H.; Räike, A.; Hartikainen, H.; Vähätalo, A.V. Iron as a source of color in river waters. *Sci. Total Environ.* **2015**, *536*, 914–923. [CrossRef] [PubMed]
2. Breneman, I.V.J.; Blasinski, H.; Farrell, J. The Color of Water: Using Underwater Photography to Estimate Water Quality. *Proc. SPIE* **2014**, 90230R. [CrossRef]
3. Wang, Y.; Kidger, T.E.; Tatsuno, K. The water colority measurement based on HSV chromaticity. *Proc. SPIE* **2014**, 1002124. [CrossRef]
4. Jin, X.; Xiang, C.; Qi, D. *A on Industrial Water Treatment Technology*, 3rd ed.; Chemical and Chemical Press: Beijing, China, 2003; pp. 37–38.
5. Grabas, M. Organic matter removal from meat processing wastewater using moving bed biofilm reactors. *Environ. Prot. Eng.* **2000**, *26*, 55–62.
6. ISO. *Clear Liquids—Estimation of Colour by the Platinum-Cobalt Scale—Part 1: Visual Method, ISO 6271-1-2004*; ISO: Geneva, Switzerland, 2004.
7. Wang, A. Study on the determination of water colority. *Environ. Monit. China* **2000**, *16*, 37–40.
8. Zhao, X.; Shen, W. Standard for water color determination based on three-wavelength luminous transmittance. *Chin. J. Environ. Eng.* **2013**, *7*, 4766–4772.
9. Funt, B.; Xiong, W. Estimating Illumination Chromaticity via Support Vector Regression. In Proceedings of the Color and Imaging Conference, Scottsdale, AZ, USA, 1 January 2004.
10. Zeng, F.; Luo, X. Determination of the colority of water samples by spectrophotometry. *Ind. Water Treat.* **2006**, *26*, 69.
11. Peng, P.; Huang, H.; Ren, H.; Ma, H.; Lin, Y.; Geng, J. Exogenous N-acyl homoserine lactones facilitate microbial adhesion of high ammonia nitrogen wastewater on biocarrier surfaces. *Sci. Total Environ.* **2018**, *624*, 1013–1022. [CrossRef]
12. Camargo, J.A.; Alonso, Á.; Puente, M.D.L. Eutrophication downstream from small reservoirs in Mountain Rivers of Central Spain. *Water Res.* **2005**, *39*, 3376–3384. [CrossRef]

13. Bian, X.; Fan, J.; Duan, H. Measurement of cyanogen chloride in drinking water with isonicotinic acid-pyrazolone spectrophotometry. *Mod. Prev. Med.* **2009**, *36*, 3328–3329.
14. Li, G.; Tong, Y.; Li, J. Research on ultraviolet-visible absorption spectrum preprocessing for water quality contamination detection. *Optik* **2018**, *164*, 277–288. [CrossRef]
15. Brkić, B.; Giannoukos, S.; Taylor, S. Mobile mass spectrometry for water quality monitoring of organic species present in nuclear waste ponds. *Anal. Methods UK* **2018**, *10*, 5827–5833. [CrossRef]
16. Charulatha, G.; Srinivasalu, S.; Maheswari, O.U. Evaluation of ground water quality contaminants using linear regression and artificial neural network models. *Arab. J. Geosci.* **2017**, *10*, 128. [CrossRef]
17. Mori, Y.; Hopmans, J.W.; Mortensen, A.P. MultiFunctional Heat Pulse Probe for the Simultaneous Measurement of Soil Water Content, Solute Concentration, and Heat Transport Parameters. *Vadose. Zone J.* **2003**, *2*, 561–571. [CrossRef]
18. Domask, W.G.; Kobe, K.A. Mercurimetric Determination of Chlorides and Water-Soluble Chlorohydrins. *Anal. Chem.* **2002**, *24*, 989–991. [CrossRef]
19. Kónya, J.; Nagy, N.M. Determination of water-soluble phosphate content of soil using heterogeneous exchange reaction with 32 P radioactive tracer. *Soil Till. Res.* **2015**, *150*, 171–179. [CrossRef]
20. Bogoslovskiy, V.V.; Chernova, N.S. Potentiometric determination of the content of water-soluble organic amides. *Fibre Chem.* **2007**, *39*, 73–75. [CrossRef]
21. Gil, S.; Reisin, H.D.; Rodríguez, E.E. Using a digital camera as a measuring device. *Am. J. Phys.* **2006**, *74*, 768–775. [CrossRef]
22. Levin, N.; Benâ Dor, E.; Singer, A. A digital camera as a tool to measure colour indices and related properties of sandy soils in semi-arid environments. *Int. J. Remote Sens.* **2005**, *26*, 5475–5492. [CrossRef]
23. Rossel, R.V.; Fouad, Y.; Walter, C. Using a digital camera to measure soil organic carbon and iron contents. *Biosyst. Eng.* **2008**, *100*, 149–159. [CrossRef]
24. Suzuki, Y.; Endo, M.; Jin, J.; Iwase, K.; Iwatsuki, M. Tristimulus colorimetry using a digital still camera and its application to determination of iron and residual chlorine in water samples. *Anal. Sci.* **2006**, *22*, 411–414. [CrossRef] [PubMed]
25. Hoguane, A.M.; Green, C.L.; Bowers, D.G.; Nordez, S. A note on using a digital camera to measure suspended sediment load in maputo bay, mozambique. *Remote Sens. Lett.* **2012**, *3*, 259–266. [CrossRef]
26. Tang, B.; Sapiro, G.; Caselles, V. Color image enhancement via chromaticity diffusion. *IEEE Trans. Image Process.* **2001**, *10*, 701–707. [CrossRef] [PubMed]
27. Cao, P.; Zhao, W.; Liu, S.; Shi, L.; Gao, H. Using a Digital Camera Combined with Fitting Algorithm and T-S Fuzzy Neural Network to Determine the Turbidity in Water. *IEEE Access* **2019**, *7*, 83589–83599. [CrossRef]
28. Bao, X.; Liu, S.; Song, W. Using a PC camera to determine the concentration of Nitrite, Ammonia Nitrogen, Sulfide, Phosphate, and Copper in Water. *Anal. Methods UK* **2018**. [CrossRef]
29. Scatena, F.N. Drinking water quality. *Drink. Water For.* **2000**, *2000*, 7.
30. White, M.; Feighery, L.; Bowers, D.; O'Riain, G.; Bowyer, P. Using digital cameras for river plume and water quality measurements. *Int. J. Remote Sens.* **2005**, *26*, 4405–4419. [CrossRef]
31. Goddijn, L.M.; White, M. Using a digital camera for water quality measurements in Galway Bay. *Estuar. Coast. Shelf. Sci.* **2006**, *66*, 429–436. [CrossRef]
32. San, L.H.; Jafri, M.M.; Abdullah, K. Aerial Photogrammetry Method for Water Quality Monitoring Using Digital Camera. *Open Environ. Sci.* **2009**, *3*, 20–25. [CrossRef]
33. Leeuw, T.; Boss, E. The HydroColor App: Above Water Measurements of Remote Sensing Reflectance and Turbidity Using a Smartphone Camera. *Sensors* **2018**, *18*, 256. [CrossRef]

Article

Monitoring of Cyanobacteria in Water Using Spectrophotometry and First Derivative of Absorbance

Adogbeji Valentine Agberien [1] and Banu Örmeci [2,*]

[1] Institute of Environmental Science and Interdisciplinary Science, Carleton University, Ottawa, ON K1S 5B6, Canada; valentineagberien@cmail.carleton.ca

[2] Department of Civil and Environmental Engineering, Carleton University, Ottawa, ON K1S 5B6, Canada

* Correspondence: banu.ormeci@carleton.ca

Received: 19 November 2019; Accepted: 20 December 2019; Published: 30 December 2019

Abstract: Management of cyanobacteria blooms and their negative impact on human and ecosystem health requires effective tools for monitoring their concentration in water bodies. This research investigated the potential of derivative spectrophotometry in detection and monitoring of cyanobacteria using toxigenic and non-toxigenic strains of *Microcystis aeruginosa*. *Microcystis aeruginosa* was quantified in deionized water and surface water using traditional spectrophotometry and the first derivative of absorbance. The first derivative of absorbance was effective in improving the signal of traditional spectrophotometry; however, it was not adequate in differentiating between signal and noise at low concentrations. Savitzky-Golay coefficients for first derivative were used to smooth the derivative spectra and improve the correlation between concentration and noise at low concentrations. Derivative spectrophotometry improved the detection limit as much as eight times in deionized water and as much as four times in surface water. The lowest detection limit measured in surface water with traditional spectrophotometry was 392,982 cells/mL, and the Savitzky-Golay first derivative of absorbance was 90,231 cells/mL. The method provided herein provides a promising tool in real-time monitoring of cyanobacteria concentrations and spectrophotometry offers the ability to measure water quality parameters together with cyanobacteria concentrations.

Keywords: cyanobacteria; *Microcystis aeruginosa*; water; monitoring; spectrophotometry; derivative absorbance

1. Introduction

Cyanobacteria have become the recent focus of scientific research due to their impacts on environmental and human health [1,2]. Some species of cyanobacteria form blooms and produce taste and odour compounds, such as Geosmin and 2-methylisoborneol (2-MIB), which negatively affect water quality. A more pressing issue is the capability of these organisms to produce toxins (cyanotoxins) which are detrimental to living organisms [3]. Over the years, there has been an increase in literature on cyanobacteria, cyanobacterial blooms, and cyanotoxins, among which *Microcystis* and its associated hepatotoxin, microcystin, have been the focus. *M. aeruginosa* is the most common toxin-producing species in cyanobacterial blooms, and microcystins are the most widespread toxins [4,5]. Other genera capable of producing microcystins include *Anabaena* (*Dolichospermum*), *Anabaenopsis*, *Nostoc*, and *Oscillatoria* [6].

To ensure public health safety, World Health Organization (WHO) recommends critical control points to reduce exposure to cyanotoxin hazards [7]. The critical control points depend on monitoring schemes that are capable of detecting cyanobacteria proliferation and require sufficient management protocol to mitigate potential environment and human health impacts. Standard monitoring programs involve visual inspection of the water body, monitoring of nutrient concentration, and monitoring of mass developments of cyanobacteria and their toxins in the respective waterbody [7]. The WHO guidelines also state the probability of adverse health effects based on exposure to specific cyanobacteria concentrations in recreational waters (Table 1).

Table 1. World Health Organization guideline values for cyanobacteria in recreational waters.

Cyanobacteria Cell Concentration (cells/ml)	Probability of Adverse Health Effects
<20,000	Relatively mild and/or low
20,000–100,000	Moderate
100,000–10,000,000	High
>10,000,000	Very high

Continuous, real-time information on cyanobacteria concentration and other water quality parameters are of utmost importance to water quality managers. This information facilitates understanding of the impact of point and non-point source pollution on a water body, captures spatial and temporal variability in cyanobacteria concentration, and aids in determining subsequent treatment protocol to be implemented [7]. Real-time information also ensures blooms are managed in their early stages, and as such prevents extensive damage to the water body [3,7–10].

There are several methods for qualitative and quantitative analysis of cyanobacteria in surface waters [11–13]. These methods employ biological, biochemical or physicochemical approaches [11,12]. The challenges include spatial and temporal variability of cyanobacteria in the water body, sampling and analysis time, cost of analytical instrument, efficacy of the analytical technique, and interferences due to both water and cyanobacteria/algal characteristics.

Identification and quantification of cyanobacteria under a microscope is an established method that provides information regarding the species of cyanobacteria present, which can also infer on the possible presence of toxins in the waterbody [12]. However, there are several limitations to this laborious and time-consuming technique, such as the large quantities of variable types of cells, sample preparation and transport, high degree of expertise required for accurate quantification, and more importantly it is not suitable for real-time monitoring.

Fluorometry is the most commonly used method for real-time monitoring of cyanobacteria and typically relies on the detection and quantification of chlorophyll-a [14]. Cyanobacteria species contain light-absorbing pigment, chlorophyll-a, together with variable concentrations of accessory phycobiliprotein pigments phycocyanin (PC) and phycoerythrin (PE) [15]. These pigments play a major role in photosynthesis and determine the external colour of the species [16]. Chlorophyll-a contributes to 0.5–1% of the ash-free dry weight of phytoplanktonic organisms, and as such has set the standard for representation of phytoplankton biomass. Chlorophyll-a does, however, vary depending on the physiological state of the organism, species composition, and growth conditions such as nutrient and sunlight availability [7,17].

Spectrophotometry is used in laboratory studies to quantify the growth of cyanobacteria cultures in broth. Compared to fluorometry, spectrophotometry has a much higher detection limit (lower sensitivity) and therefore it is not employed for monitoring of cyanobacteria in water supplies. Recently, [18] developed a method to determine the concentration of microalgae in water and wastewater using spectrophotometry and the first derivative of absorbance. The detection limits established using the first derivative of absorbance for *C. vulgaris* were 0.47, 0.56, and 1.96 mg TVS (total volatile solids)/L in distilled water, surface water and wastewater, respectively. The results indicated that the first derivative of absorbance can successfully be used for monitoring of microalgae in bioreactors for industrial applications and microalgae-based wastewater treatment systems and potentially for surface water monitoring if detection limits can be improved.

Derivative spectrophotometry is capable of separating overlapping peaks and reducing background noise caused by the presence of other compounds in a sample and can improve the selectivity and sensitivity of classic spectrophotometry [19,20]. Digital (mathematical), electronic, or optical methods can be used to obtain first order or higher order derivatives from zero-order derivatives (original absorbance spectra). Digital methods, however, are more useful than electronic or optical methods because of the ease with which they calculate the derivative spectra, and the flexibility of the method for reanalysis of data (i.e., the derivative spectra can be recalculated with different parameters) [20]. If the zero-order derivative follows Beer-Lambert Law, all orders of derivative will reflect the linear relationship between concentration and absorbance. The improved selectivity and sensitivity associated with derivative spectrophotometry has increased its use in clinical, environmental, and pharmaceutical fields of analysis [20]. It does, however, seem that no research has explored the utilization of derivative spectrophotometry for the detection and quantification of cyanobacteria in water.

Even though spectrophotometry has a lower sensitivity compared to fluorometry for cyanobacteria, it offers other advantages. In-line and real-time UV-vis spectrophotometers are available in the market and are widely used by drinking water treatment plants to measure organic carbon at 254 nm. New generation real-time spectrophotometers have the capability to scan the whole UV-vis range (190–800 nm), measure the concentrations of a wide range of compounds simultaneously, apply algorithms and analyze the data, and transfer the data wirelessly. Furthermore, using chemometrics and advanced statistics, absorbance characteristics of a water sample at different wavelengths can be correlated to water quality parameters that are of interest to decision makers and regulators. These water quality parameters include UV254, dissolved organic carbon (DOC), biochemical oxygen demand (BOD), chemical oxygen demand (COD), total suspended solids, nitrate and nitrite, and color, which are all important in monitoring cyanobacteria blooms. The ability to measure the water quality parameters is important as they directly impact the growth conditions, competition and abundance of cyanobacteria. For example, it has been reported that increases in nitrate loads not only increases the growth rate and toxicity of toxigenic *M. aeruginosa*, but that higher nitrate concentrations favor toxigenic strains over non-toxigenic strains [21,22]. Therefore, if a spectrophotometry-based method that is sensitive enough for monitoring cyanobacteria in surface waters can be developed, only one instrument would be adequate to measure cyanobacteria and the complementary water quality parameters instead of the bundles of instruments that are currently used for monitoring applications. This would reduce the complexity and cost of real-time monitoring, which are the main roadblocks to the wide use of real-time monitoring.

The main objective of this research was to investigate the use of derivative spectrophotometry for monitoring of cyanobacteria and determine the detection limits that can be achieved using first derivative of absorbance and complementary statistical tools, and whether spectrophotometry would be a viable method for cyanobacteria monitoring.

2. Materials and Methods

2.1. Cultivation of Cyanobacteria

Pure cultures of *Microcystis aeruginosa* CPCC 299 (toxigenic strain) and *Microcystis aeruginosa* CPCC 632 (non-toxigenic strain) were used in this study. The strains and the sterile growth media BG-11 and 3N-BBM were purchased from the Canadian Phycological Culture Centre (CPCC) at the University of Waterloo (Waterloo, ON, Canada).

M. aeruginosa CPCC 299 was cultured in BG-11 while *M. aeruginosa* CPCC 632 was cultured in 3N-BBM. Each growth medium was inoculated with 1ml of its respective strain such that the dilution ratio was 1:100. To avoid contamination of cultures, both 250 mL Erlenmeyer flasks (250 mL) used to culture each strain of *M. aeruginosa* were rinsed with deionized (D.I.) water, after which the mouth of each flask was covered in foil paper and autoclaved at 15 psi and 121 °C for 30 min. The flasks were then let to cool at room temperature and labelled according to strain and growth medium before culturing as per dilution ratio aforementioned. The cultures were incubated in a temperature-controlled incubator and maintained at 25 °C under 24-h light intensity of 1800–2500 lux. *M. aeruginosa* CPCC 299 was cultured for 10 days and *M. aeruginosa* CPCC 632 was cultured for 11 days. Each flask was stirred manually twice a day at eight-hour time intervals or more.

2.2. Preparation of Cyanobacteria Samples for Spectrophotometric Analysis

M. aeruginosa cultures were enumerated under a light microscope using a Neubauer chamber (hemocytometer), after which dilutions of the sample were prepared in D.I. water and surface water. The same dilution ratios were prepared in both D.I. water and surface water to make comparisons when quantifying in each water sample. Surface water was taken from the Rideau River (Ottawa, Ontario). After dilution, concentration of samples ranged from 129 cells/mL to 10,288,000 cells/mL for *M. aeruginosa* CPCC 299, and from 88 cells/mL to 7,064,000 cells/mL for *M. aeruginosa* CPCC 632.

2.3. Absorbance Measurements

Spectrophotometric analysis was done using a Cary 100 Bio UV-Vis Spectrophotometer and a 1 cm quartz cuvette. A spectral scan was carried out between 200 nm to 800 nm, however, analysis focused on absorption of light between 400 nm to 800 nm due to the presence of significant peaks. The spectrophotometer was set to baseline subtraction to account for absorption by blanks (water samples), after which absorption spectra for respective growth media were taken and subtracted from final absorbance measurements (growth media and cyanobacteria). Standard calibration curves were plotted at low, medium, and high concentration ranges to validate whether spectral analysis was in agreement with the Beer-Lambert Law. For CPCC 299, the low, medium, and high concentration ranges were (129–803,750 cells/mL), (1,286,000–5,144,000 cells/mL), and (5,144,000–10,288,000 cells/mL), respectively.

2.4. First Derivative Of Absorbance

The first order derivative of absorbance refers to the rate of change in absorbance with respect to wavelength ($\partial A/\partial \lambda$). The first derivative was plotted using changes in absorbance values at 1 nm bandwidth.

2.5. Savitzky-Golay First Derivative of Absorbance

Savitzky-Golay coefficients for first derivative of absorbance were used to smooth the plot of the first derivative [23].

Savitzky-Golay first derivative of absorbance, $A_j = \frac{\sum_{i=-\frac{m-1}{2}}^{\frac{m-1}{2}} C_i A_{j+i}}{N}$ $\left(\frac{m+1}{2} \le j \le n - \frac{m-1}{2}\right)$, where m = number of data points used for smoothing; C_i = filter coefficient; A = observed absorbance value at a specific wavelength; i = the relative position of the convolution coefficient from the smoothed (central) data point; j = the smoothed data point; n = number of data points in absorbance spectra for respective concentrations; N = standardization factor.

Twenty-three data points were used for smoothing such that $m = 23$; $i = -11, -10, \ldots .10, 11$; $N = 1012$ (each m has its respective standardization factor, N, and can be readily obtained from a Savitzky-Golay table of coefficients for first derivate); C_i were fitted according to Savitzky-Golay coefficients for first derivative of absorbance.

2.6. Method Detection Limit

Method Detection Limit (MDL) is the minimum concentration of a compound that can be measured with 99% confidence that the concentration of then compound is greater than zero. The MDLs were calculated using Chemiasoft MDL calculator (www.chemiasoft.com), and a minimum of 7–10 spike replicates were used for the calculation of each MDL.

3. Results

3.1. Absorbance Measurements in Deionized Water

Initial experiments were carried out in contaminant free D.I. water. Absorbance spectra of *M. aeruginosa* CPCC 299 at three concentration ranges resulted in peaks at approximately 445 nm, 620 nm, and 682 nm, with the most obvious peaks at 682 nm. This peak (682 nm) is indicative of the presence of chlorophyll-a. Chlorophyll-a absorbs light at both 440 and 680 nm and phycocyanin absorbs light at 620 nm [15], which explains all the significant peaks observed for *M. aeruginosa* CPCC 299. The absorbance spectra for low (129–803,750 cells/mL) and medium (1,286,000–5,144,000 cells/mL) concentrations are illustrated in Figure 1a,b. The absorbance spectra for high concentrations are not shown, as the absorbance spectra at medium concentrations show that the method can easily detect these concentrations.

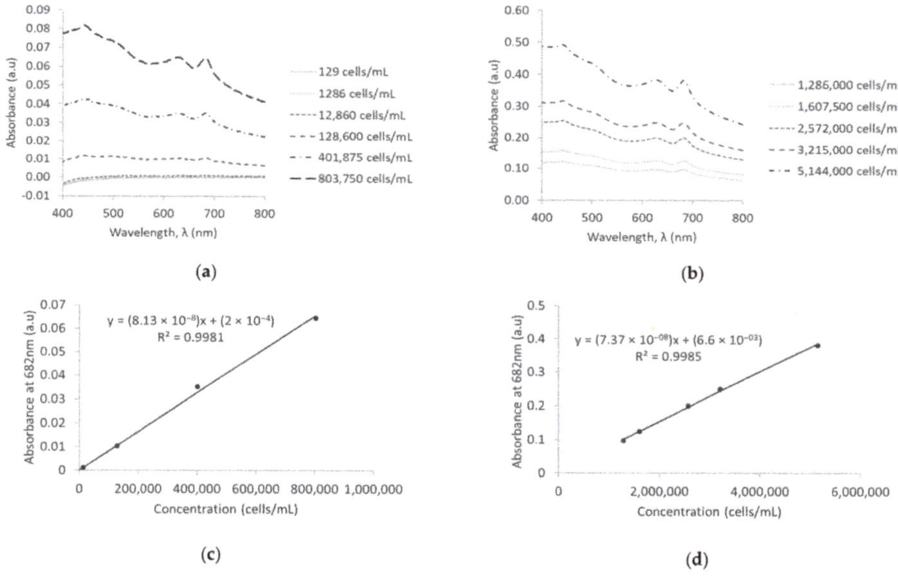

Figure 1. Absorbance spectra of *M. aeruginosa* CPCC 299 at low concentrations (**a**) and medium concentrations (**b**) in D.I. water, and the corresponding calibration curves at 682 nm (**c**,**d**).

Comparison of both spectra indicates an increase in absorbance with cyanobacteria concentration. To validate that spectral analysis was coherent with the Beer-Lambert Law, a standard calibration curve at 682 nm was generated. There was a positive linear trend, and strong correlation between cyanobacteria concentration and absorbance ($R^2 = 0.9981$ at low concentrations, $R^2 = 0.9985$ at medium concentrations) (Figure 1c,d). Absorbance measurements increased linearly from −0.0002 au to 0.7438 au for concentrations ranging from 129 cells/mL to 10,288,000 cells/mL. The presence of negative absorbance values is due to the subtraction to account for the growth medium during the spectral scan. There was also a strong correlation between absorbance and concentration at 620 nm ($R^2 > 0.99$), and the slopes at 620 nm and 682 nm were approximately the same. At 620 nm, the slope for low and medium concentrations were $8.128 \times 10^{-8} \pm 1.532 \times 10^{-9}$ au/(cells/mL) and $7.366 \times 10^{-8} \pm 1.482 \times 10^{-9}$ au/(cells/mL), respectively; at 682 nm, the slope for low and medium concentrations were $8.132 \times 10^{-8} \pm 1.753 \times 10^{-9}$ au/(cells/mL) and $7.369 \times 10^{-8} \pm 1.675 \times 10^{-9}$ au/(cells/mL) respectively. However, as illustrated in Figure 1a,b, the peaks at 620 nm are not readily distinguished from background noise (broad and blunt peaks), and at concentrations ≤128,600 cells/mL, the peaks at 682 nm are not readily observed. The method detection limit for the zero-order derivative at 682 nm was calculated to be 338,950 cells/mL in D.I. water based on the concentration increments tested.

3.2. First Derivative of Absorbance in D.I. Water

A plot of the first derivative of absorbance (y-axis) against concentration (x-axis) reveals over twice the quantity of peaks indicated on the zero-order absorbance plot. This is a characteristic of first order derivatives and is comprehensively explained by [19]. Simply, a first order derivative passes through zero at the original point where the maximum absorbance value was observed. Flanking this zero point are positive and negative bands with maximum and minimum at the same wavelengths as the inflection point (location of rapid change in direction). As illustrated in Figure 2a, the signal is not readily detected at low concentrations due to the presence of significant noise. This unwanted effect as explained by reference [20] is due to rapid, random changes of noise amplitude in the spectrum. Due to the aforementioned effect of noise, there is a relatively good correlation between the first derivative of absorbance and low concentrations of *M. aeruginosa* (R^2 = 0.8601 at λ = 694 nm) (Figure 2c). However, at medium concentrations, the peaks are readily detected, despite the presence of noise through the spectral scan (Figure 2b). There is also a stronger correlation between the first derivative of absorbance and concentration at medium concentrations than at low concentrations (R^2 = 0.9782 at λ = 694 nm) (Figure 2d). There is a negative linear relationship between the first derivative of absorbance at 694 nm and concentration, a trend which is due to the utilization of negative bands. By taking the absolute value of absorbance at locations where peaks are observed, λ_{max} is observed at 694 nm (a region of negative bands). Comparison of the zero-order derivative absorbance spectra and first derivative of absorbance spectra indicates that response measurements (absorbance values) are greater for zero-order derivatives than for first derivative of absorbance, yet the peaks for first derivative of absorbance are more easily observed than that of zero-order derivative of absorbance. This characteristic exists because the first derivative of absorbance is a plot of the rate of change of absorbance, as opposed to absorbance, against wavelength.

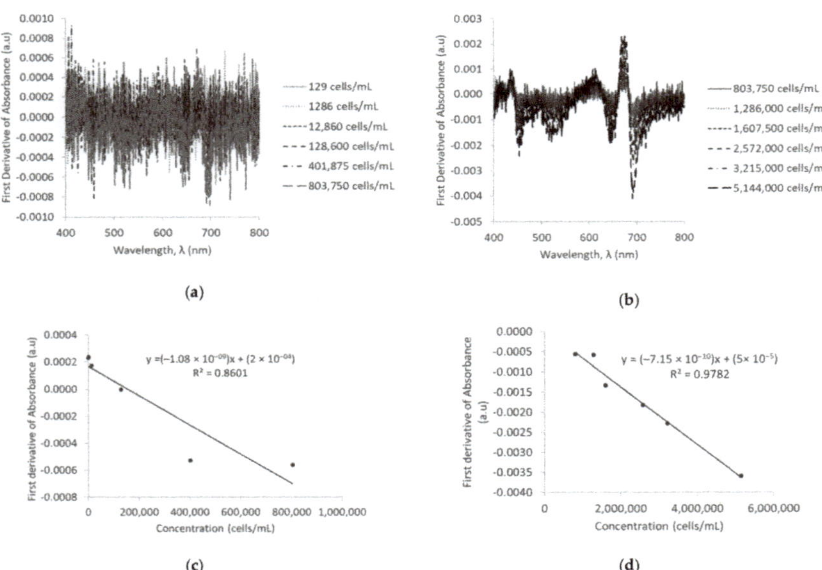

Figure 2. First derivative of absorbance spectra of *M. aeruginosa* CPCC 299 at low concentrations (**a**) and medium concentrations (**b**) in D.I. water, and the corresponding calibration curves at 694 nm (**c**,**d**).

3.3. Savitzky-Golay First Derivative of Absorbance in D.I. Water

Due to the presence of significant noise when utilizing the first derivative of absorbance, Savitzky-Golay coefficients for first derivate were used to smoothen the spectra of the first derivative. Utilization of a digital filter such as the Savitzky-Golay polynomial smoothening algorithm [23] can substantially improve the sensitivity and detection limit of the method.

Smoothening of the data using Savitzky-Golay coefficients for the first derivative resulted in sharp and distinct peaks at low, medium, and high concentrations. However, similar to the zero-order derivative spectra, the peaks for 129 cells/mL, 1286 cells/mL, and 12,860 cells/mL were not identified on the spectral scan. The Savitzky-Golay first derivative of absorbance spectra for low and medium concentrations are illustrated in Figure 3a,b, respectively. The standard calibration curve of concentration versus Savitzky-Golay first derivative of absorbance revealed a linear trend and strong positive correlation between the Savitzky-Golay first derivative of absorbance and concentration (R^2 = 0.9948 at low concentrations; R^2 = 0.9994 at medium concentrations; R^2 = 0.9999 at high concentrations). The standard calibration curves of low and medium concentrations are illustrated in Figure 3c,d. The method detection limit using the Savitzky-Golay first derivative of absorbance was calculated to be as low as 41,802 cells/mL. This improvement in detection limit was approximately eight times that of zero-order derivative of absorbance.

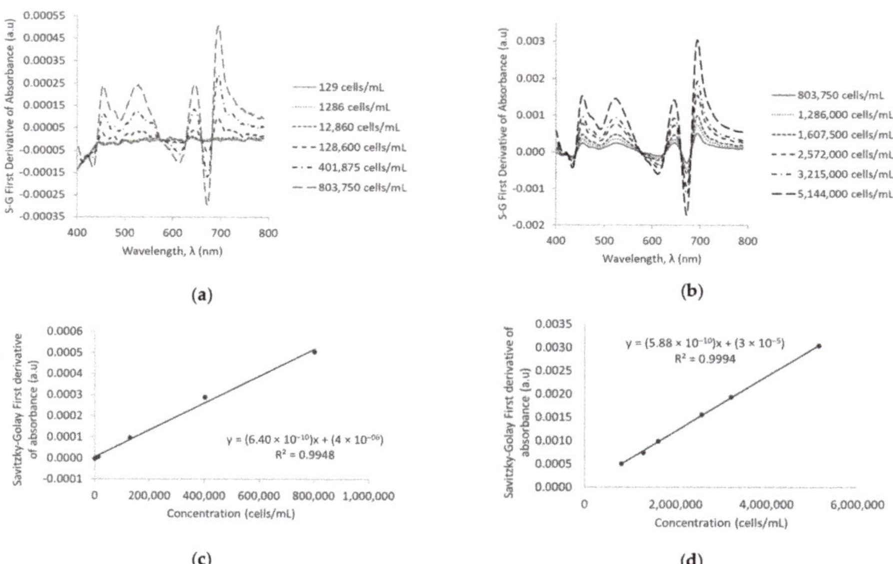

Figure 3. Savitzky-Golay first derivative of absorbance spectra of *M. aeruginosa* CPCC 299 at low concentrations (**a**) and medium concentrations (**b**) in D.I. water, and the corresponding calibration curves at 694 nm (**c**,**d**).

3.4. Absorbance Measurements in Surface Water

Initial experiments were carried out in D.I. water to evaluate the capability of the method in a contaminant-free water sample. The overall goal, however, was to determine the viability of derivative spectrophotometry in surface water, which is more challenging due to the presence of organic and inorganic contaminants. Therefore, measurements were repeated in surface water to determine the effectiveness of the method under more realistic conditions. The objective was not to study the impact of each water quality parameter on the performance of the method and the detection limits. The same *M.*

aerugionsa culture and dilutions were used in both the D.I. and surface water experiments, and as a result water characteristics were the primary factors resulting in differences in absorbance measurements.

Similar to measurements in D.I. water, the absorbance spectra of all concentration ranges of *M. aeruginosa* CPCC 299 resulted in peaks at 445 nm, 620 nm, and 682 nm (Figure 4). The sharpest peaks were observed at 682 nm; however, peaks were not readily identified at concentrations < 401,875 cells/mL. The absorbance spectra of low and medium concentrations are illustrated in Figure 4a,b, respectively. Analysis at 682 nm revealed a linear trend and strong correlation between absorbance and concentration at all concentration ranges (R^2 = 0.9994 at low concentrations; R^2 = 0.9993 at medium concentrations; R^2 = 0.9980 at high concentrations). The standard calibration curve of low and medium concentrations is illustrated in Figure 4c,d, respectively. The detection limit in surface water was poorer than in deionized water (i.e., 392,982 cells/mL in surface water compared to 338,950 cells/mL in D.I.), a response which was expected due to interferences from impurities.

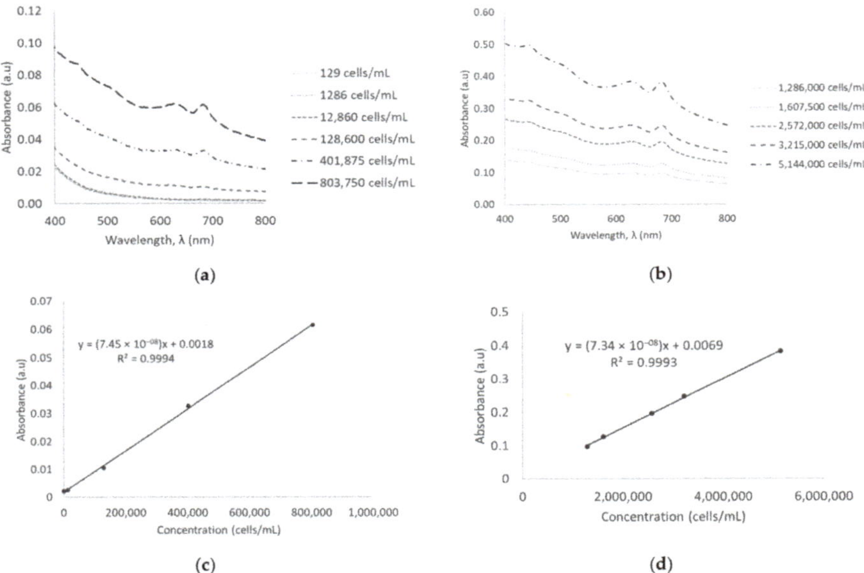

Figure 4. Absorbance spectra of *M. aeruginosa* CPCC 299 at low concentrations (**a**) and medium concentrations (**b**) in surface water, and the corresponding calibration curves at 682 nm (**c**,**d**).

3.5. First Derivative of Absorbance and Savitzky-Golay First Derivative of Absorbance in Surface Water

As was expected, due to the sharpness and rapid change in noise, the peaks were not readily identified using the first derivative of absorbance at low concentrations (Figure 5a). However, the Savitzky-Golay coefficients for first derivative of absorbance were effective in improving the signal-to-noise ratio and facilitated distinct identification of peaks (Figure 5b). The location of peaks was the same for both quantification in D.I. water and surface water, with the most obvious peak still observed at 694 nm. As was the case in D.I. water, there was still a strong linear relationship between the Savitzky-Golay first derivative of absorbance at 694 nm and all concentration ranges of *M. aeruginosa* (R^2 = 0.9989 at low concentrations; R^2 = 0.9990 at medium concentrations; R^2 = 0.9977 at high concentrations). The standard calibration curves of low concentrations versus first derivative and Savitzky-Golay first derivative of absorbance are illustrated in Figure 5c,d, respectively. The method detection limit was calculated to be as low as 90,231 cells/mL and is approximately four times an improvement in detection compared to zero-order derivative measurements.

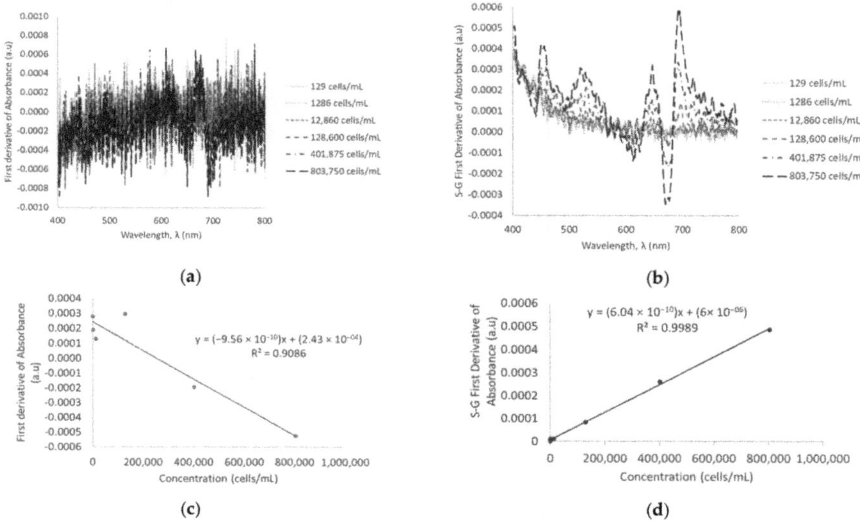

Figure 5. First derivative (**a**) and Savitzky-Golay first derivative (**b**) of absorbance spectra at medium concentrations in surface water (Rideau River), and the corresponding calibration curves at 694 nm (**c**,**d**).

A summary of all results obtained from the above experiments including the detection limits are presented in Table 2.

Table 2. Critical data for *Microcystis aeruginosa* CPCC 299 in both deionized water and surface water in low (129–803,750 cells/mL), medium (1,286,000–5,144,000 cells/mL), and high (5,144,000–10,288,000 cells/mL) concentration ranges.

Water Sample	Test	Concentration Range (cells/mL)	Slope	R^2	Detection Limit (cells/mL)
Deionized water	Absorbance	129–803,750	$8.13208 \times 10^{-08} \pm 1.75287 \times 10^{-09}$	0.9981	338,950
		1,286,000–5,144,000	$7.36864 \times 10^{-08} \pm 1.67453 \times 10^{-09}$	0.9985	
		5,144,000–10,288,000	$6.99701 \times 10^{-08} \pm 8.81363 \times 10^{-10}$	0.9997	
	First derivative of absorbance	129–803,750	$-1.07539 \times 10^{-09} \pm 2.16885 \times 10^{-10}$	0.8601	
		1,286,000–5,144,000	$-7.14544 \times 10^{-10} \pm 5.33754 \times 10^{-11}$	0.9782	
		5,144,000–10,288,000	$-7.50473 \times 10^{-10} \pm 1.23489 \times 10^{-11}$	0.9995	
	Savitzky-Golay first derivative of absorbance	129–803,750	$6.39754 \times 10^{-10} \pm 2.32216 \times 10^{-11}$	0.9948	41,802
		1,286,000–5,144,000	$5.88213 \times 10^{-10} \pm 7.20283 \times 10^{-12}$	0.9994	
		5,144,000–10,288,000	$6.26434 \times 10^{-10} \pm 3.44903 \times 10^{-12}$	0.9999	

Table 2. *Cont.*

Water Sample	Test	Concentration Range (cells/mL)	Slope	R^2	Detection Limit (cells/mL)
Surface water	Absorbance	129–803,750	$7.45411 \times 10^{-08} \pm 9.27447 \times 10^{-10}$	0.9994	392,982
		1,286,000–5,144,000	$7.33732 \times 10^{-08} \pm 1.1411 \times 10^{-09}$	0.9993	
		5,144,000–10,288,000	$7.37948 \times 10^{-08} \pm 2.32234 \times 10^{-09}$	0.9980	
	First derivative of absorbance	129–803,750	$-9.5609 \times 10^{-10} \pm 1.51611 \times 10^{-10}$	0.9086	
		1,286,000–5,144,000	$-7.3683 \times 10^{-10} \pm 3.83158 \times 10^{-11}$	0.9920	
		5,144,000–10,288,000	$-7.3868 \times 10^{-10} \pm 2.82227 \times 10^{-11}$	0.9971	
	Savitzky-Golay first derivative of absorbance	129–803,750	$6.04214 \times 10^{-10} \pm 1.03161 \times 10^{-11}$	0.9989	90,231
		1,286,000–5,144,000	$5.8526 \times 10^{-10} \pm 1.0753 \times 10^{-11}$	0.9990	
		5,144,000–10,288,000	$6.20429 \times 10^{-10} \pm 2.0855 \times 10^{-11}$	0.9977	

3.6. Comparison of *M. aeruginosa* CPCC 299 and *M. aeruginosa* CPCC 632 Spectra

The next phase investigated potential differences in the absorbance spectra of toxigenic (*M. aeruginosa* CPCC 299) and non-toxigenic (*M. aeruginosa* CPCC 632) strains to determine whether there would be any significant differences in their absorbance spectra and whether they could be differentiated based on their absorbance characteristics. *M. aeruginosa* CPCC 632 and *M. aeruginosa* CPCC 299 have identical morphologies and there are no specific morphological differences between the toxigenic and non-toxigenic strains of *Microcystis* spp. [24]. However, the presence of toxins may impact the absorbance measurements due to the physical and chemical changes they induce in the water matrix. In addition, [25] reported that the presence of environmentally relevant concentrations of microcystin increased the size of *Microcystis* spp. colonies 2.7 times in six days, and this aggregation behavior can also impact the absorbance spectra. Results indicated that there was no significant differences in the absorbance spectra of *M. aeruginosa* CPCC 632 and *M. aeruginosa* CPCC 299 at similar concentrations (Figure 6a,b), indicating that spectrophotometry would not be able to differentiate them.

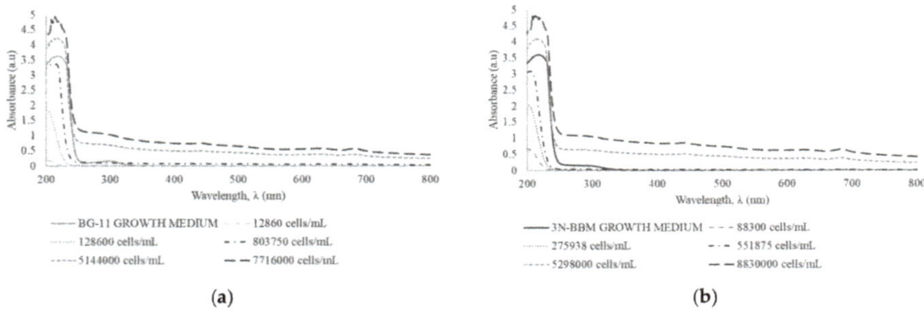

Figure 6. Absorbance spectra of toxigenic *M. aeruginosa* CPCC 299 (**a**) and non-toxigenic CPCC 632 (**b**) in D.I water.

4. Discussion

The results presented in this study show that spectrophotometry can be a valuable tool for monitoring of cyanobacteria in surface waters. The detection limit of spectrophotometry alone is not very sensitive, but the detection limit can be greatly improved by employing mathematical and statistical tools such as the Savitzky-Golay first derivative of absorbance as shown in this study. The detection limit for *M. aeruginosa* CPCC 299 in surface water was approximately 393,000 cells/mL using absorbance measurements and was reduced to approximately 90,000 cells/mL with the application of Savitzky-Golay first derivative of absorbance.

Although the method presented in this paper sufficiently improved the method detection limit both in D.I. water and surface water, it was not able to improve the detection limit to a range of low probability of adverse health effects (<20,000 cells/mL) established by the WHO guidelines (Table 1). However, the detection limit can further be improved using different pathlengths for absorbance measurements, employing signal processing and mathematical tools that should be investigated in future research. The main advantage of using spectrophotometry is its simplicity, real-time capability, and ability to measure a wide range of water quality parameters (e.g., UV254, DOC, BOD, COD, turbidity, nitrate, and nitrite) simultaneously together with the cyanobacteria concentrations. This is particularly important since several studies have reported the impact of water quality and nutrient concentrations on the growth and numbers of cyanobacteria and on the competition and dominance of the toxigenic and non-toxigenic strains. No sample processing, treatment with reagents, or pigment extraction are required for the method and there is a well-established market (e.g., RealTech Inc. UV-VIS sensors) for real-time spectrophotometers that can scan the complete visible and ultraviolet range and analyze results in real-time.

Spectrophotometry neither has the low detection limits nor the specificity of fluorometry that are its main disadvantages. Fluorometric scans have both an excitation and emission spectra, and each compound has its unique fluorometric spectra commonly referred to as its fluorescence signature. Because of these reasons, fluorometry is a more sensitive tool with higher specificity in monitoring of cyanobacteria compared to spectrophotometry. Fluorometry is capable of monitoring cyanobacteria concentrations within WHO management thresholds, however, the maximum attainable concentration for all fluorometric probe technology is 200,000 cells/mL, above this threshold, this tool is insufficient for quantification of cyanobacteria [13]. The more concentrated the water body, as is the case with highly eutrophic lakes, the more likely fluorometric probes are to underestimate measurements. In comparison, spectrophotometry is sufficient for quantification over a wider concentration range [26].

Many of the interferences that hinder the applicability of fluorometry also interfere with spectrophotometric measurements. For example, spectrophotometry and fluorometry were both subject to interference from pigments such as chlorophyll b, c, and pheophytin [27]. This is due to the overlapping spectra caused by other pigments that absorb light and fluoresce at the same wavelength as chlorophyll-a. In addition, the presence of other phytoplankton communities would impact and interfere with the measurements as well as water quality parameters such as turbidity and light scattering particles [13].

5. Conclusions

This study presents a spectrophotometry-based method as a promising tool for real-time detection and monitoring of cyanobacteria. The first derivative of absorbance improved the differentiation between signal and noise; however, it was not effective in improving the signal-to-noise ratio at cyanobacteria concentrations lower than 800,000 cells/mL. Smoothing of the first derivative of absorbance using Savitzky-Golay first derivative of absorbance resulted in narrower, sharper, and more distinct peaks and substantially improved the detection limits both in deionized water and surface water. The lowest detection limit measured in surface water with traditional spectrophotometry was 392,982 cells/mL and with the Savitzky-Golay first derivative of absorbance was 90,231 cells/mL. Detection limits were lower in deionized water, and 338,950 cells/mL and 41,802 cells/mL were achieved with traditional spectrophotometry and the Savitzky-Golay first derivative of absorbance, respectively. The main advantage of a spectrophotometry-based method is its simplicity, real-time capability, and ability to measure a wide range of water quality parameters (e.g., UV254, DOC, BOD, COD, turbidity, nitrate, and nitrite) together with the cyanobacteria concentrations without the need for reagents or processing times. Further research is recommended to bring down the detection limits to lower than 20,000 cells/mL based on the WHO guideline for low probability of adverse health effects.

Author Contributions: Conceptualization, B.Ö.; methodology, B.Ö. and A.V.A.; validation, A.V.A.; formal analysis, A.V.A.; investigation, B.Ö. and A.V.A.; writing—original draft preparation, A.V.A.; writing—review and editing, B.Ö.; supervision, B.Ö.; funding acquisition, B.Ö. All authors have read and agreed to the published version of the manuscript.

Funding: This research was funded by the Natural Sciences and Engineering Research Council (NSERC) under the Strategic Partnership Grants program (STPGP 463663-14).

Conflicts of Interest: The authors declare no conflict of interest.

References

1. Falconer, I.R. An Overview of problems caused by toxic blue–green algae (cyanobacteria) in drinking and recreational water. *Environ. Toxicol.* **1999**, *14*, 5–12. [CrossRef]
2. Carmichael, W.W.; Boyer, G.L. Health impacts from cyanobacteria harmful algae blooms: Implications for the North American Great Lakes. *Harmful Algae* **2016**, *54*, 194–212. [CrossRef]
3. Kenefick, S.L.; Hrudey, S.E.; Prepas, E.E.; Motkosky, N.; Peterson, H.G. Odorous Substances and Cyanobacterial Toxins in Prairie Drinking Water Sources. *Water Sci. Technol.* **1992**, *25*, 147–154. [CrossRef]
4. Sivonen, K.; Niemelä, S.I.; Niemi, R.M.; Lepistö, T.; Luoma, T.H.; Räsänen, L.A. Toxic cyanobacteria (blue-green algae) in Finnish fresh and coastal waters. *Hydrobiologia* **1990**, *190*, 267–275. [CrossRef]
5. Bláha, L.; Babica, P.; Maršalek, B. Toxins produced in cyanobacterial water blooms-toxicity and risks. *Interdiscip. Toxicol.* **2009**, *2*, 36–41. [CrossRef] [PubMed]
6. Kaebernick, M.; Neilan, B.A. Ecological and molecular investigations of cyanotoxin production. *FEMS Microbiol. Ecol.* **2001**, *35*, 1–9. [CrossRef]
7. Chorus, I.; Bartram, J. *Toxic Cyanobacteria in Water: A Guide to Their Public Health Consequences, Monitoring and Management*; E & FN Spon, Published on Behalf of the World Health Organization: New York, NY, USA, 1999.
8. Bullerjahn, G.S.; McKay, R.M.; Davis, T.W.; Baker, D.B.; Boyer, G.L.; D'Anglada, L.V.; Doucette, G.J.; Ho, J.C.; Irwin, E.G.; Kling, C.L.; et al. Global solutions to regional problems: Collecting global expertise to address the problem of harmful cyanobacterial blooms. A Lake Erie case study. *Harmful Algae* **2016**, *54*, 223–238. [CrossRef]
9. Chaffin, J.D.; Kane, D.D.; Stanislawczyk, K.; Parker, E.M. Accuracy of data buoys for measurement of cyanobacteria, chlorophyll, and turbidity in a large lake (Lake Erie, North America): Implications for estimati.on of cyanobacterial bloom parameters from water quality sonde measurements. *Environ. Sci. Pollut. Res.* **2018**, *25*, 25175–25189. [CrossRef]
10. Read, J.; Klump, V.; Johengen, T.; Schwab, D.; Paige, K.; Eddy, S.; Anderson, E.; Manninen, C. Working in freshwater: The Great Lakes observing system contributions to regional and national observations, data infrastructure, and decision support. *Mar. Technol. Soc. J.* **2010**, *44*, 84–98. [CrossRef]
11. Srivastava, A.; Singh, S.; Ahn, C.-Y.; Oh, H.-M.; Asthana, R.K. Monitoring Approaches for a Toxic Cyanobacterial Bloom. *Environ. Sci. Technol.* **2013**, *47*, 8999–9013. [CrossRef]
12. Moreira, C.; Ramos, V.; Azevedo, J.; Vasconcelos, V. Methods to detect cyanobacteria and their toxins in the environment. *Appl. Microbiol. Biotechnol.* **2014**, *98*, 8073–8082. [CrossRef] [PubMed]
13. Zamyadi, A.; Choo, F.; Newcombe, G.; Stuetz, R. A review of monitoring technologies for real-time management of cyanobacteria: Recent advances and future direction. *Trends Anal. Chem.* **2016**, *85*, 83–96. [CrossRef]
14. Falkowski, P.; Kiefer, D.A. Chlorophyll a fluorescence in phytoplankton: Relationship to photosynthesis and biomass. *J. Plankton Res.* **1985**, *7*, 715–731. [CrossRef]
15. Gan, F.; Shen, G.; Bryant, D.A. Occurrence of Far-Red Light Photoacclimation (FaRLiP) in Diverse Cyanobacteria. *Life* **2015**, *5*, 4–24. [CrossRef] [PubMed]
16. Kulasooriya, S.A. Cyanobacteria: Pioneers of Planet Earth. *Ceylon J. Sci. (Bio. Sci.)* **2011**, *40*, 71–88. [CrossRef]

17. Conroy, J.D.; Kane, D.D.; Dolan, D.M.; Edwards, W.J.; Charlton, M.N.; Culver, D.A. Temporal trends in Lake Erie plankton biomass: Roles of external phosphorus loading and Dreissenid mussels. *J. Great Lakes Res.* **2005**, *31* (Suppl. S2), 89–110. [CrossRef]
18. Almomani, F.A.; Örmeci, B. Monitoring and measurement of microalgae using the first derivative of absorbance and comparison with chlorophyll extraction method. *Environ. Monit. Assess.* **2018**, *190*, 90. [CrossRef]
19. Kuś, S.; Marczenko, Z.; Obarski, N. Derivative UV-VIS Spectrophotometry in Analytical Chemistry. *Chem. Anal. (Wars.)* **1996**, *41*, 899–927.
20. Patel, K.N.; Patel, J.K.; Rajput, G.C.; Rajgor, N.B. Derivative spectrophotometry method for chemical analysis: A review. *Sch. Res. Libr.* **2010**, *2*, 139–150.
21. Beversdorf, L.J.; Miller, T.R.; McMahon, K.D. The Role of Nitrogen Fixation in Cyanobacterial Bloom Toxicity in a Temperate, Eutrophic Lake. *PLoS ONE* **2013**, *8*, e56103. [CrossRef]
22. Davis, T.W.; Berry, D.L.; Boyer, G.L.; Gobler, C.J. The effects of temperature and nutrients on the growth and dynamics of toxic and non-toxic strains of microcystis during cyanobacteria blooms. *Harmful Algae* **2009**, *8*, 715–725. [CrossRef]
23. Savitzky, A.; Golay, M.J.E. Smoothing and Differentiation of Data by Simplified Least Squares. *Anal. Chem.* **1964**, *36*, 1627–1639. [CrossRef]
24. Fastner, J.; Erhard, M.; von Döhren, H. Determination of oligopeptide diversity within a natural population of Microcystis spp. (Cyanobacteria) by typing single colonies by matrix-assisted laser desorption ionization-time of flight mass spectrometry. *Appl. Environ. Microbiol.* **2001**, *67*, 5069–5076. [CrossRef] [PubMed]
25. Gan, N.; Xiao, Y.; Zhu, L.; Wu, Z.; Liu, J.; Hu, C.; Song, L. The role of microcystins in maintaining colonies of bloom forming Microcystis spp. *Environ. Microbiol.* **2012**, *14*, 730–742. [CrossRef] [PubMed]
26. Sobiechowska-Sasim, M.; Stoń-Egiert, J.; Kosakowska, A. Quantitative analysis of extracted phycobilin pigments in cyanobacteria-an assessment of spectrophotometric and spectrofluorometric methods. *J. Appl. Phycol.* **2014**, *26*, 2065–2074. [CrossRef]
27. Dos Santos, A.C.; Calijuri, M.C.; Moraes, E.M.; Adorno, M.A.; Falco, P.B.; Carvalho, D.P.; Deberdt, G.L.; Benassi, S.F. Comparison of three methods for Chlorophyll determination: Spectrophotometry and Fluorimetry in samples containing pigment mixtures and spectrophotometry in samples with separate pigments through High Performance Liquid Chromatography. *Acta Limnol. Bras.* **2003**, *15*, 7–18.

Article

Intelligent Wide-Area Water Quality Monitoring and Analysis System Exploiting Unmanned Surface Vehicles and Ensemble Learning

Huiru Cao [1], Zhongwei Guo [2], Shian Wang [1,*], Haixiu Cheng [2,3] and Choujun Zhan [4]

[1] Department of Information Engineering, Guangzhou Institute of Technology, Guangzhou 510725, China; xiaocao0924@163.com

[2] School of Electronical and Computer Engineering, Nanfang College of Sun Yat-sen University, Guangzhou 510970, China; bombax0613@163.com (Z.G.); cshx.cheng@mail.scut.edu.cn (H.C.)

[3] School of Computer Science and Engineering, South China University of Technology, Guangzhou 510641, China

[4] School of Computer, South China Normal University, Guangzhou 510631, China; choujun_zhan@163.com

* Correspondence: wangshian2004@163.com

Received: 27 December 2019; Accepted: 25 February 2020; Published: 2 March 2020

Abstract: Water environment pollution is an acute problem, especially in developing countries, so water quality monitoring is crucial for water protection. This paper presents an intelligent three-dimensional wide-area water quality monitoring and online analysis system. The proposed system is composed of an automatic cruise intelligent unmanned surface vehicle (USV), a water quality monitoring system (WQMS), and a water quality analysis algorithm. An automatic positioning cruising system is constructed for the USV. The WQMS consists of a series of low-power water quality detecting sensors and a lifting device that can collect the water quality monitoring data at different water depths. These data are analyzed by the proposed water quality analysis algorithm based on the ensemble learning method to estimate the water quality level. Then, a real experiment is conducted in a lake to verify the feasibility of the proposed design. The experimental results obtained in real application demonstrate good performance and feasibility of the proposed monitoring system.

Keywords: unmanned surface vehicle; water monitoring; ensemble learning; dynamic power management

1. Introduction

High-quality water supply is essential for human survival [1,2]. Therefore, water protection has been a hot topic in academic and industrial domains [3,4]. With the rapid development of industry and urbanization, industrial and sanitary sewage has severely affected fresh-water sources worldwide, especially in developing countries, and thus has significantly influenced the living conditions of human beings [5]. In water protection, water quality monitoring is a key task. Therefore, in smart cities, it is extremely important to monitor the quality of water resources effectively [6,7].

In recent years, the evolution of high-resolution sensors and Internet of Things (IoT) technologies has significantly improved water quality monitoring technologies [8–10]. Various WQMSs have been designed to solve the problems related to the water quality monitoring of lakes [11,12], rivers [13,14], and groundwater [15,16]. Generally, water quality can be accurately determined by laboratory analysis, but spatial resolution and rapid assessment cannot be obtained with efficiency, simultaneously.

Significant progress has been made in the water quality monitoring field by the introduction of the IoT [17] and unmanned techniques [18,19] that can reduce the monitoring cost and improve the intelligence of the WQMS [20,21]. On the one hand, the new measurement technologies, such as ship-borne measurements [22] and IoT system [23], have been used to monitor water quality. Particularly, [24,25], an unmanned surface vehicle (USV) was used to monitor water quality. On the other hand, various water analysis methods have been used to improve water quality estimation systems [26–28]. However, there are still many challenges, such as those related to low cost, high efficiency, and good real-time performance. Motivated by these studies, in our work we integrate the USVs into water quality monitoring at different depths.

In this work, an intelligent three-dimensional wide-area water monitoring and analysis system and a dynamic energy-saving system are proposed. For the purpose of convenient reference, acronyms used in this paper are given in Table 1. The conceptual diagram of the proposed design is shown in Figure 1. Most of the traditional water quality monitoring devices can obtain only the water quality parameters related to the water surface [29,30]. However, due to the sewage spreading problem, the traditional methods have the shortcoming of collecting water quality monitoring data with limited precision; instead, water quality should preferably be monitored at different depth levels. Therefore, a device that can realize three-dimensional water quality monitoring at different depth levels is designed in this work. In the proposed system, a sensor node collects the water quality data and uploads it to the cloud, and then a water quality monitoring model based on ensemble learning is used to determine the water quality level.

Table 1. Acronyms used in this paper.

Acronym	Description	Acronym	Description
ADC	Analog digital converter	PVC	Polyvinyl chloride
APCS	Automatic positioning cruise system	ROC	Receiver operating characteristic
AUC	Area under the curve	SPI	Serial Peripheral Interface
DC	Direct current	SVM	Support Vector Machine
GPRS	General packet radio service	TDS	Total dissolved solids
GPS	Global position system	UART	Universal asynchronous receiver/transmitter
I/O	Input/output	USV	Unmanned surface vehicle
IoT	Internet of Things	WQMM	Water quality monitoring module
MCU	Micro control unit	WQMS	Water quality monitoring system

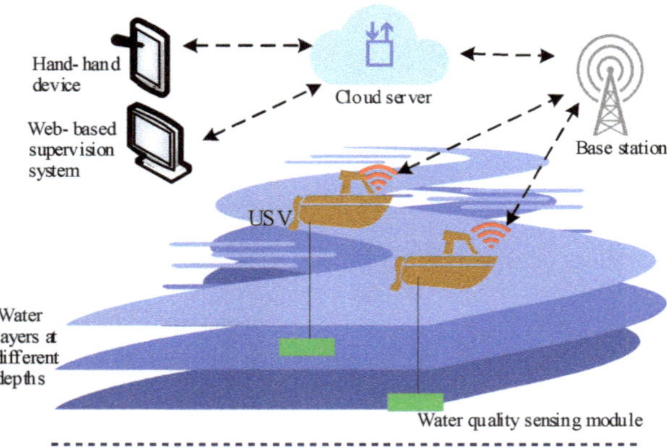

Figure 1. The overview of the proposed three-dimensional water quality monitoring system. USV, unmanned surface vehicle.

The rest of the paper is organized as follows. Section 2 presents the overall design of the USV and automatic cruising strategy. Section 3 describes the proposed water quality monitoring system. Section 4 introduces the water quality analysis algorithm. The experimental results of the proposed system are presented in Section 5. Section 6 concludes the paper.

2. USV Design

This section introduces the architecture design of the proposed USV and an automatic positioning cruise system.

2.1. USV Architecture

The overall architecture design of the USV is shown in Figure 2. The USV weighs 5 kg, and its length, width, and height are 0.8 m, 1.8 m, and 0.6 m, respectively. The draft of the vehicle is 0.1 m. Meanwhile, the catamaran is equipped with two brushless DC (direct current) motors, which can provide thrust from 1 kg to 2 kg. The hull structure of the USV adopts the structure based on the catamaran. Both hulls adopt a pontoon made of a PVC (polyvinyl chloride) mesh cloth, and the connecting bridge has a PVC composite frame. The frame of the connecting bridge is covered with a plexiglass panel to prevent damages to the system components caused by external factors. The entire USV system mainly consists of three parts: an MCU (micro control unit), an automatic positioning cruise system, and a wireless communication transmission module.

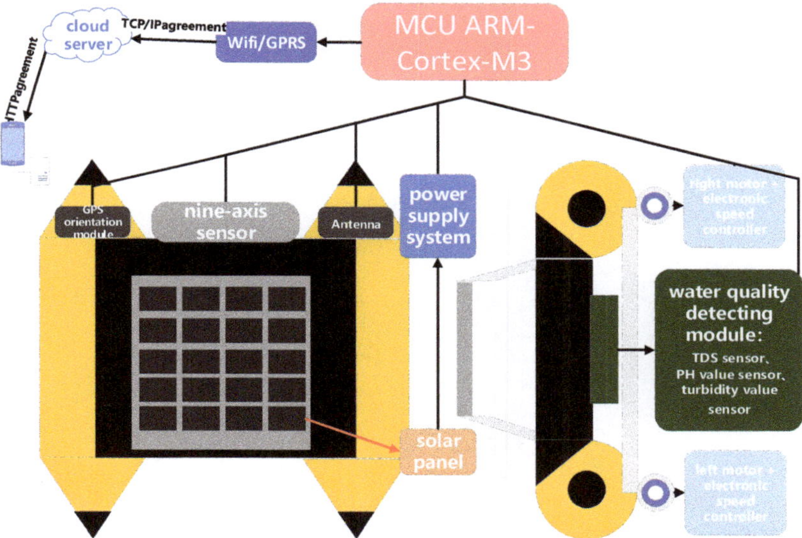

Figure 2. The overall architecture design of the unmanned service vehicle (USV). GPRS, general packet radio service; MCU, micro control unit; GPS, global position system; TDS, total dissolved solids; ARM, advanced reduced instruction-set computer machine.

2.2. USV Embedded Control Unit

The USV control unit is based on a 32-bit ARM microcontroller. An image of the USV is shown in Figure 3. The USV MCU is directly connected to the cruise system, power control module, and two wireless communication modules, as demonstrated in Figure 3. The USV MCU can establish wireless links to the communication module. The MCU of the USV is responsible for power control, cruising, and other tasks. Since the USV is sailing, an alterable power consumption mode is adopted according to the distance between the current and target locations. For instance, a USV is in the working and position modifying mode when the distance from the target location is less than or equal to 50 m, and when the distance is greater than 50 m the USV will go to the standby mode.

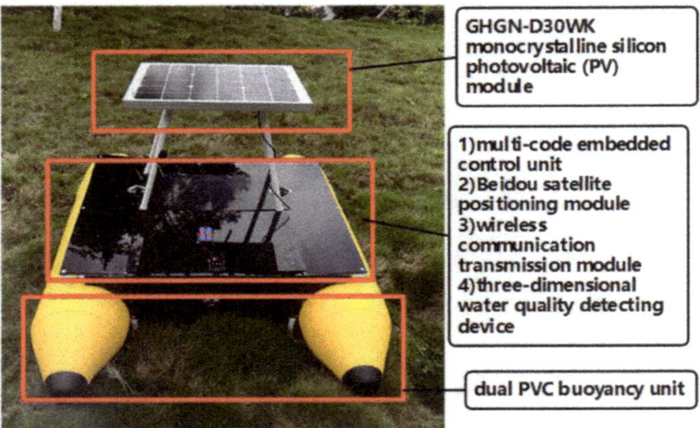

Figure 3. Structural distribution map of the USV. PVC, polyvinyl chloride.

2.3. Automatic Positioning Cruise System

The simplified block diagram of an automatic positioning cruise system of a USV is presented in Figure 4. The positioning cruise system includes a nine-axis sensor MPU9250 (three-axis gyroscope, three-axis acceleration, and three-axis magnetometer) and a GPS positioning module, which are used for positioning and USV moving direction determination. The nine-axis sensor and the positioning module are connected to the MCU (STM32F103ZET6) via SPI and UART interference, as shown in Figure 4.

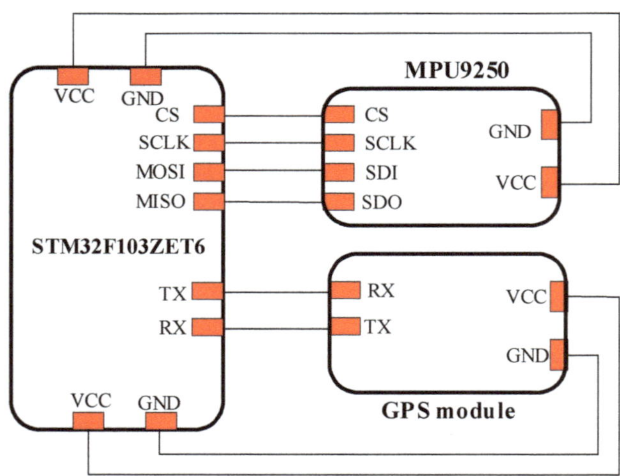

Figure 4. The architecture of the automatic positioning cruise system. GPS, global positioning system; VCC, volt current condenser; GND, ground; TX, transport; RX, receive; CS, chip select; SCLK, serial clock; MOSI, master out slave in; MISO, master in slave out; SDI, serial data input; SDO, serial data output.

In some of the previous studies [31,32], an automatic positioning achieved by a nine-axis sensor and a GPS (global position system) module was presented. Therefore, based on the circuit architecture displayed in Figure 4, an APCS (automatic positioning cruise system) is proposed. The flowchart of the APCS working principle is shown in Figure 5.

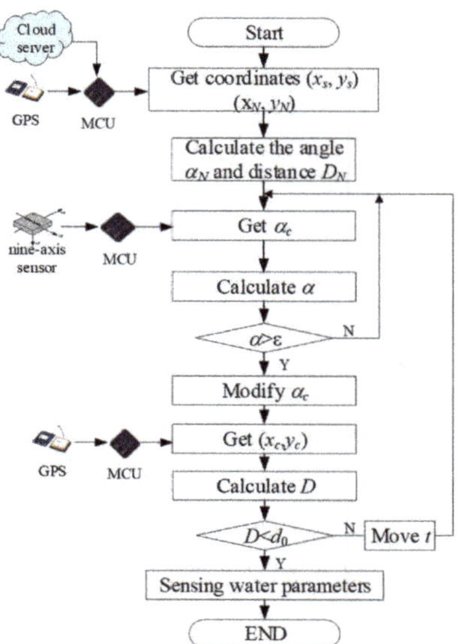

Figure 5. Flowchart of the APCS. (x_s, y_s), the starting position; (x_N, y_N), the next position; D_N, the distance between the next and starting positions; α_N, the angle between the next and starting positions; α_C, the current heading angle; α, the angle deviation;ε, the angle resolution; (x_C, y_C), the current position; D, the distance deviation; d_0, the distance resolution; t, moving time.

The main APCS working steps are as follows. First, the coordinate system is set up such that the starting point of the monitoring denotes the origin, the east-west axis represents the x-axis, and the north-south axis represents the y-axis. The MCU of a USV obtains the starting position by a GPS module and uses a wireless communication module to obtain the information on the next position, which are respectively denoted as (x_s, y_s) and (x_N, y_N), from the cloud server for the water monitoring task. Then, the MCU calculates distance D_N and angle α_N between the next and starting positions by:

$$D_N = \sqrt{(x_s - x_N)^2 + (y_s - y_N)^2} \tag{1}$$

$$\alpha_N = \arctan\frac{y_N - y_s}{x_N - x_s} \tag{2}$$

After obtaining the distance and angle of the next position, the MCU makes the USV move to the next position. During the USV's movement, the APCS periodically drives the sensor of MPU9250 and GPS module weighted fusion algorithm and sends the current heading angle α_C to the MCU by weighted fusion algorithm. Then, the MCU calculates the angle deviation between α_C and α_N, and makes the USV change direction until the angle deviation becomes smaller than a predefined value ε, which is given by:

$$\alpha = |\alpha_C - \alpha_N| \leq \varepsilon \tag{3}$$

Then, before the USV reaches the next position, the MCU periodically acquires the information on USV current position (x_C, y_C) and calculates distance deviation D of the next position by Equation (4). The presented steps are repeated until distance deviation D satisfies the condition given by Equation (4), i.e., until the USV reaches the target position.

$$D = |D_C - D_N| \leq d_0 \tag{4}$$

Finally, when the USV reaches the next position, the water quality parameters are collected by the WQMM (water quality monitoring module).

In practice, a USV is placed on a lake, and its starting point is set at a specified location. Then, water quality monitoring is completed in the cruise mode. The USV stops at each measurement point until the water quality monitoring at different depths is completed. In the above automatic cruise steps, there may be a deviation at the position level. Therefore, by comparing the position information from a GPS module and that of the nine-axis sensor with α_N and D_N intermittently, the USV's position is modified by adjusting the angle α and distance D.

2.4. Wireless Communication Transmission Modules

Two different types of wireless modules are adopted to connect to the cloud server and the WQMS, respectively. The first module uses the GPRS (general packet radio service) to realize wireless communication transmission between USV and cloud server. This technology is built based on a cellular network with a long communication distance, and it can be directly used in GSM (global system for mobile communications) or LTE (long term evolution) network. The other wireless communication transmission system is a Wi-Fi module intended for the USV and WQMM, and it uses the IEEE (institute of electrical and electronic engineers) 802.11n communication protocol.

3. Water Quality Monitoring System

The WQMS represents the key component in this study. This section introduces the main modules of the WQMS, which are the water quality monitoring module, a lifting device, and data transmission and storage system.

3.1. Water Quality Monitoring Module

In the proposed system, the WQMM consists of an MCU, power, sensors, wireless communication module, and plastic enclosure, as demonstrated in Figure 6a. As already mentioned, the MCU of the WQMM is based on a 32-bit ARM microcontroller. The main water quality values, including the pH value, total dissolved solids (TDS) value, and turbidity value, are monitored [33]. The parameters of sensors used to measure these values, such as sensing range, accuracy, and working voltage, are presented in Table 2. The MCU of the WQMM mainly controls the water quality monitoring sensor. After obtaining the relevant water quality data, the collected data are transmitted to the main control board via the wireless module.

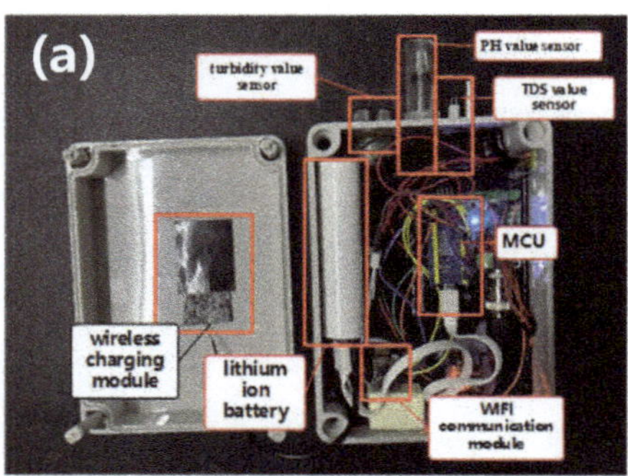

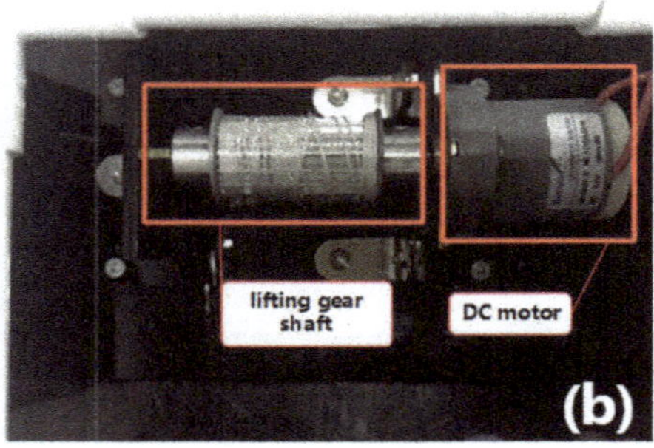

Figure 6. (**a**) Arrangement of sensors of water quality monitoring system (WQMM), (**b**) Lifting device.

Table 2. Parameters of sensors.

Sensor Type	Image	Range	Working Voltage (V)	Accuracy (%)	Producer
pH		0~14	5	±0.7	TELESKY [1]
TDS		0~1000 ppm	3.3~5	±5	WAAAX [2]
Turbidity		0~4000 NTU (nephelometric turbidity unit)	5	±0.75	EIXPSY [3]

Note: [1] Shengzhen, China; [2] Luoyang, China; [3] Shengzhen, China.

pH sensor: The pH sensor is used to monitor the pH value of the water. Meanwhile, the determination coefficient R^2 of the calibration model of this sensor reaches a value of 0.999. The pH sensor detects the hydrogen ion concentration in a solution using a hydrogen ion glass electrode and a reference electrode that form a primary battery. During the ion exchange process, the potential difference between the electrodes is measured between the glass membrane and the hydrogen solution to detect the hydrogen ion concentration in the solution and to determine the pH value of the solution.

TDS sensor: The working voltage of the TDS sensor is 3.3~5 V, while its measurement range and accuracy are 0~1000 ppm and ±5% F.S (full scale) (25 °C), respectively. There is a difference between the measured value and the true value of the TDS sensor, which is closely related to the water temperature. Therefore, the TDS sensor should be modified during practical application.

Turbidity sensor: The turbidity value is monitored based on the light transmittance and scattering rate of a liquid solution. The turbidity information is obtained by the AD (analog to digital) conversion interface in the dynamic monitoring environment. The working voltage and accuracy of turbidity sensor are 5 V and 0.75%, respectively. The response time is shorter than 500 ms.

3.2. WQMM Lifting Device

In order to realize a three-dimensional water quality monitoring system, the proposed design uses the ZS-RE81.3i DC (direct current) motor (ZHENG KE ELECTROMOTOR, Wenzhou, China) to lift and lower the water quality monitoring device. As shown in Figure 6b, the WQMM lifting device consists of a lifting gear shaft and DC motor. The voltage of the motor is 12 V, and its maximum value of revolutions per minute is 60, which indicates that the number of revolutions per minute can reach the value of 60. The PID (proportion integration differentiation) algorithm is used to obtain the desired rotating speed while the device is running.

The upgrade device can reach different depths in the range of 0~2 m. Meanwhile, the number of revolutions (n_r) of the motor can be calculated using the number of pulses. Assume R denotes the radius of the lifting gear shaft. In addition, the system can obtain the descending depth l_d of the probe box, which is given by:

$$l_d = 2n_r\pi R \tag{5}$$

Because the WQMM is initially flush with the bottom of a USV, the starting position is determined by the USV's draft. The USV MCU stops the WQMM descent by stopping the motor working. When the USV arrives at a predefined location, the USV MCU computes the sinking time and the standing time of the WQMM. The water quality parameters are transmitted before the water monitoring.

3.3. Sensing Data Transmission and Storage

The water sensing data are stored in the local memory of a monitoring module. After the probe box leaves the water body, the WQMM MCU activates the wireless module to establish a wireless connection with the UVS. Then, the sensing data are transmitted to the USV via wireless links. The USV transmits the sensing data to the remote cloud server via the GPRS. The diagram of data transmission is shown in Figure 7.

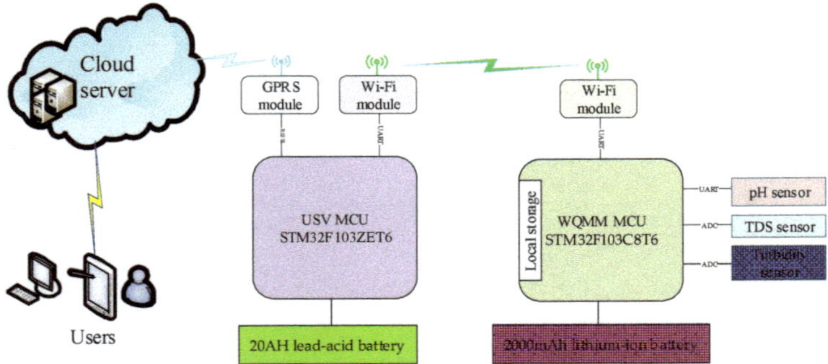

Figure 7. Diagram of the data transmission process. WQMM, water quality monitoring module.

On the open-source cloud platform, the unique data access JSON (JavaScript object notation) format message can be obtained, which is then sent to the cloud platform by the wireless module. The cloud platform parses the data and stores it. The protocol format of data access is different for each device. When a device account is created, the system automatically assigns a device ID (identifier) and an interface key (APIKEY) to the device. When data need to be transmitted to the cloud platform, a device is required to store the data. As mentioned, each device has a fixed ID and APIKEY. When sending a data request message to the cloud platform, the device ID number and APIKEY are used to access and transfer the data.

4. Water Quality Analysis Algorithm

The water quality level can be determined relatively straightforward use of various parameters, including the pH value, turbidity, the total number of bacteria, oxygen content, TDS value, and others. The traditional water quality monitoring methods mainly conduct water quality testing based on water quality samples with a large number of parameters, which lacks the features of automation and high efficiency. Therefore, water quality evaluation using a small number of parameters can be an effective alternative. It was shown [34,35] that pH value was significantly positively correlated with the dissolved oxygen value, electrical conductivity, and other parameters. Besides, the turbidity promotes the growth and reproduction of bacteria and adsorption of harmful toxic inorganic and organic substances. The turbidity particles have a certain impact on human and fish health. In the water without electrolysis or acid-base treatment, the salt cations are mainly calcium and magnesium, which coincides with the definition of water hardness. Therefore, water hardness can be indirectly expressed by the TDS value. Namely, when this value changes, the water quality also changes. Therefore, in the water quality analysis system, the three parameters (pH, TDS, and turbidity) are used to assess the water quality.

The proposed system uses the ensemble learning method [36,37] to predict subsequent changes in water quality by analyzing the collected water quality data in order to determine the correlation between the pH, turbidity, and TDS values. The Random Forest algorithm represents a concrete implementation of the bagging method. This algorithm trains multiple decision trees and combines the results of these trees to obtain the final result. The Random Forest can be used for splitting and regression, which are used to find the best fitting parameters. Its performance mainly depends on the decision tree type. The decision tree type is selected according to the specific task. For instance, in machine learning, if a set of objects can be classified into multiple categories, then the information on a certain class (x_i) can be defined as follows:

$$I(X = x_i) = -\log_2 P(x_i) \tag{6}$$

where $I(x)$ represents the information on a random variable, and $p(x_i)$ refers to the probability that x_i occurs.

In Figure 8, the flowchart of the water quality analysis algorithm based on the Random Forest is presented. The steps of this algorithm are as follows.

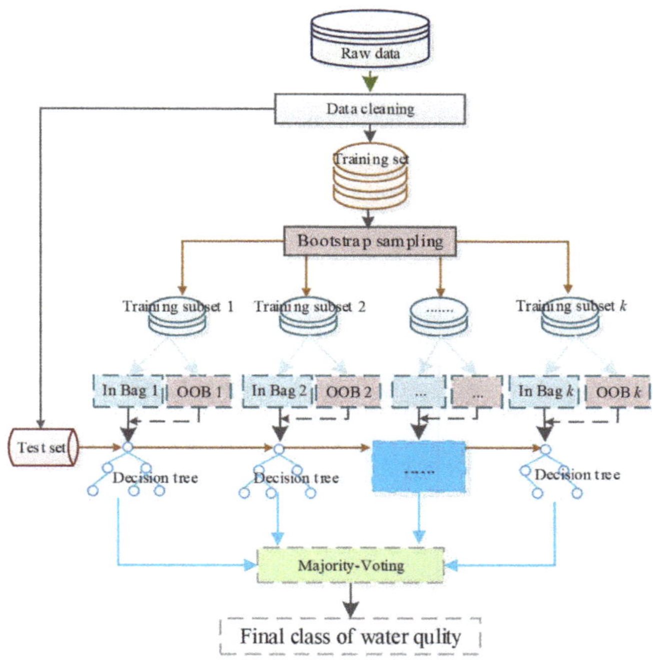

Figure 8. Flowchart of the water quality analysis algorithm. k, the number of trees; OOB, out-of-bag.

Step 1: Raw dataset is obtained by extracting the three features, namely the pH, turbidity, and TDS values. Then, these values are combined with the water quality evaluation results obtained from the historical dataset. The historical data consists of water quality parameters at different depths that are manually collected, and water quality evaluation value obtained by the average method.

Step 2: The data obtained in Step 1 is cleaned by retrieving and processing abnormal values of the three parameters and water quality evaluation value.

Step 3: The sub training set and test samples are generated for each decision tree by the bootstrap sampling technique.

Step 4: Steps 2 and 3 are repeated k times to construct k decision trees to generate random forest.

Step 5: The classification results of each decision tree for the test samples are summarized, and the class with the maximum number of votes is the final classification result.

5. Results and Analysis

This section presents and analyzes experimental sensing results, the performance of the water quality analysis algorithm, and water quality monitoring time.

5.1. Sensing Results

The proposed design was evaluated in the aquaculture zone, and the coordinates of the position where the water quality was measured were at Lat. 21.881131, Long. 110.842761, as shown in Figure 9. For the purpose of accurate measurement of water quality parameters, the water area was gridded. Meanwhile, 100 measuring points were selected, as shown in Figure 9, and the designed USV was used to measure water parameters at three different depths, which were 10 cm, 50 cm, and 100 cm.

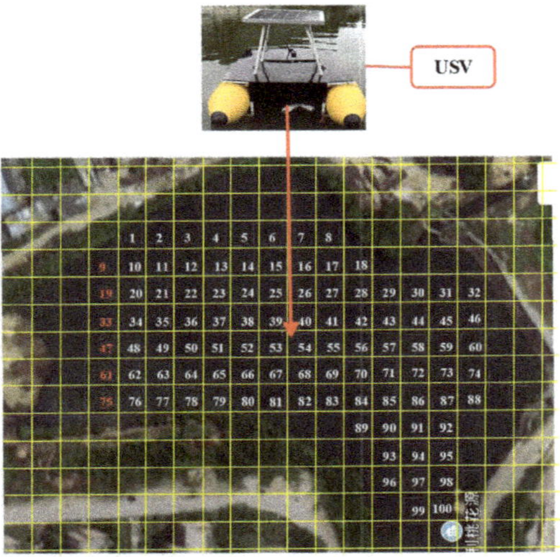

Figure 9. Locations of the water sample extraction.

Meanwhile, for displaying the measurement part results at different depths, we measured the water parameters at points (9, 19, 33, 47, 61, 75) at three different depth values, as shown in Figure 9. There were 18 monitoring points in the 15-square-meter area. Due to the continuity of monitored values, a spline interpolation technique was used to generate the cross-section map of each water parameter. In other words, the cross-section map denoted an objective reflection of the actual parameter values. The cross-section maps of the pH, TDS, turbidity, are presented in Figure 10. In Figure 10, it can be seen that average pH, TDS, and turbidity values were 8, 45 ppm, and 8.5 NTU, respectively. At the width from 10 to 15 m and the depth from 0 to 1 m, the pH value was greater than 9, which was beyond the safe limit. By using the proposed evaluation algorithm, it was found that the water quality in this region was slightly polluted.

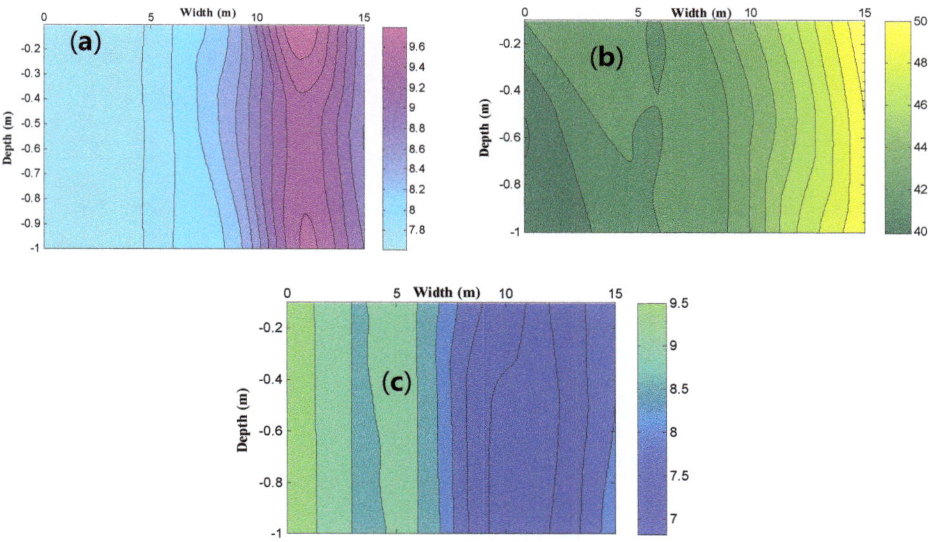

Figure 10. Cross-section maps of: (**a**) pH, (**b**) TDS, and (**c**) turbidity.

5.2. Intelligent Water Quality Analysis Algorithm

Model evaluation results: the Random Forest model was tested using Python 3.6 (Python Software Foundation, Wilmington, DE, USA) programming language. A total of 2870 samples corresponding to five different water quality levels were used in the experiment. The parameters of the proposed algorithm are presented in Table 3. In the learning process of the Random Forest model, 90% of the available data were selected as the training set, whereas the remaining 10% was used to test the developed model. The logistic regression and SVM methods were also used to analyze water quality, and their results were compared with the results obtained by the proposed method. In comparison, the precision, recall, and F1 measure were used because these measures are usually adopted to evaluate the performances of diffident algorithms.

Table 3. Random Forest parameters for water quality analysis.

Parameter	Value	Description
k	50	The number of trees in random forest.
max_features	2	The maximum number of features.
max_depth	10	The maximum depth of the decision tree.
min_sample_split	2	The minimum number of samples required for a split.
min_sample_leaf	1	The minimum number of samples required to form a leaf node.

The static random model was trained off-line using different values of pH, TDS, and turbidity at different depth levels as input parameters to estimate the water quality, which represented the output parameter. The evaluation indexes of different methods on the test dataset are presented in Table 4, where it can be seen that Random Forest was superior to the other two methods regarding the precision, recall, and F1 measure (H-mean). The precisions of the Random Forest, logistic regression, and SVM (support vector machine) were 92%, 39%, and 40%, respectively. In other words, among the tested methods, the proposed algorithm best evaluated the water quality on the test dataset.

Table 4. Model evaluation indexes on the test dataset.

Assessment Index	Random Forest	Logistic Regression	SVM
Precision	92%	39%	40%
Recall	0.7	0.5	0.5
F1 measure (H-means)	0.74	0.43	0.4

The ROC (receiver operating characteristic) curves of different methods are presented in Figure 11, where it can be seen that the values of the area under the curve (AUC) of the Random Forest, logistic regression, and SVM were 0.6, 0.7 and 0.5, respectively. Furthermore, the AUC value of the Random Forest algorithm was larger than those of the other two methods. Thus, the Random Forest that uses only three parameters (pH, TDS, and turbidity) represents a good water quality classifier.

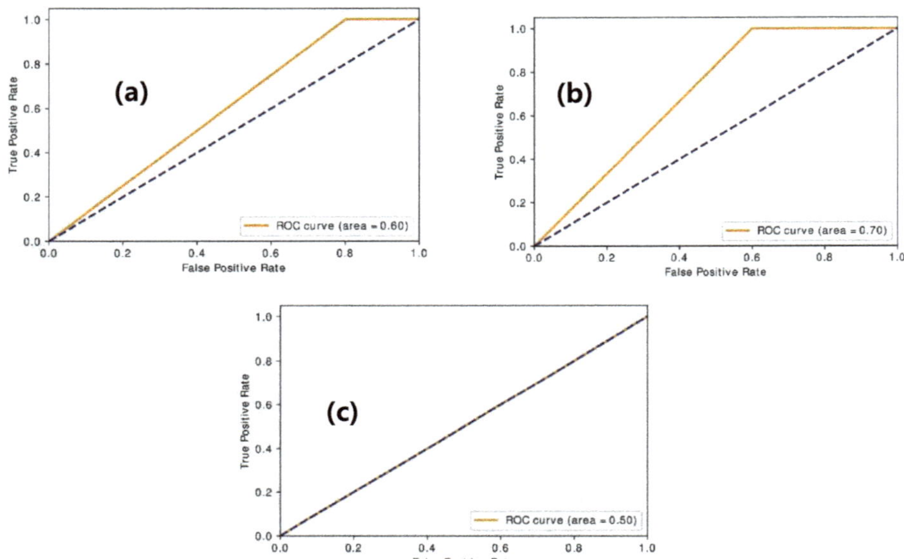

Figure 11. The ROC curves of: (**a**) logistic regression, (**b**) Random Forest, and (**c**) SVM. ROC, receiver operating characteristic; SVM, support vector machine.

Water quality classification precision: a total of two water sample sets denoted as I and II were taken at each position in different time portions, as shown in Figure 9. The water quality of these samples was extracted by laboratory analysis and then compared with the values obtained by the Random Forest, SVM, and logistic regression methods.

The comparison of actual water quality value and water quality values predicted by the SVM, logistic regression, and Random Forest methods is presented in Figure 12. For the sample set I, the precision of the proposed Random Forest algorithm was higher than 92%, and those of the SVM and logistic regression methods were less than 40% and 39.8%, respectively. For the sample set II, the proposed algorithm had a precision of more than 95%, and it outperformed the SVM and logistic regression methods. In Figure 12, it can be seen that the Random Forest algorithm was superior to the other methods regarding the precision rate. The experimental results proved the feasibility of the proposed method. Therefore, the water quality level can be accurately predicted by the Random Forest based on the pH, TDS, and turbidity values.

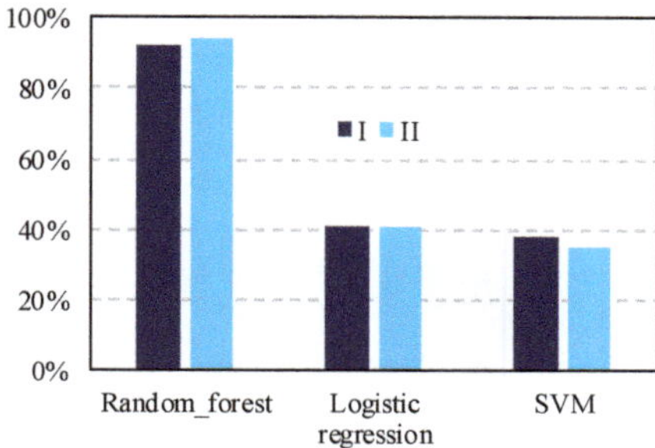

Figure 12. Precision rate of different methods in the real experiment.

5.3. *Water Quality Monitoring Time*

Under normal working conditions, the USV's moving speed could reach 0.5 m/s, and the WQMM's descending speed could reach 0.01 m/s. Therefore, these values were used in the experiment on water quality monitoring. At the above parameters, the experiment of water quality monitoring was constructed. The continuous monitoring was conducted at different depth values at 1 point (donates as C I), 5 points (donates as C II), and points (donates as C III) that were 2 m apart. The average monitoring times were obtained after the experiment was repeated three times under the same conditions. The average monitoring times of C I, C II, and C III were 400 s, 2100 s, and 4100 s, respectively. It only took three minutes from uploading the monitoring data to getting the evaluation results. The proposed system saved more than 60% of time compared with the manual approach.

6. Conclusions

In this paper, an intelligent wide-area water quality monitoring and analysis system is proposed, which represents a combination of intelligent USV, water quality monitoring module, and online water quality analysis. An unmanned system is designed to control a USV cruise automatically. By integrating the water quality sensor and lifting devices, the WQMM is designed. The ensemble learning method is proposed to analyze water quality, which provides a scientific basis for wide-area water quality testing. The experimental results demonstrate and validate that the proposed system can satisfy the requirements for water quality monitoring while improving the overall work efficiency. In the future, we will study the USV drifting control and the working performance of accuracy under different environmental conditions.

Author Contributions: Conceptualization, H.C. and S.W.; methodology, C.Z.; software, H.C.; validation, Z.G. All authors have read and agreed to the published version of the manuscript.

Funding: This work is supported by Guangzhou Science and Technology program (No. 201804010427), China.

Conflicts of Interest: The authors declare no conflict of interest.

References

1. Jury, W.A.; Vaux, H. The role of science in solving the world's emerging water problems. *Proc. Natl. Acad. Sci. USA* **2005**, *102*, 15715–15720. [CrossRef]
2. Moe, C.L.; Rheingans, R.D. Global challenges in water, sanitation and health. *J. Water Health* **2006**, *4*, 41–57. [CrossRef]
3. Benjamin, J.; Smith, E.; Kearns, M.; Rosen, J.; Stevens, K. Improving water utilities' access to source water protection and emergency response data. *J. Am. Water Work Assoc.* **2018**, *110*, E33–E44. [CrossRef]
4. Han, B.; Meng, N.; Zhang, J.; Cai, W.; Wu, T.; Kong, L.; Ouyang, Z. Assessment and management of pressure on water quality protection along the middle route of the south-to-north water diversion project. *Sustainability* **2019**, *11*, 3087. [CrossRef]

5. Zessner, M.; Lindtner, S. Estimations of municipal point source pollution in the context of river basin management. *Water Sci. Technol.* **2005**, *52*, 175–182. [CrossRef]
6. Zessner, M.; Schönhart, M.; Parajka, J.; Trautvetter, H.; Mitter, H.; Kirchner, M.; Hepp, G.; Blaschke, A.P.; Strenn, B.; Schmid, E. A novel integrated modelling framework to assess the impacts of climate and socio-economic drivers on land use and water quality. *Sci. Total Environ.* **2017**, *579*, 1137–1151. [CrossRef] [PubMed]
7. Smith, R.A.; Schwarz, G.E.; Alexander, R.B. Regional interpretation of water-quality monitoring data. *Water Resour. Res.* **1997**, *33*, 2781–2798. [CrossRef]
8. Vijayakumar, N.; Ramya, A.R. The Real Time Monitoring of Water Quality in IoT Environment. In Proceedings of the International Conference on Innovations in Information, Embedded and Communication Systems (ICIIECS), Amsterdam, The Netherlands, 19–20 March 2015; pp. 1–5.
9. Prasad, A.N.; Mamun, K.A.; Islam, F.R.; Haqva, H. Smart Water Quality Monitoring System. In Proceedings of the 2015 2nd Asia-Pacific World Congress on Computer Science and Engineering (APWC on CSE), Nadi, Fiji, 2–4 December 2015; pp. 1–6.
10. Chi, Q.; Yan, H.; Zhang, C.; Pang, Z.; Da Xu, L. A reconfigurable smart sensor interface for industrial WSN in IoT environment. *IEEE Trans. Ind. Inform.* **2014**, *10*, 1417–1425.
11. Jácome, G.; Valarezo, C.; Yoo, C. Assessment of water quality monitoring for the optimal sensor placement in lake Yahuarcocha using pattern recognition techniques and geographical information systems. *Environ. Monit. Assess.* **2018**, *190*, 259.
12. Wu, Z.; Wang, X.; Chen, Y.; Cai, Y.; Deng, J. Assessing river water quality using water quality index in Lake Taihu Basin, China. *Sci. Total Environ.* **2018**, *612*, 914–922. [CrossRef]
13. Engel, F.; Attermeyer, K.; Ayala, A.I.; Fischer, H.; Kirchesch, V.; Pierson, D.C.; Weyhenmeyer, G.A. Phytoplankton gross primary production increases along cascading impoundments in a temperate, low-discharge river: Insights from high frequency water quality monitoring. *Sci. Rep.* **2019**, *9*, 6701. [CrossRef] [PubMed]
14. Wilson, N.J.; Mutter, E.; Inkster, J.; Satterfield, T. Community-Based Monitoring as the practice of Indigenous governance: A case study of Indigenous-led water quality monitoring in the Yukon River Basin. *J. Environ. Manag.* **2018**, *210*, 290–298. [CrossRef] [PubMed]
15. Van Driezum, I.; Saracevic, E.; Scheibz, J.; Zessner, M.; Kirschner, A.; Sommer, R.; Farnleitner, A.; Blaschke, A.P. Finding an optimal strategy for measuring the quality of groundwater as a source for drinking water. *EGU Gen. Assem. Conf. Abstr.* **2015**, *17*.
16. Hatvani, I.G.; Magyar, N.; Zessner, M.; Kovács, J.; Blaschke, A.P. The water framework directive: Can more information be extracted from groundwater data? A case study of Seewinkel, Burgenland, eastern Austria. *Hydrogeol. J.* **2014**, *22*, 779–794. [CrossRef]
17. Mukta, M.; Islam, S.; Barman, S.D.; Reza, A.W.; Khan, M.S.H. Iot based Smart Water Quality Monitoring System. In Proceedings of the IEEE 4th International Conference on Computer and Communication Systems (ICCCS), Singapore, 23–25 February 2019; pp. 669–673.
18. Totsuka, S.; Kageyama, Y.; Ishikawa, M.; Kobori, B.; Nagamoto, D. Noise removal method for unmanned aerial vehicle data to estimate water Quality of miharu dam reservoir, Japan. *J. Adv. Comput. Intell. Intell. Inform.* **2019**, *23*, 34–41. [CrossRef]
19. Caccia, M.; Bruzzone, G.; Bono, R. A practical approach to modeling and identification of small autonomous surface craft. *IEEE J. Ocean. Eng.* **2008**, *33*, 133–145. [CrossRef]
20. Alilou, H.; Nia, A.M.; Keshtkar, H.; Han, D.; Bray, M. A cost-effective and efficient framework to determine water quality monitoring network locations. *Sci. Total Environ.* **2018**, *624*, 283–293. [CrossRef]
21. Behmel, S.; Damour, M.; Ludwig, R.; Rodriguez, M.J. Water quality monitoring strategies—A review and future perspectives. *Sci. Total Environ.* **2016**, *571*, 1312–1329. [CrossRef]
22. Stadler, P.; Loken, L.C.; Crawford, J.T.; Schramm, P.J.; Sorsa, K.; Kuhn, C.; Savio, D.; Striegl, R.G.; Butman, D.; Stanley, E.H. Spatial patterns of enzymatic activity in large water bodies: Ship-borne measurements of β-D-glucuronidase activity as a rapid indicator of microbial water quality. *Sci. Total Environ.* **2019**, *651*, 1742–1752. [CrossRef]
23. Wang, H.; Li, H.; Zhang, Y. Design of green water quality monitoring vessel based on dual operation mode. *IOP Conf. Ser. Earth Environ. Sci.* **2019**, *242*, 3. [CrossRef]

24. Ferri, G.; Manzi, A.; Fornai, F.; Ciuchi, F.; Laschi, C. The HydroNet ASV, a small-sized autonomous catamaran for real-time monitoring of water quality: From design to missions at sea. *IEEE J. Ocean. Eng.* **2014**, *40*, 710–726. [CrossRef]
25. Odetti, A.; Altosole, M.; Bruzzone, G.; Caccia, M.; Viviani, M. Design and construction of a modular pump-jet thruster for autonomous surface vehicle operations in extremely shallow water. *J. Mar. Sci. Eng.* **2019**, *7*, 222. [CrossRef]
26. Rajakumar, A.G.; Mohan Kumar, M.S.; Amrutur, B.; Kapelan, Z. Real-time water quality modeling with ensemble kalman filter for state and parameter estimation in water distribution networks. *J. Water Res. Plan. Manag.* **2019**, *145*, 04019049. [CrossRef]
27. Folorunso, T.A.; Aibinu, M.A.; Kolo, J.G.; Sadiku, S.O.E.; Orire, A.M. Water quality index estimation model for aquaculture system using artificial neural network. *J. Adv. Comput. Eng. Technol.* **2019**, *5*, 179–190.
28. Bercu, B.; Capderou, S.; Durrieu, G. Nonparametric recursive estimation of the derivative of the regression function with application to sea shores water quality. *Stat. Inference Stoch. Process.* **2019**, *22*, 17–40. [CrossRef]
29. Neale, P.A.; Altenburger, R.; Aït-Aïssa, S.; Brion, F.; Busch, W.; de Aragão Umbuzeiro, G.; Denison, M.S.; Du Pasquier, D.; Hilscherová, K.; Hollert, H. Development of a bioanalytical test battery for water quality monitoring: Fingerprinting identified micropollutants and their contribution to effects in surface water. *Water Res.* **2017**, *123*, 734–750. [CrossRef]
30. Yang, T.H.; Hsiung, S.H.; Kuo, C.H.; Tsai, Y.D.; Peng, K.C.; Hsieh, Y.C.; Shen, Z.J.; Feng, J.; Kuo, C. Development of Unmanned Surface Vehicle for Water Quality Monitoring and Measurement. In Proceedings of the IEEE International Conference on Applied System Invention (ICASI), Chiba, Japan, 13–17 April 2018; pp. 566–569.
31. Das, S.S. Simple, Inexpensive, Accurate Calibration of 9 Axis Inertial Motion Unit. In Proceedings of the 28th IEEE International Conference on Robot and Human Interactive Communication (RO-MAN), New Delhi, India, 14–18 October 2019; pp. 1–6.
32. Skvortzov, V.Y.; Lee, H.; Bang, S.; Lee, Y. Application of Electronic Compass for Mobile Robot in an Indoor Environment. In Proceedings of the IEEE International Conference on Robotics and Automation, Roma, Italy, 10–14 April 2007; pp. 2963–2970.
33. Awomeso, J.A.; Taiwo, A.M.; Idowu, O.A.; Gbadebo, A.M.; Oyetunde, O.A. Assessment of water quality of Ogun river in southwestern Nigeria. *IFE J. Sci.* **2019**, *21*, 375–388. [CrossRef]
34. Trivedi, P.; Bajpai, A.; Thareja, S. Evaluation of water quality: Physico–chemical characteristics of Ganga river at Kanpur by using correlation study. *Nat. Sci.* **2009**, *1*, 91–94.
35. Mei, Y.; Bai, Y.; Wang, L. Effect of pH on binding of pyrene to hydrophobic fractions of dissolved organic matter (DOM) isolated from lake water. *Acta Geochim.* **2016**, *35*, 288–293. [CrossRef]
36. Zhou, Z. Ensemble learning. *Encycl. Biom.* **2015**, 411–416.
37. Krawczyk, B.; Minku, L.L.; Gama, J.; Stefanowski, J.; Woźniak, M. Ensemble learning for data stream analysis: A survey. *Inf. Fusion* **2017**, *37*, 132–156. [CrossRef]

Article

Sentinel-2 Satellites Provide Near-Real Time Evaluation of Catastrophic Floods in the West Mediterranean

Isabel Caballero *, Javier Ruiz and Gabriel Navarro

Instituto de Ciencias Marinas de Andalucía (ICMAN), Consejo Superior de Investigaciones Científicas (CSIC), Avenida República Saharaui, 11519 Puerto Real, Spain; javier.ruiz@icman.csic.es (J.R.); gabriel.navarro@icman.csic.es (G.N.)

* Correspondence: isabel.caballero@icman.csic.es

Received: 18 October 2019; Accepted: 25 November 2019; Published: 27 November 2019

Abstract: Flooding is among the most common natural disasters in our planet and one of the main causes of economic and human life loss worldwide. Evidence suggests the increase of floods at European scale with the Mediterranean coast being critically vulnerable to this risk. The devastating event in the West Mediterranean during the second week of September 2019 is a clear case of this risk crystallization, when a record-breaking flood (locally called the "Cold Drop" (Gota Fría)) has swollen into a catastrophe to the southeast of Spain surpassing previous all-time records. By using a straightforward approach with the Sentinel-2 twin satellites from the Copernicus Programme and the ACOLITE atmospheric correction processor, an initial approximation of the delineated flooded zones, including agriculture and urban areas, was accomplished in quasi-real time. The robust and flexible approach requires no ancillary data for rapid implementation. A composite of pre- and post-flood images was obtained to identify change detection and mask water pixels. Sentinel-2 identifies not only impacts on land but also on water ecosystem and its services, providing information on water quality deterioration and concentration of suspended matter in highly sensitive environments. Subsequent water quality deterioration occurred in large portions of Mar Menor, the largest coastal lagoon in the Mediterranean. The present study demonstrates the potentials brought by the free and open-data policy of Sentinel-2, a valuable source of rapid synoptic spatio-temporal information at the local or regional scale to support scientists, managers, stakeholders, and society in general during and after the emergency.

Keywords: Copernicus Programme; ACOLITE; flooding; quasi-real time monitoring; inundation mapping; suspended matter; Spain

1. Introduction

Flooding is among the most common natural disasters in our planet and one of the main causes of economic and human life loss worldwide [1]. During the period 1980–2013, flood losses exceeded $1 trillion globally and resulted in ca. 220,000 fatalities [2]. In addition, the frequency and scale of floods are likely to increase in next decades not only due to climate-related extremes and natural hazards but also because ongoing socio-economic development [3]. Several studies of river floods and storms in Europe suggested that the observed increases in losses were principally due to economic wealth, increases in populations, and developments in hazard-prone areas, but the increase in heavy precipitation in some regions of Europe may have also influenced these [4,5]. Human activity is undoubtedly contributing to an increase in the likelihood and adverse impacts of extreme flood events. The population inhabiting risky zones is growing and, at the same time, construction in flood plains and inappropriate river management is reducing rivers capacity to absorb floodwaters. Coastal areas are also at risk of flooding; according to the European Environment Agency State of the Environment and Eurosion, the total value of European economic assets located within 500 m of the coastline, including agricultural, beaches, land and industrial facilities, is currently estimated at €500 to €1000 billion [6]. In 2007, the Floods Directive for the assessment and management of flood risk was adopted by the European Commission (EC), and worth mentioning is the operative monitoring by the EC Copernicus Emergency Management Service (Copernicus EMS) [7]. In addition to social and economic damage, floods may cause severe environmental impacts, as well as reduce biodiversity and destroy wetland areas. In Europe, a climatic-change signal in flood discharges has been demonstrated in relation to changes in the timing of floods within the year [8]. However, the regional scale diagnose of these events is hampered by lack of a coherent signal mainly due to the restricted spatial coverage and density of hydrometric stations [9]. A recent study analysing co-occurring heavy precipitation and high sea level (compound flooding) in Europe showed that the Mediterranean coasts are at present experiencing the highest compound flooding probability, being critically vulnerable to this risk [10].

The devastating event in the west Mediterranean during the second week of September 2019 is a clear case of this risk crystallization, when a record-breaking flood (locally called the "Cold Drop" (Gota Fría)) has swollen into a catastrophe to the southeast of Spain (Figure 1). This weather event typical of the fall season, where a sudden fall in temperatures appears along the east coast caused by the arrival of very cold polar air [11], is already being described as one of the worst. Wide areas have dramatically broken historical records of daily rainfall and flash floods, created chaos on roads, cut public transport, blocked roads, destroyed homes, closed schools, collapsed rivers, and caused six human fatalities in Valencia, Alicante, Murcia, and eastern Andalusia (Figure 1). These provinces remained on red alert during several days. Spain's weather agency continued to issue severe weather warnings during various days as storms bringing torrential rain in excess of 300 L/m^2, hail, winds up to 100 km/h, huge waves, and flood risk swept across the Southeast of the Iberian Peninsula. A clear example of the extreme rain event could be observed at a meteorological station in the Orihuela region (red star in Figure 1), where the daily precipitation rate during 12–13 September surpassed that of previous years (Figure 2). The government informed the material and economic damages caused by this extraordinary meteorological phenomenon are still unquantifiable owing to its magnitude, declaring the region a disaster area. On 13 September, the authorities reported that the Segura River had burst its banks in Orihuela, a town of around 80,000 residents in Alicante province. Thousands of residents from different municipalities were evacuated due to the extreme weather conditions of one of the most spectacular floods in Spain, where current catastrophic flooding has far surpassed previous all-time records in the West Mediterranean.

Understanding the spatiotemporal and physical characteristics of risk drivers (exposure, hazard, and vulnerability) is needed in order to implement effective flood mitigation measures [12]. As a means of emergency response after a flooding or inland inundation, flood mapping helps to identify the extent of the event on a large scale as well as the affected infrastructure such as roads and settlements and impaired regions of interest such as agricultural areas. This information can be used by disaster management agencies and other stakeholders to undertake the rescue in affected areas, coordinating appropriate recovery activities and prevention measures for possible upcoming events. Currently, the catastrophic impacts of flooding can be evaluated by means of remote sensing technologies, which provide a unique source of data for the implementation of European Union (EU) directives [13]. Remote sensing technologies are one of the most widely used platforms for emergency management and extended area mapping. Satellite images can be applied to assess the extent of flooded areas and their impact on human, economic, environment, and infrastructure offering a synoptic view of large areas, frequent observations, and historical archives. Several studies have focused on obtaining information using remote sensing imagery with the application of different methodologies [14–18]. Particularly, when optical satellite data are available before and after a flooding event, identification of flooded regions and water quality deterioration in the coastal areas would be much easier by means of change detection approaches. These approaches frequently require the application of pre-and post-flood image pairs to distinguish pixels that have been modified from non-water to water between image dates and have been implemented using post classification image differencing [19]. Whereas these techniques can offer useful and accurate results, the combined challenges of storm-related cloud cover and low temporal resolution remain important limitations. For this evaluation, high temporal and fine spatial resolution images are required in order to comprehensively generate accurate maps.

Figure 1. Location of the Spanish regions impacted by the flooding in September 2019 from south to north: Almeria, Murcia, Alicante, and Valencia. The red star corresponds to the location of the meteorological station in Orihuela (Figure 2). Photos showing the record-breaking flooding during the Cold Drop (Gota Fría) in Alicante (**a**) and (**c**), Murcia (**b**) and (**d**), and Valencia (**e**), and (**f**) during 11–14 September 2019. The photos are courtesy of Levante, Efe, España Diario, REUTERS, and El Confidencial.

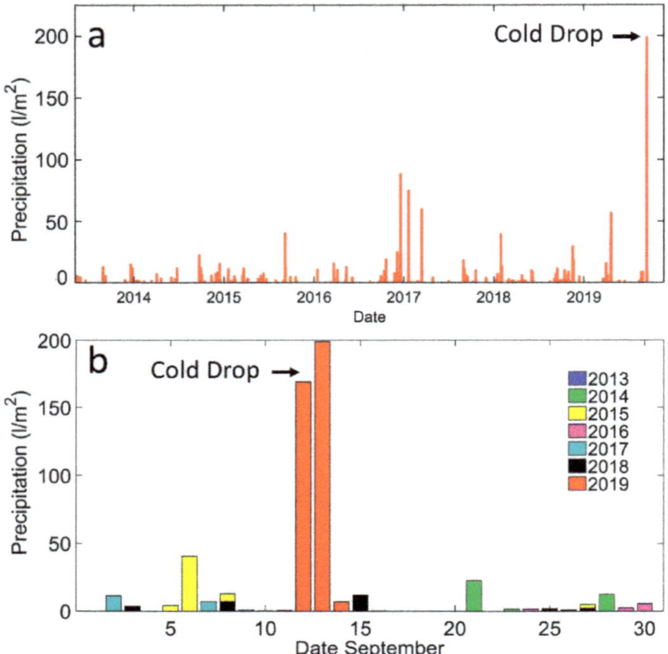

Figure 2. (**a**) Historical daily rainfall (litre per square metre, L/m^2) in a meteorological station in the Orihuela region (Alicante, red star in Figure 1) and (**b**) daily precipitation during September 2013–2019, where the extreme precipitation during the Cold Drop event on 12–13 September 2019 is highlighted in red.

In this regard, the European Space Agency (ESA), in partnership with the EC, is developing a new family of missions called the Sentinels for the operational needs of the Copernicus Programme. Each Sentinel is based on a constellation of two twin satellites to fulfil revisit and coverage requirements, providing robust datasets for Earth Observation Services, beginning a new era in disaster monitoring and emergency management [20]. These missions carry a range of technologies, such as multi-spectral imaging and radar instruments for ocean, land, and atmospheric monitoring, and are designed to deliver the wealth of data and imagery that are central to improve the management of the environment, safeguarding lives every day. Specifically, the Copernicus Sentinel-2 mission comprises a constellation of two polar-orbiting satellites placed in the same sun-synchronous orbit, phased at 180° from each other. It aims at monitoring variability in land and water surface conditions to provide imagery of vegetation, soil, water cover, and inland waterways. The 5-day revisit of the pair of Multi Spectral Imagers (MSI) onboard Sentinel-2 with 10 m spatial resolution in visible bands, combined with freely available data, is unprecedented, providing new insights into the observation of nearly any coastal place on Earth [21]. This enables for new endeavours in monitoring, change detection, and mapping of coastal zones.

In the present contribution, a demonstration of a satellite-based tool as an effective solution to address and delimit flood-affected areas and to evaluate the effects of the flood on the water quality (suspended sediments) of the Spanish coastal regions following the catastrophic flood during September 2019 was accomplished. A straightforward approach has been applied using a recently released software suited for atmospheric correction (ACOLITE processor) with Sentinel-2A/B images before and after the event, indicating the potentials brought by the twin satellite's wide coverage and high revisit for change detection. The findings are an example of addressing near-real time flooding monitoring in large areas providing a preliminary valuable source of synoptic information to support managers, scientists, stakeholders, and society during and after the emergency.

2. Materials and Methods

In this study, the Sentinel-2A and 2B twin polar-orbiting satellites were used. Both Multispectral Instruments (MSI) on-board are now operational: Sentinel-2A was launched on 23 June 2015 and Sentinel-2B followed on 7 March 2017. A free, full, and open data policy has been adopted by the EU for the Copernicus Programme, which foresees access by all users to the Copernicus Sentinels core products. The radiometric, spectral, and spatial characteristics of the bands used in this study are specified in the User Handbook [21]. A temporal examination provided Level-1C Sentinel-2 images that were typically geo-located within two pixels of each other (20 m) which is within the stated quality requirements for absolute geo-location [22]. Sentinel-2 Level 1 datasets (zone 30, sub-tile SXG and SXH) were downloaded from the ESA official website Copernicus Open Access Hub [23]. The Copernicus Services Data Hub is an access to Copernicus Sentinels data dedicated to Copernicus Services and European Institutions. Copernicus Sentinel-2 data are systematically processed to L1C products and made available online between 2 and 12 h from sensing (on average, 7 h after sensing) in the Copernicus Open Access Hub and Copernicus Services Data Hub. Only images with low cloud coverage (<20% over the region of study) were considered and downloading time was generally around 0.5 h. At least two images with the same orbit track and the same coverage are required for change detection, namely, the reference image (pre-event) and the target image (co-event), respectively. The reference image was selected as the latest available image prior to the event with minimum cloud coverage. The scenes were selected carefully with two Sentinel-2A/B satellite images before (19 August 2B, 3 September 2A) and two scene-captures during or immediately following (13 September 2A, 18 September 2B) the event (Figure 2).

A straightforward approach has been accomplished by means of the ACOLITE software developed by the Royal Belgian Institute of Natural Sciences RBINS (version 20190326.0). ACOLITE bundles the processing software and atmospheric correction algorithms implemented at RBINS for aquatic applications of Sentinel-2 (A/B) and Landsat (5/7/8) satellite information. ACOLITE performs the atmospheric correction and can provide several parameters derived from water reflectance. ACOLITE was originally developed in IDL (2014–2017) and has been translated into Python (2018–2019). The Dark Spectrum Fitting (DSF) atmospheric correction was used [24]. This model was originally implemented for water applications of fine resolution metre-scale optical satellites but proved capacities for use with MSI due to their finer spectral coverage [25]. The DSF computes atmospheric path reflectance based on multiple dark targets in the scene or subscene, with no a priori defined dark band. For each band, the darkest object is estimated from the offset fit to the first thousand pixels in the histogram and a "dark spectrum". A Continental or Maritime aerosol model is chosen based on the lowest RMSD between the observed reflectance and the retrieved path reflectance for the two closest fitting bands. The settings selected for the DSF were a tiled path reflectance option, which divides the full scene in approximately 6×6 km tiles and interpolates retrieved path reflectance. In this study, optional image-based sun glint correction of the surface reflectance was also incorporated. Total time for ACOLITE processing was ~2.5 h (see workflow diagram in Figure 3). Processing speed and time were associated to a Windows 10 Pro desktop machine with Inter(R) Core™ i7.

MSI has spectral bands at different spatial resolutions, 10, 20, and 60 m and ACOLITE converts internally the bands to the same (user-specified) resolution. For bands at lower resolution than the processing resolution, values are replicated by nearest neighbour resampling, i.e., no new pixel values are computed, and for bands at higher resolution, pixels are spatially mean averaged. By default, the 10 m grid is used, which means the values from the 20 and 60 m bands are replicated 4 and 36 times to form a 10 m grid. In this study, ACOLITE products resampled to 10 m pixel size corresponded to Remote Sensing Reflectance (Rrs, 1/sr) in all visible and Near-Infrared (NIR) bands. The concentration of total suspended material (TSM, g/m^3) was also calculated using a standard approach [26], a model which is suitable for any ocean colour sensor. Theory indicates that use of a single band, if chosen adequately, offers a robust and TSM-sensitive algorithm. Retrieving suspended matter in optically shallow water requires the use of spectral bands that have limited depth penetration in order to avoid substantial interference from the bottom but at the same time are sensitive to turbidity. Red bands are useful for estimation of suspended matter but can have a strong bottom signal in shallow waters [27]. The absorption by water increases rapidly from red to the "red edge" NIR (700–780 nm). This absorption limits the light received from the bottom, while still returning light scattered from materials in the water. In this sense, the red-edge NIR band 704 nm has sufficient water absorption to be a good compromise between detecting turbidity with limited detection of the bottom, so it was selected for the suspended matter algorithm [26] within the ACOLITE processor. Several studies have already indicated these red-edge spectral bands are appropriate for turbidity or suspended solids monitoring in optically shallow regions [27,28].

Non-water pixels were masked on the maps using a combination of thresholds and spectral tests within ACOLITE. Masking is performed after the atmospheric correction, using the retrieved surface reflectance. Rrsmin and Rrsmax represent the minimum and maximum reflectance in the sensor bands, respectively. The data were masked if any one of the four spectral tests used was true: "Bright" spectral test, "NIR peak" spectral test, "Non-water" spectral test, and "White" spectral test (see more details on [24]):

"Bright" spectral test: Water pixels have a maximum reflectance much less than 20%, and hence pixels should be masked where:

$$Rrs_{max} > 0.2 \tag{1}$$

"NIR peak" spectral test: Water pixels have a visible band reflectance higher than NIR reflectance (except for extremely turbid waters), and hence, pixels should be masked where:

$$Max\left(Rrs_{blue}, Rrs_{green}, Rrs_{red}\right) < Rrs_{NIR} \tag{2}$$

"Non-water" spectral test: The NIR to red reflectance ratio is limited to a given ratio for water pixels (except for extremely turbid waters), and hence, pixels should be masked where:

$$\frac{Rrs_{NIR}}{Rrs_{red}} > 0.9 \tag{3}$$

$$Rrs_{NIR} > 0.05 \tag{4}$$

"White" spectral test: Water pixels typically have a large spectral variability in the visible and NIR, and hence, pixels should be masked where:

$$\frac{Rrs_{max} - Rrs_{min}}{Rrs_{max}} < 0.2 \tag{5}$$

Based on the later and while computing thresholds, the two satellite imageries on 13 and 18 September after change detection were clarified into damaged and un-damaged areas. These data were processed within few hours, saving on time and resources and enabling the near-real time quantification of the flooded areas and water quality issues (see workflow diagram in Figure 3). Moreover, information on the historical daily rainfall of a meteorological station in Orihuela, Alicante (red star in Figure 1, 38°04′04″ N, 0°58′53″ W) was obtained from Aemet Open Data [29].

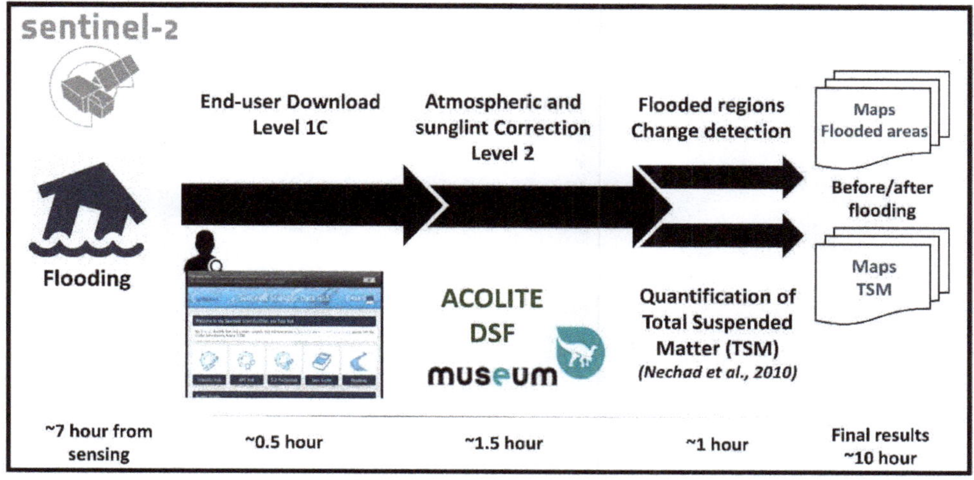

Figure 3. Diagram with workflow and time processing for each step using Sentinel-2 images and ACOLITE processor.

3. Results

The Sentinel-2 images as true colour composite Red–green–blue from the EO Browser [30] at 10 m spatial resolution show the before on 3 September 2019 (Figure 4a) and after on 13 September 2019 (Figure 4b) of the flooding event in Murcia province (Spain). The affected and non-affected zones were delineated, resulting in a flooded area of 1307 ha (Figure 5) over the Region of Interest on 13 September (ROI detailed in Figure 4b). This phenomenon has damaged the cultivated lands, mainly in Los Alcázares region (Figure 4b), where the exceptional rainfall caused widespread flooding and damage to agricultural lands, much of them farmed under agri-environment agreements. While flooding in winter is a common event in this region, flooding in late summer or early autumn (associated to the Gota Fría) is unusual and much more damaging. The information retrieved can be used to evaluate the impacts on flooded areas and help inform strategies to deal with changes in flood risk in areas of agricultural and environmental interest.

Moreover, Sentinel-2 identifies not only impacts on land but also on water ecosystem and its services. Due to the large quantity of sediments brought by the flood, the affected regions show an extreme increase of total suspended material and severe turbidity levels, visible in the RGB image (Figure 4b), leading to a contamination and deterioration of the rivers and the coastal adjacent region (Figure 2b). Large portions of Mar Menor (the largest coastal lagoon in the Mediterranean) showed up to 200 mg/m^3 of suspended solids on 13 September (Figure 4d) compared to the standard situation ten days before on 3 September with minimum concentration <10 mg/m^3 (Figure 4c). There were no in situ data to validate the TSM algorithm, but realistic information can be observed for the mapping of suspended matter using the TSM model [26] with the 704 nm band. A recent study has confirmed that Sentinel-2 red-edge bands achieved good accuracies when compared with suspended solid concentrations [31], in addition to the red or NIR bands [32,33]. One of the main advantages of Sentinel-2 over Landsat-8 is the inclusion of these three red-edge and NIR bands, allowing for the determination of chlorophyll in turbid and productive waters and the retrieval of turbidity or suspended particulate matter concentration, even in narrow inlets and ports, providing an invaluable dataset for several scientific and management purposes [34]. The TSM model was not able to perform accurately over the extremely turbid region close to the coast and part of the turbid intrusion was masked out by ACOLITE processor (similar results were found using other TSM algorithms within ACOLITE software).

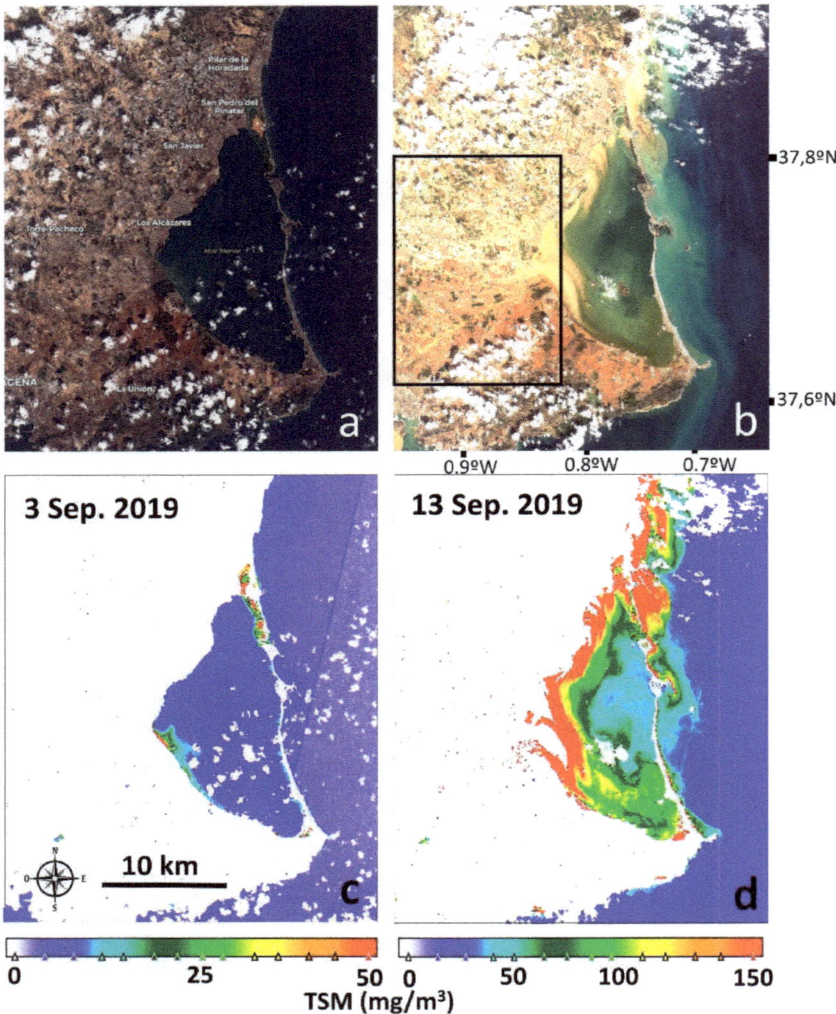

Figure 4. Sentinel-2 images at 10 m spatial resolution as true colour composite (red–green–blue) from the EOBroswer showing the (**a**) before (3 September 2019) and (**b**) after (13 September 2019) of the flooding event in Murcia province (Spain). The high concentration of suspended material (TSM, mg/m^3) in Mar Menor can be observed in the map (**d**) on 13 September with TSS > 200 mg/m^3 compared to the normal situation (**c**) on 3 September 2019 with minimum concentration (10 mg/m^3). White areas correspond to land, clouds, or cloud shadows masked after ACOLITE. Black rectangle shows the Region of Interest (ROI) for flooded area calculation. The TSS scales for each scene are different.

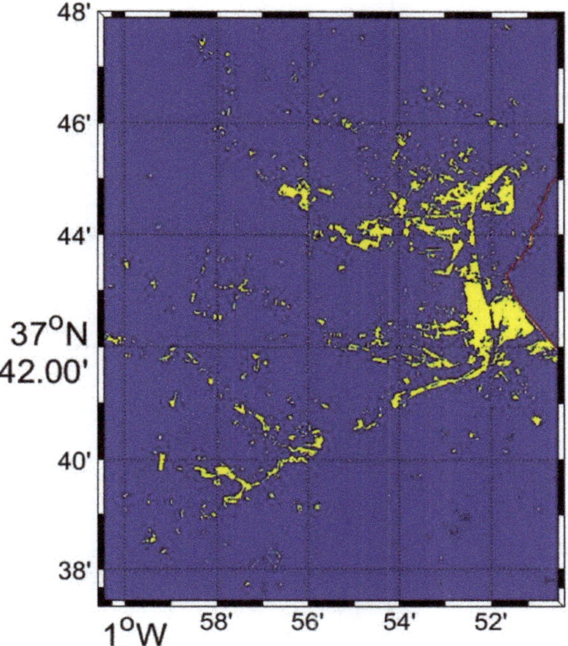

Figure 5. The flood-affected areas (yellow) in Murcia on 13 September 2019 within the Region of Interest (ROI) defined in Figure 4b (black rectangle). The red line represents the coastal line.

In the Dolores region (Alicante), pre (19 August 2019, Figure 6) and post (18 September 2019, Figure 6b) Sentinel-2 false colour composite (near infrared–green–blue) from the EO Browser indicated the affected regions, with an estimated flooded area of 3150 ha on 18 Sep (Figure 7) within the ROI (rectangle in Figure 6b). This scene corresponds to some days following the event, thus some flood areas cloud not be well documented (scene 13 September over Alicante region has severe cloud contamination). The delineation of the flooded zones, including agriculture and urban areas impacted as well as identification of the affected infrastructures can be observed in the false colour image (Figure 6b). Water quality in reservoirs servicing human consumption and agriculture was also deteriorated as can be observed in the turbidity plume associated to the Segura River on 18 Sep with concentration of suspended solids up to 150 mg/m^3 (Figure 6d) compared to the normal situation on 19 August 2019 with minimum concentration <10 mg/m^3 (Figure 6c).

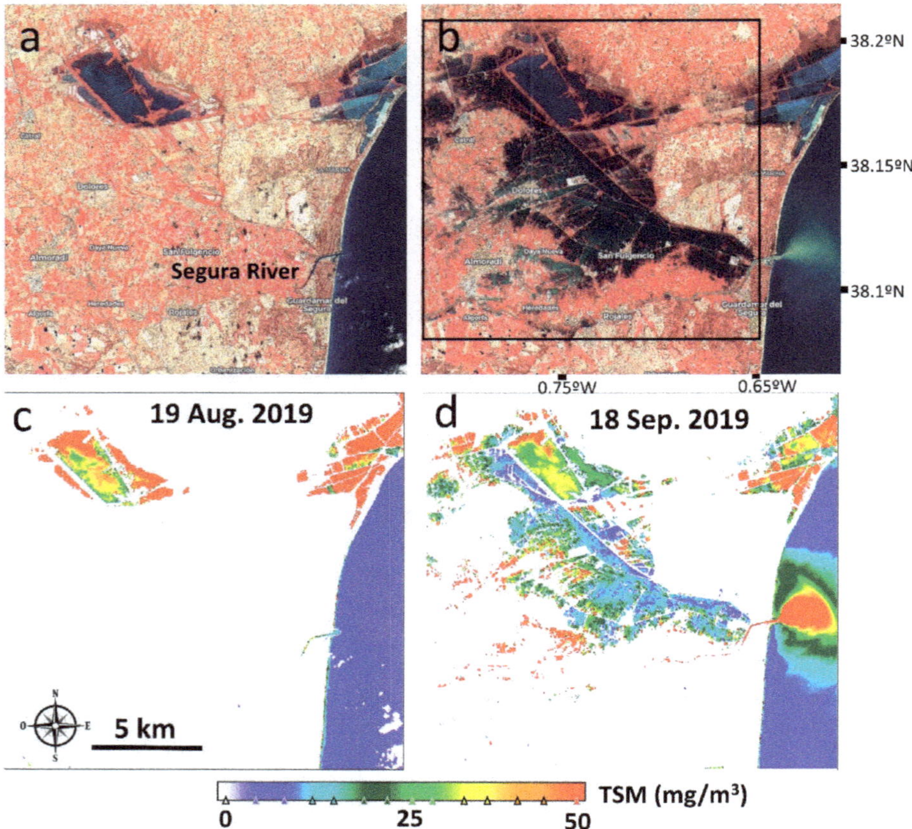

Figure 6. Sentinel-2 images at 10 m spatial resolution as false colour composites (near infrared–green–blue) from the EO Browser showing the (**a**) before (19 August 2019) and (**b**) after (18 September 2019) of the flooding event in the Dolores region, Alicante province (Spain). The high concentration of suspended material (TSM, mg/m^3) in the Segura River coastal region >50 mg/m^3 can be observed in the map (**d**) on 18 September compared to the normal situation (**c**) on 19 August 2019 with minimum concentration (<10 mg/m^3). White areas correspond to land, clouds or clouds shadows masked after ACOLITE. Black rectangle shows the Region of Interest (ROI) for flooded area calculation.

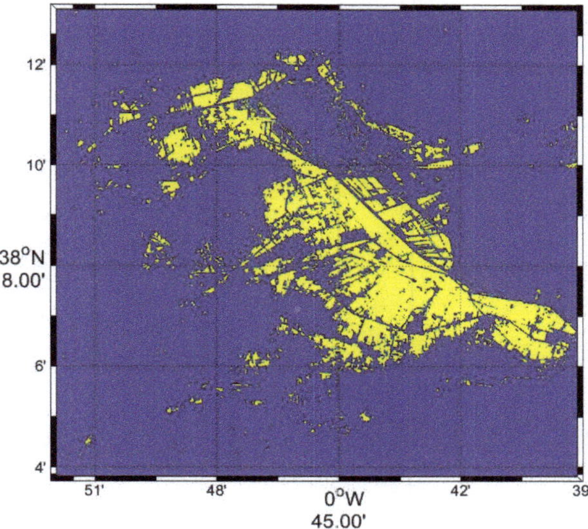

Figure 7. The flood-affected areas (yellow) in Alicante on 18 September 2019 within the Region of Interest (ROI) defined in Figure 6b (black rectangle).

4. Discussion

In this study, an operative procedure after the flooding event was applied to automatically map flooded areas and suspended matter from multispectral Sentinel-2 images with a novel ACOLITE processor based on the DSF model. Up to our knowledge, it is the first time ACOLITE tool has been tested for quick assessment of flooded-affected areas. This framework is proposed as a way to rapidly obtain key preliminary information while potentially automating the image processing, allowing accurate initial approximation maps. The ability to retrieve products for quasi-real-time monitoring is proved, with final products obtained on average after 10 h from sensing: 7 h to process and made available the data by Copernicus, 0.5 h for downloading, and 2.5 h for ACOLITE processing and generation of final maps (Figure 3). The promising employment of advanced software such as ACOLITE combined with the fine-scale Sentinel-2 satellite imagery to optimize mapping in short times might benefit operational and scientific monitoring and management, especially in data-poor regions of the world.

Subsequent hypoxic and other undesirable consequences occurred in the highly sensitive environment of Mar Menor after the flooding. The intrusion of turbid waters to the Mediterranean (Figure 4) also played its role in the massive mortality of bluefin tuna and seahorses landing in the beaches during the event. In addition, during the first week of October 2019, the lack of oxygen killed thousands of fish and shellfish in this basin, and the Minister of Environment in the Murcia Government confirmed that the water quality in the lagoon was much worse than before the Cold Drop. The information provided by Sentinel-2 might assist the coordination of efforts by all bodies involved in the administration of the Mar Menor in order to resolve the problems exacerbated by the Cold Drop.

The method establishes a plausible high-quality strategy with sufficient sensitivity to support interannual and pre-post hazards, which may benefit change detection analysis from multiple images to discriminate variability. Change detection plays a crucial role in flood monitoring using remote sensing data, while the selection of a reference image is crucial in order to obtain an accurate thematic map. We needed to acquire suitable cloud-free scenes without sun-glint and other noise-inducing environmental variables, but this work was not straightforward. However, the high revisit time of

Sentinel-2 provided increased likelihood of obtaining cloud-free scenes delivering time-sensitive data. The availability of the Sentinel-2A/B images each 5 days made possible to select pre/post-flooding or co-occurring images. It is also important to highlight that this study benefited from the availability of the post-flood image captured days after the extreme flooding (18 September) in Dolores region, when many inundated areas were still saturated at the surface.

The flooded-affected areas estimated with Sentinel-2 data collected for post-event assessment have been visually validated with datasets from Copernicus [7] as well as based on dwellers information. Activation of the Copernicus Emergency Management Service (EMS) provided mapping products to support emergency management activities immediately following a disaster, in this case screened to identify flood events with available clear sky post-event images using the very fine resolution WorldView-2 and Pléiades-1A satellites. Visual assessment against EMS data indicated close agreement with scoured riverbanks and high-water marks, and the performance in this study is consistent and comparable to standard approaches focused on the Copernicus Programme. Similar flooded affected zones have been delineated with Sentinel-2 in Dolores region, although differences might appear due to the lower resolution of MSI (10 m) compared to the very fine resolution of both WorldView-2 and Pléiades-1A (~2m). A more exhaustive assessment of the flooding mapping could be considered through quantitative comparison with independent data, such as Copernicus Emergency maps, for a more comprehensive evaluation of methodology and performance. However, the aim of this research was to explore the suitability of the atmospheric correction method applied to Sentinel-2 data for quick mapping of flooded areas and suspended matter. Potential timely uses of the retrieved information can be the basis for civil protection or for prompt delimitation of areas or districts hit by the flood. This information could be considered without requiring very high quantitative performances of the proposed methodology in terms of accuracy in flooded area delimitation or in the mapping of the real levels of suspended matter in the coastal waters. The application of the ACOLITE processor for coastal mapping has been recently reported in other studies [32,33].

Sentinel-2 might provide a way to use other satellites more effectively, assisting as a reference data set that can be used to inform mapping conducted with the various commercial very high-resolution sensors such as the World-View fleet [35] or the novel CubeSats from Planet [36]. On the other scale, Landsat has a four-decade record of Thematic Mapper and has been shown useful even with 30 m pixels [37,38]. Comparisons of Sentinel-2 with Landsat-8 may ultimately lead to results that could expand the utility of the entire Landsat data record for change detection. The development of new cloud environments and standardizing methodologies (e.g.,: Google Earth Engine or the Coastal Thematic Exploitation Platform Coastal-TEP by ESA) may aid in the routine application of Sentinel-2. Accordingly, for further avenues of research with Sentinel-2, we intend to upscale the study results to more environments as well as implement the approach in cloud-based computing platforms such as Google Earth Engine.

Recent studies have already suggested the potential of Sentinel-2 for automated mapping of flooded areas [39]. The state of the art of the methods for water detection from multispectral images is well described in the literature with approaches based on the Normalized Difference Water Index-NDWI [40], the modified NDWI [41], the Water Ratio Index [42], or the Water Index [43]. From an end user perspective, ACOLITE allows a quick, easy, and automated processing to generate preliminary information for change detection based on a well-described thresholding approach. Compared to flood mapping with a single image, change detection-based methods have an advantage in masking out the permanent water bodies (rivers, lakes) and some water look-alike objects through the identification of inundated soil and land from post-flood imagery [44]. The robust and flexible approach was able to identify flood-related water pre-existing water bodies in Dolores (Alicante province). The flooding maps based on Sentinel-2 data may be operative as a preliminary emergency map, as it allows an accurate initial approximation of flooded-affected areas and water quality issues. Additionally, this information might be useful for defining the economic losses for the insurance sector and evaluating public compensation schemes [45], as this region of the Mediterranean is extremely vulnerable to climate change impacts [10].

Generally, Synthetic Aperture Radar (SAR) sensors have emerged as the main source of operational techniques for mapping flood extent and flood-prone areas by planning emergency response agencies [46,47]. SAR tools provide advantages to monitor flooding as they have unique capabilities compared with optical data; radar can work at night and with clouds due to their all-weather capabilities, so the use of radar images has increased considerably [48,49]. In this work, it is demonstrated that Sentinel-2A/B 5-day revisit, its rapid acquisition time, and the methodology adapted provided a robust product for emergency services of flood mapping and immediate damage assessment in disaster monitoring. Moreover, the straightforward nature of the strategy and its independence from ancillary training data makes it applicable and accessible for a wide variety of stakeholders, managers, and end users, especially in data-poor or remote areas. The Mar Menor is suffering several human pressures, such as sediment and nutrient inputs from agriculture and other activities, so implementation of operational strategies to monitor its evolution is critical to protect its fragile natural ecosystem [50]. In this regard, the open data policy and long-term mission commitment of Sentinel-2 opens future promising time series evaluation over years and even decades that can be an important tool to provide crucial missing information at the local or regional scale in a rapid and non-intrusive manner.

5. Conclusions

It is demonstrated that, in the event of a natural disaster, as the recent catastrophic flooding event occurred in Spain during the second week of September 2019, the Sentinel-2 fleet is a valuable rapid-response source of synoptic spatio-temporal information, and its benefit has been proven here. With their fine spatial and high temporal resolutions, these twin open-source satellites, particularly when coupled, have the potential to generate rapid information of the affected flooded parcels and of the water quality, data that is key for both the local government body as well as for the habitats to better assist policies plans and to deliver compensations based on it. This scheme can be established operationally for monitoring procedures bringing a considerable benefit to support managers, stakeholders and society during the emergency. On a longer term, the scientific rigor of the derived diagnoses can guide research and help to assist in objectively accounting losses and compensation needs as well as to design mitigation plans for future events. They can also provide a solid base of information to determine whether or not these events have a climatic component as well as to aid in the establishment of flood warning systems. It is an exciting avenue for future research how these services provided by science to society can be delivered with the Sentinel-2 mission, allowing new endeavours in high-resolution mapping and monitoring which will significantly increase the availability of crucial information after flooding events.

Author Contributions: Conceptualization, I.C., J.R., and G.N.; methodology, I.C.; formal analysis, I.C.; investigation, I.C., J.R., and G.N.; Writing—original draft preparation, I.C., J.R., and G.N.

Funding: This research was funded by the Spanish Ministry of Science, Innovation and Universities (MCIU), the State Research Agency (AEI), and the European Regional Development Fund (ERDF) in the frame of the Sen2Coast Project "Improved assessment of algorithms for bathymetry and water quality parameters with Sentinel-2A/B high-resolution satellites for advancement of coastal applications" (RTI2018-098784-J-I00).

Acknowledgments: We acknowledge the European Commission and the European Space Agency for the Sentinel-2 datasets. The authors acknowledge the three anonymous reviewers, whose comments helped to improve this manuscript.

Conflicts of Interest: The authors declare no conflict of interest. The funders had no role in the design of the study; in the collection, analyses, or interpretation of data; in the writing of the manuscript; or in the decision to publish the results.

References

1. European Commission. Directive 2007/60/EC of the European Parliament and of the Council of 23 October 2007 on the assessment and management of flood risks. *Off. J. Eur. Union L.* **2007**, *288*, 27–34.
2. ReNatCat. SERVICE Database Munich RE. Munich. 2014. Available online: https://www.munichre.com/en/solutions/for-industry-clients/natcatservice.html (accessed on 3 October 2019).
3. Jongman, B.; Ward, P.J.; Aerts, J.C. Global exposure to river and coastal flooding: Long term trends and changes. *Glob. Environ. Change* **2012**, *22*, 823–835. [CrossRef]
4. Barredo, J.I. Normalised flood losses in Europe: 1970–2006. *Nat. Hazards Earth Syst. Sci.* **2009**, *9*, 97–104. [CrossRef]
5. Feyen, L.; Dankers, R.; Bódis, K.; Salamon, P.; Barredo, J.I. Fluvial flood risk in Europe in present and future climates. *Clim. Change* **2012**, *112*, 47–62. [CrossRef]
6. Eurosion. Available online: http://www.eurosion.org/ (accessed on 25 September 2019).
7. European Commission Copernicus Emergency Management Service. Available online: http://emergency.copernicus.eu (accessed on 25 September 2019).
8. Blöschl, G.; Hall, J.; Parajka, J.; Perdigão, R.A.; Merz, B.; Arheimer, B.; Aronica, G.T.; Bilibashi, A.; Bonacci, O.; Borga, M. Changing climate shifts timing of European floods. *Science* **2017**, *357*, 588–590. [CrossRef]
9. Hall, J.; Arheimer, B.; Borga, M.; Brázdil, R.; Claps, P.; Kiss, A.; Llasat, M.C. Understanding flood regime changes in Europe: A state of the art assessment. *Hydrol. Earth Syst. Sci.* **2014**, *18*, 2735–2772. [CrossRef]
10. Bevacqua, E.; Maraun, D.; Vousdoukas, M.I.; Voukouvalas, E.; Vrac, M.; Mentaschi, L.; Widmann, M. Higher potential compound flood risk in Northern Europe under anthropogenic climate change. *Sci. Adv.* **2019**, *5*. [CrossRef]
11. Diez, J.J.; Esteban, M.D.; Silvestre, J.M. Understanding Extreme Spanish Coastal Flood Events. In *EGU General Assembly Conference Abstracts*; EGU: München, Germany, 2013; Volume 15, Available online: https://ui.adsabs.harvard.edu/abs/2013EGUGA..15.1957D/abstract (accessed on 7 October 2019).
12. Tanoue, M.; Hirabayashi, Y.; Ikeuchi, H. Global-scale river flood vulnerability in the last 50 years. *Sci. Rep.* **2016**, *6*, 36021. [CrossRef]
13. Bresciani, M.; Stroppiana, D.; Odermatt, D.; Morabito, G.; Giardino, C. Assessing remotely sensed chlorophyll-a for the implementation of the Water Framework Directive in European perialpine lakes. *Sci. Total Environ.* **2011**, *409*, 3083–3091. [CrossRef]
14. Sanyal, J.; Lu, X.X. Application of remote sensing in flood management with special reference to monsoon Asia: A review. *Nat. Hazards* **2004**, *33*, 283–301. [CrossRef]
15. Joyce, K.E.; Belliss, S.E.; Samsonov, S.V.; McNeill, S.J.; Glassey, P.J. A review of the status of satellite remote sensing and image processing techniques for mapping natural hazards and disasters. *Prog. Phys. Geogr.* **2009**, *33*, 183–207. [CrossRef]
16. Schnebele, E.; Cervone, G. Improving remote sensing flood assessment using volunteered geographical data. *Nat. Hazards Earth Syst. Sci.* **2013**, *13*, 669–677. [CrossRef]
17. Li, Y.; Martinis, S.; Plank, S.; Ludwig, R. An automatic change detection approach for rapid flood mapping in Sentinel-1 SAR data. *Int. J. Appl. Earth Obs. Geoinf.* **2018**, *73*, 123–135. [CrossRef]
18. Ovando, A.; Martinez, J.M.; Tomasella, J.; Rodriguez, D.A.; von Randow, C. Multi-temporal flood mapping and satellite altimetry used to evaluate the flood dynamics of the Bolivian Amazon wetlands. *Int. J. Appl. Earth Obs. Geoinf.* **2018**, *69*, 27–40. [CrossRef]
19. Wang, Y. Using Landsat 7 TM data acquired days after a flood event to delineate the maximum flood extent on a coastal floodplain. *Int. J. Remote Sens.* **2004**, *25*, 959–974. [CrossRef]
20. Sentinel Constellation. Available online: https://sentinel.esa.int/web/sentinel/missions (accessed on 25 September 2019).
21. European Space Agency. Sentinel-2 User Handbook. ESA Standard Document Paris, France 2015. Available online: https://sentinel.esa.int/documents/247904/685211/Sentinel-2_User_Handbook (accessed on 1 October 2019).
22. European Space Agency. Sentinel-2 MSI Technical Guide 2017. Available online: https://earth.esa.int/web/sentinel/technical-guides/sentinel-2-msi (accessed on 1 October 2019).
23. Sentinel's Scientific Data Hub. Available online: https://scihub.copernicus.eu/ (accessed on 19 September 2019).

24. Vanhellemont, Q.; Ruddick, K. Atmospheric correction of metre-scale optical satellite data for inland and coastal water applications. *Remote Sens. Environ.* **2018**, *216*, 586–597. [CrossRef]
25. Vanhellemont, Q. Adaptation of the dark spectrum fitting atmospheric correction for aquatic applications of the Landsat and Sentinel-2 archives. *Remote Sens. Environ.* **2019**, *225*, 175–192. [CrossRef]
26. Nechad, B.; Ruddick, K.G.; Park, Y. Calibration and validation of a generic multisensor algorithm for mapping of total suspended matter in turbid waters. *Remote Sens. Environ.* **2010**, *114*, 854–866. [CrossRef]
27. International Ocean-Colour Coordinating Group (IOCCG). *Remote Sensing of Ocean Colour in Coastal, and Other Optically-Complex, Waters*; IOCCG: Dartmouth, NS, Canada, 2000.
28. Kutser, T.; Paavel, B.; Verpoorter, C.; Ligi, M.; Soomets, T.; Toming, K.; Casal, G. Remote sensing of black lakes and using 810 nm reflectance peak for retrieving water quality parameters of optically complex waters. *Remote Sens.* **2016**, *8*, 497. [CrossRef]
29. Aemet Open Data. Available online: https://datosclima.es (accessed on 27 September 2019).
30. EO Browser. Available online: https://www.sentinel-hub.com/explore/eobrowser (accessed on 23 September 2019).
31. Liu, H.; Li, Q.; Shi, T.; Hu, S.; Wu, G.; Zhou, Q. Application of sentinel 2 MSI images to retrieve suspended particulate matter concentrations in Poyang Lake. *Remote Sens.* **2017**, *9*, 761. [CrossRef]
32. Caballero, I.; Steinmetz, F.; Navarro, G. Evaluation of the first year of operational Sentinel-2A data for retrieval of suspended solids in medium-to high-turbidity waters. *Remote Sens.* **2018**, *10*, 982. [CrossRef]
33. Caballero, I.; Navarro, G. Retrieval of Suspended Solids from Landsat-8 and Sentinel-2: A Tool for Coastal Monitoring in Extremely Turbid Waters. In Proceedings of the IGARSS 2018–2018 IEEE International Geoscience and Remote Sensing Symposium 2018, Valencia, Spain, 22–27 July 2018; pp. 7874–7877. [CrossRef]
34. Vanhellemont, Q.; Ruddick, K. Acolite for Sentinel-2: Aquatic applications of MSI imagery. In Proceedings of the 2016 ESA Living Planet Symposium, Prague, Czech Republic, 9–13 May 2016.
35. DigitalGlobe. Available online: https://www.digitalglobe.com/company/ (accessed on 25 September 2019).
36. Planet Labs. Available online: https://www.planet.com/ (accessed on 28 September 2019).
37. Amarnath, G. An algorithm for rapid flood inundation mapping from optical data using a reflectance differencing technique. *J. Flood Risk Manag.* **2014**, *7*, 239–250. [CrossRef]
38. Chignell, S.; Anderson, R.; Evangelista, P.; Laituri, M.; Merritt, D. Multi-temporal independent component analysis and Landsat 8 for delineating maximum extent of the 2013 Colorado front range flood. *Remote Sens.* **2015**, *7*, 9822–9843. [CrossRef]
39. Goffi, A.; Stroppiana, D.; Brivio, P.A.; Bordogna, G.; Boschetti, M. Towards an automated approach to map flooded areas from Sentinel-2 MSI data and soft integration of water spectral features. *Int. J. Appl. Earth Obs. Geoinf.* **2020**, *84*, 101951. [CrossRef]
40. McFeeters, S. The use of the Normalized Difference Water Index (NDWI) in the delineation of open water features. *Int. J. Remote Sens.* **1996**, *17*, 1425–1432. [CrossRef]
41. Xu, H. Modification of normalised differrence water index (NDWI) to enhance open water features in remotely sensed imagery. *Int. J. Remote Sens.* **2006**, *27*, 3025–3033. [CrossRef]
42. Shen, L.; Li, C. Water Body Extraction from Landsat ETM+ Imagery Using Adaboost Algorithm. In Proceedings of 18th International Conference on Geoinformatics 2010, Beijing, China, 18–20 June 2010; pp. 1–4.
43. Fisher, A.; Flood, N.; Danaher, T. Comparing Landsat water index methods for automated water classification in eastern Australia. *Remote Sens. Environ.* **2016**, *175*, 167–182. [CrossRef]
44. Schumann, G.; Bates, P.D.; Horritt, M.S.; Matgen, P.; Pappenberger, F. Progress in integration of remote sensing-derived flood extent and stage data and hydraulic models. *Rev. Geophys.* **2009**, *47*. [CrossRef]
45. Arnell, N. Flood insurance. *Floods* **2000**, *1*, 412–424.
46. Brown, K.M.; Hambidge, C.H.; Brownett, J.M. Progress in operational flood mapping using satellite synthetic aperture radar (SAR) and airborne light detection and ranging (LiDAR) data. *Prog. Phys. Geogr.* **2016**, *40*, 196–214. [CrossRef]
47. Dasgupta, A.; Grimaldi, S.; Ramsankaran, R.A.A.J.; Pauwels, V.R.; Walker, J.P.; Chini, M.; Matgen, P. Flood mapping using synthetic aperture radar sensors from local to global scales. In *Global Flood Hazard: Applications in Modeling, Mapping, and Forecasting*; John Wiley & Sons: Hoboken, NJ, USA, 2018; pp. 55–77. Available online: https://agupubs.onlinelibrary.wiley.com/doi/10.1002/9781119217886.ch4 (accessed on 27 November 2019). [CrossRef]
48. Pritchard, M.E.; Yun, S.H. Satellite Radar Imaging and Its Application to Natural Hazards. In *Natural Hazards*; Singh, R., Bartlett, D., Eds.; CRC Press: Boca Raton, FL, USA, 2018; pp. 95–114.

49. Clement, M.A.; Kilsby, C.G.; Moore, P. Multi-temporal synthetic aperture radar flood mapping using change detection. *J. Flood Risk Manag.* **2018**, *11*, 152–168. [CrossRef]
50. Erena, M.; Domínguez, J.A.; Aguado-Giménez, F.; Soria, J.; García-Galiano, S. Monitoring Coastal Lagoon Water Quality through Remote Sensing: The Mar Menor as a Case Study. *Water* **2019**, *11*, 1468. [CrossRef]

Article

An Observational Process Ontology-Based Modeling Approach for Water Quality Monitoring

Xiaolei Wang [1], Haitao Wei [1,*], Nengcheng Chen [2], Xiaohui He [1] and Zhihui Tian [1]

1 School of Geo-Science and Technology, Zhengzhou University, Zhengzhou 450052, China; xiaolei8788@zzu.edu.cn (X.W.); hexh@zzu.edu.cn (X.H.); iezhtian@zzu.edu.cn (Z.T.)

2 State Key Laboratory of Information Engineering in Surveying, Mapping and Remote Sensing, Wuhan University, Wuhan 430079, China; cnc@whu.edu.cn

* Correspondence: zzu_wei@zzu.edu.cn

Received: 31 January 2020; Accepted: 2 March 2020; Published: 5 March 2020

Abstract: The increasing deterioration of aquatic environments has attracted more attention to water quality monitoring techniques, with most researchers focusing on the acquisition and assessment of water quality data, but seldom on the discovery and tracing of pollution sources. In this study, a semantic-enhanced modeling method for ontology modeling and rules building is proposed, which can be used for river water quality monitoring and relevant data observation processing. The observational process ontology (OPO) method can describe the semantic properties of water resources and observation data. In addition, it can provide the semantic relevance among the different concepts involved in the observational process of water quality monitoring. A pollution alert can be achieved using the reasoning rules for the water quality monitoring stations. In this study, a case is made for the usability testing of the OPO models and reasoning rules by utilizing a water quality monitoring system. The system contributes to the water quality observational monitoring process and traces the source of pollutants using sensors, observation data, process models, and observation products that users can access in a timely manner.

Keywords: observational process ontology; water quality monitoring; water pollution alert; semantic discovery

---

1. Introduction

Water quality monitoring using in-situ environmental sensors presents many challenges for data discovery, particularly for the warning and querying of water pollution. The growth in freshwater demand has forced water managers and decision makers to contemplate comprehensive strategies for water quality monitoring [1]. Different large sensory data collection methods have the problems of too much data but not enough useful information, a lack of integration of different database sources, and poor interoperability between these resources [2].

Moreover, monitoring practices are further complicated due to the increasing variety of sensors and multiple user needs. It is therefore important to consider approaches for using ontology in combination with rules to enhance the discovery of pollution events using water quality monitoring [3]. Semantic technology has already been identified as a challenging research topic for various disciplinary and domain applications [4]. Sheth and Sahoo proposed the semantic sensor web (SSW) [5]. Related sensor ontology models describe sensors and their observations, the studied features of interest, the observed properties, as well as actuators [6,7].

Semantic techniques could enhance the quantity and quality of information available for water quality management [8–10]. The use of semantics in sensor networks enables the discovery and analysis of sensor data based on spatial, temporal, and thematic information [11]. To enable the encoding of semantics using the observation data in the Sensor Web, well-known technologies, such as web ontology language (OWL) [12] and semantic web rule language (SWRL) [13], have been used. The OWL is an ontological language for the semantic web with formally defined meanings that provide the description of classes, properties, individuals, and data values, and these are stored as Semantic Web documents. SWRL includes a high-level abstract syntax for rules in both the OWL DL and OWL Lite sublanguages of OWL.

These tools provide two promising areas for water quality monitoring research: (a) the semantic modeling of water observation data and (b) how to use the semantic information of observation data in a water quality monitoring system.

(a) With regard to the first research area, the ontology models can be used for the description of water environmental resources. A hydrological sensor web ontology based on the Semantic Sensor Network (SSN) ontology has been proposed to describe the heterogeneous hydrological sensor web resources [14]. An ontology model has been specially designed for river water quality monitoring [15]. The related ontology primarily focuses on sensor devices and observation data for water management, but it lacks a description of the observational process and models. It has been increasingly applied to observational resources from various sensors to provide data management and query capabilities [10].
(b) The second research area, smart water quality monitoring, which is regarded as the future of water quality monitoring technology, promotes progress in data collection capabilities, communication, data analysis, and early warning strategies [16]. Ontology and rules have been applied to manage water sources. For example, lake water quality monitoring is based on an integration of a semantic framework [17]. Mihaela et al. [18] a knowledge modeling framework for intelligent environmental decision support systems integrated an ontological approach and data analysis approach. This was then applied for the generation of a knowledge base for decision making.

In summary, there has been a growing awareness that ontology and rules should be utilized. However, there exist two limitations in the related work: (1) Only using the observation data values from one type of monitoring station could easily cause an initial warning misjudgment. (2) The results of the query focus on observation data, and lacks a complete display of observational process. For example, only the data value of the water quality at the location where the accident occurred is shown on the map. This method cannot trace the data and models associated with water pollution.

Successful water quality monitoring requires an understanding of the observational processes of sensors, as well as understanding of the pollution warning process. In this study, the objective is to is to achieve a semantic modeling of the observational process in Sensor Web by integrating water quality monitoring and alerts. To achieve this goal, an observational process ontology modeling method is described, which consists of observation acquisition, observation data, observation processing, and observation products. In addition, some rules are defined that can be used in combination for tracing the entire process of pollution events.

This study is organized as follows: In Section 2, a brief overview of study area is presented. In Section 3, an ontology of the water quality observational process and the reasoning rules for the discovery of water quality are described. The discovery service performed and discussions are described in Section 4. Finally, Section 5 outlines the conclusions and directions for future work.

2. Study Area

Qingyi River passes through the city center of Xuchang and has a drainage area of 2362 km^2 in the Henan Province of China. The total length of the Qingyi River is 149 km, and it empties into the Ying River. The river basin has a warm temperate humid monsoon climate, four distinct seasons, rich heat energy, concentrated rainfall, and abundant sunshine.

The Qing Yi River is an important water body for flood control and pollution prevention for the region and the landscape of Xuchang City. Hence, the water quality is of great significance for the people and to provide healthy urban development. The primary pollution source of the Qingyi Basin pollutant emissions originates from the leather, wig products, and paper industries in the area. Xuchang is China's largest hair products distribution center and export production base. The export value of hair products accounts for 49% of China's exports of similar goods. Xuchang City's COD and NH_3 emissions from wig products, papermaking, and leather industries accounted for more than 80% of the total industrial COD emissions in the basin, respectively [19]. Therefore, it is important to improve water quality monitoring (especially COD and NH_3N monitoring) to reduce industrial effluent pollution.

This study utilized two types of monitoring stations: 16 fixed monitoring stations (the red points in Figure 1) and 21 monitoring stations near the polluting enterprises (the blue points in Figure 1). Various sensors were deployed on these monitoring stations, and the observations made by these sensors were utilized as data. The observational parameters that were acquired by these sensors included the chemical oxygen demand (COD), NH_3-N, and other water quality factors of the Qingyi River Basin (see Section 3.1.3 for details). Data from the Qingyi River were collected from 1 June 2017 to 31 December 2019. The data were acquired using in-situ sensors located in the Qingyi River. The data of NH_3N, COD, P, N, PH, DO, and turbidity come from sensors produced by the Beijing SDL Technology Co., Ltd. (Beijing, China). The manufacturers of sensors for TOC and BOD_5 are the Beijing Security Control Technology Co., Ltd. (Beijing, China) and the Shanghai Kelan Instrument Technology Co., Ltd., (Shanghai, China) respectively.

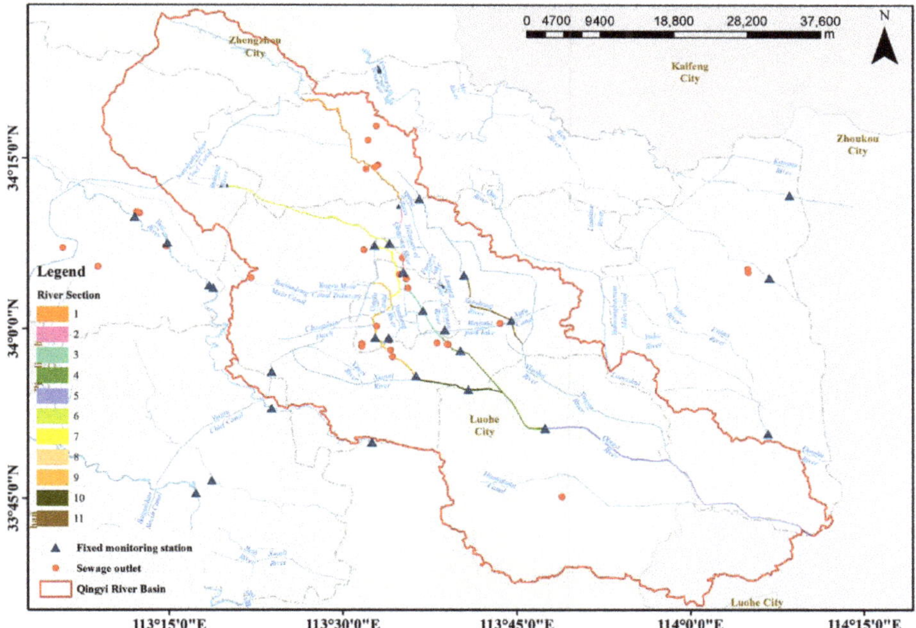

Figure 1. The Qingyi River study area.

3. Ontology and Rules

The procedure used for the ontology modeling demonstrated how to describe the observational process of observations for water quality monitoring (see Section 3.1). Based on the proposed ontology, the rules enable the development of a water quality early warning solution that could be adapted to different river sections (see Section 3.2).

3.1. The Proposed Observational Process Ontology

The construction of the ontology follows seven steps: (1) Determine the domain and scope of the ontology; (2) consider reusing existing ontologies; (3) enumerate important terms in the ontology (core classes); (4) define the classes and the class hierarchy; (5) define the properties of classes—slots; (6) define the facets of the slots; and (7) create instances (take COD and NH_3N as examples). For steps 1, 2, and 3, see Section 3.1.1. For steps 4–6, see Section 3.1.2. For the last step, see Section 3.1.3.

3.1.1. The Core Classes

The ontology revolves around the central pattern of the observational process. Several conceptual modules have been built on patterns that contain observational process concepts. The ontology modelling method refers to the Descriptive Ontology for Linguistic and Cognitive Engineering (DOLCE) Ultra-Light foundational ontology to assist users in understanding the proposed ontology. The DOLCE Ultra-Light (DUL) ontology is the most utilized ontology in the engineering cognitive domain and in linguistics, which defines 29 classes and 43 properties, including the Object, Region, Quality, and InformationEntity [20].

The observational process ontology (OPO) of water quality based on the DOLCE has the following features: (1) The general top-level definition concepts of DOLCE can be used to ensure interoperability with other resources. (2) It can implement the reuse of the existing ontology. For example, the concepts in the existing ontology can be mapped to the top-level ontology, which can then realize the integration of the domain ontology. (3) The shared terminology and definitions of the DOLECE enable the interoperability of the OPO.

Therefore, the developed ontology uses similar design patterns to those of the DUL ontology; and hence focuses on the quality (*DUL:Qaulity*) and the object (*DUL:Object*) of the entity, as well as the region (*DUL:region*) of the entity and, in particular, their resulting event (*DUL:Event*). Figure 2 shows the DUL core classes, relations, and the ontological modules.

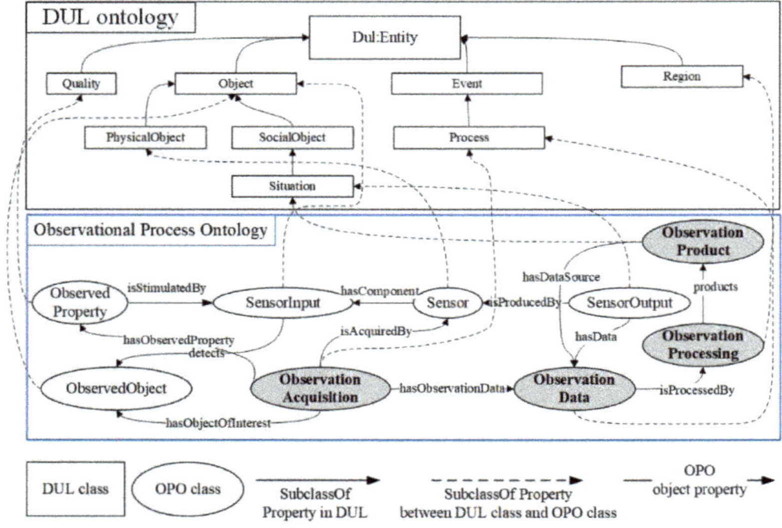

Figure 2. Core classes of the Observational Process Ontology (OPO) and DUL ontology. The top of the figure shows the upper level ontology (DUL ontology). The sub-class relationships between DUL classes are indicated by the solid arrows. The dashed arrows represent the subclass relationships between the OPO classes. The object properties of OPO are expressed by the annotated solid arrows.

The majority of the ontology is hence an extension of these classes with subclasses and additional properties, although concepts related to water quality features were also added. The OPO of the water quality modeled the concepts, relationships, properties, datatypes, and restrictions of the water quality of a river using OWL constructs.

This approach was grounded in a view of the observational process consisting of four main observational phases: observation acquisition, observation data (acquired directly by sensors), observation processing (used to describe the model), and the processed observation products. The ontology therefore captured the domain knowledge from these four core classes and across each of these four perspectives.

3.1.2. The Core Modules

The view of the ontology presented in Figure 3 describes the following:

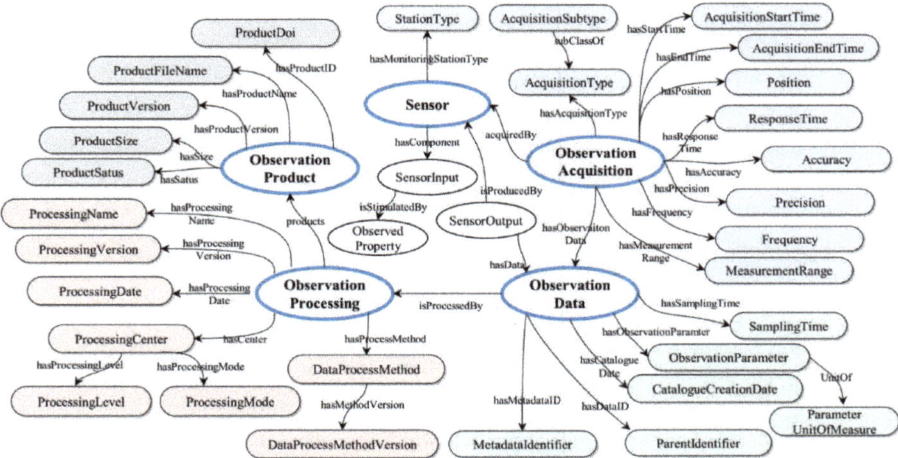

Figure 3. Core modules of the Observational Process Ontology (OPO). The classes and properties of OPO are described using four core classes (Observation Acquisition, Observation Data, Observation Processing, and Observation Product) and the Sensor class. It shows the links involved during each stage of the observational process.

Ontology classes from the 'Observation Acquisition' perspective

Water quality changes during different periods. Therefore, there are two important classes (*opo:AcquisitionStartTime* and *opo:AcquisitionEndTime*) of the observation acquisition for water quality monitoring. The *opo:ResponseTime* class expresses the time between a change in the value of an observed property and when the sensor 'settles' on an observed value.

Additionally, the following are the seven classes that describe the measurement capability of the observation acquisition.

- The *opo:Position* class is used to describe the corresponding river section from where the observation data were obtained.
- The closeness of agreement between the result of an observation and the true value of the observed property under the defined conditions is described using the *opo:Accuracy* class.
- The *opo:Precision* class represents a measure of a sensor's ability to consistently reproduce an observation of an unchanged or similar quality value.
- The *opo:Frequency* class denotes the smallest possible time between one observation or sampling and the next.

- The *opo:MeasurementRange* class is used to describe the set of values that the sensor can return as the result of an observation using the defined sensor device properties.
- The *opo:AcquisitionType* class is used to distinguish the equipped platform, for example, *in-situ*, UAV, satellite.
- The *opo:AcquisitionSubtype* class describes the means of acquisition, for example, electromagnetic, ultrasonic, laser, visual, pressure, and other methods.

Ontology classes from the 'observation data' perspective

Large amounts of data are collected by the sensing device and are processed and transmitted. The temporal characteristics of the observation data are described using two classes:

- The *opo:SamplingTime* class records when the observation data was retrieved from the observed property.
- The *opo:CatelogueCreationDate* class is used to describe the creation date for the metadata item. The creationdate is the date when the metadata item was inserted for the first time in the catalogue.

The key features of observation data are the observation parameter and the ID, which are described using four classes:

- The *opo:ObservationParameter* class can describe a single phenomenon, a constrained phenomenon, or composite phenomenon, such as height, radiance bands, or several discrete phenomena.
- The *opo:ParameterUnitOfMeasure* class denotes units of measurements for the observed phenomenon.
- The *opo:ParentIdentifier* class represents the collection identifier of observed data.
- The *opo:MetadataIdentifier* class records the identifier for metadata item.

Ontology classes from the 'observation processing' perspective

There are eight classes related to observation processing. Each class is defined as follows.

- The *opo:ProcessingDate* class is used to describe the date time of processing.
- The processor software name and version are represented using the *opo:ProcessingName* and *opo:ProcessingVersion* class separately.
- The *opo:DataProcessMethod* class represents the method used to compute the datalayer.
- The *opo:DataProcessMethodVersion* class describes the version of the method.
- The processing center code is expressed using the *opo:ProcessingCenter* class.
- The *opo:ProcessingLevel* class denotes the processing level applied to the product.
- The *opo:ProcessingMode* class records the processing mode taken from the mission specific code list.

Ontology classes from the 'observation product' perspective

The 'observation product' perspective includes five aspects.

- The *opo:ProductDoi* class is used to describe the digital object identifier of the product.
- The *opo:ProductFileName* class represents the reference to the file.
- The version of the product is described using the *opo:ProductVersion* class.
- The *opo:ProductSize* class denotes the product size (bytes), allowing the user to realize how long a download is likely to take.
- The *opo:ProductSatus* class refers to the product status, for example, archived, acquired, cancelled, failed, planned, potential, or rejected.

Properties of the OPO

The properties include object properties and data properties. The data properties include the *'opo:hasDataValue'* property (which expresses the result of an observation or actuation); the *'opo:hasSectionValue'* property (which expresses the section number of the river); and the *'opo:hasProductValue'* property (which expresses the value through the observation products). The object properties in the OPO link classes/individuals to class/individuals. The object properties can be divided into six categories:

- The subclass property: The relationships between the classes of the upper level ontology and OPO classes can be described by the opo:subClass property. The Subtype class is subclass of type class, for example, the 'AcquisitionSubtype' class is subclass of 'AcquisitionType' class.
- The identifier properties: The identifier properties include two types; one of which is the 'hasDataID', 'hasMetadataID', and 'hasProductID' properties; the other is the 'hasProcessingName' and 'hasProductName' properties.
- The time properties: The related properties refer to the 'hasStartTime', 'hasEndTime', and 'hasSamplingTime' properties.
- The position properties: The 'hasPosition' property represents the position of the observation acquisition.
- The classification property: the 'isClassifiedBy' property that describes the type of observation acquisition.
- The domain-specific properties: Other properties are related to specific attributes during the observational process; for example, the 'hasAccuracy', 'hasPrecision', 'hasFrequency', 'hasProcessinglevel', and 'hasProcessMethod' properties.

3.1.3. Ontological Implementation

The OPO based on the ontological structure described in Sections 3.1.1 and 3.1.2 was created using Protégé [21]. The individuals of the *'opo:ObservedParameter'* class in the ontology correspond to water quality indicators. The water quality indicators, include conductivity, chlorophyll content, dissolved oxygen, dissolved organic matter, pH, permanganate index, turbidity, total nitrogen, and water temperature [22]. There are nine types of indicator (NH_3N, COD, TOC, BOD_5, P, N, DO, PH, and turbidity) obtained by the fixed monitoring stations and six types of water indicators (NH_3N, COD, TOC, BOD_5, P, and N) acquired by the monitoring stations of polluting enterprises. These factors can be defined in the ontology individuals and applied to the formulation of subsequent rules.

The time series of COD and NH_3N in the Qingying River Basin are more complete than other indicators; hence, these two indicators were used as examples to set the ontology individuals and to describe the rule examples. The primary water quality monitoring-related individuals contain the 'COD' and 'NH_3N' individuals of the *'opo:ObservedParameter'* class; the 'FixedMonitoring', and the 'PollutionEnterprise' individuals of the *'opo:StationType'* class; the 'CODprocessmethod', and the 'NH_3Nprocessmethod' individuals of the *'opo:DataProcessMethod'* class. The properties are the most important elements and build the connection between the heterogeneous observational resources. The primary classes and properties in the proposed ontology are shown in Figure 4.

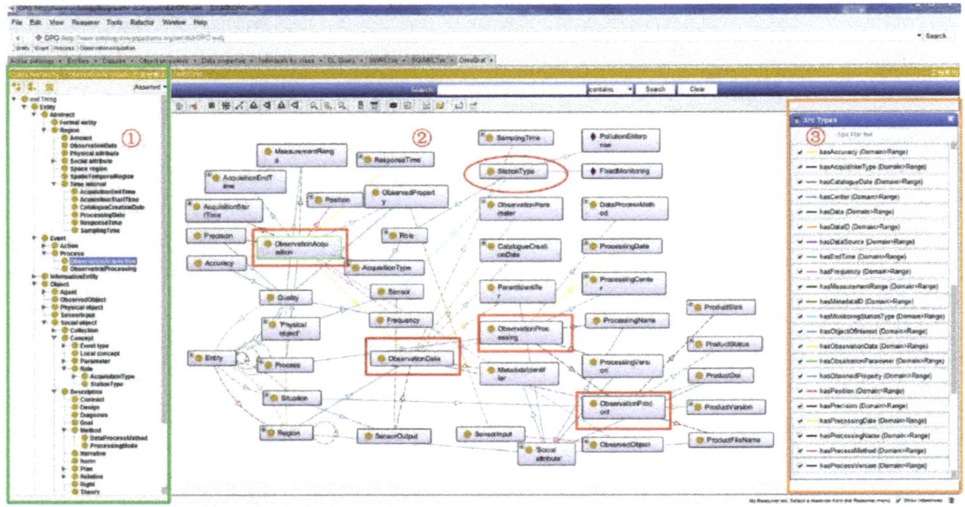

Figure 4. shows the schematic structure of the OPO in protégé. The first portion on the left side shows the subclass relationship between the classes. The third portion on the right side represents the object properties between the classes, which are shown in second portion in the middle of the figure.

3.2. *Reasoning Rules*

A water quality query is a complex task that requires intelligent inference based on specific rules. Owing to this requirement, with ontology used in memory, additional rules have been developed to extend the information inferred by the semantic framework.

SWRL can be used to write rules to reason about OWL individuals and to infer new knowledge about those individuals. Rules are in the form of an implication between an antecedent (body) and a consequent (head), that is, antecedent → consequent. The intended meaning of a rule is: if the antecedent holds (i.e., "true"), then the consequent must also hold. Both the antecedent and consequent of a rule are conjunctions of atoms. The syntax in SWRL can be used to express the judgment of water quality monitoring. The antecedent of the rule described by SWRL includes the statement of classes and properties, as well as judgment of the conditions, while the consequent represents inferred properties. The classes, object properties, and data properties of rules come from the observational process ontology. The judgment of rule is described in built-ins for comparisons (e.g., swrlb:equal, swrlb:greaterThan) and strings.

According to the types of monitoring stations, this study defines the rules for fixed monitoring stations (the blue points in Figure 1) and sewage outlets of the polluting enterprises (the red points in Figure 1) based on SWRL, which are implemented in SWRLTab of protege. The design of the rules considers the following two points: (1) Different types of monitoring stations cannot be defined uniformly, so the rules are described in Sections 3.2.1 and 3.2.2 separately. (2) The rules of fixed monitoring stations are suitable for judging a certain parameter value at a fixed monitoring position. The rules of enterprise sewage outlets could be used to determine the values after the pollution has spread.

For the Qingyi River, the upper limit of the water quality factors was determined by the Chinese GB3838 regulation, socio-economic, industrial and mining enterprises, population distribution, agricultural production, pollution capacity, and water quality status along the river. In this section, the river basin was divided into 11 river sections, which are shown in Figure 1. By using COD and NH_3N as examples, the upper limit values for each river section are shown in Table 1, which were used as the alert maximum values in the rules.

Table 1. The upper limit of COD and NH_3N for each section of the Qingyi River Basin. The section number corresponds to the legend in Figure 1.

Section Number	Upper Limit of COD (mg/L)	Upper Limit of NH_3N (mg/L)
1,2,3,11	40	2
4,5,6,7,9,10	30	1.5
8	20	1

3.2.1. The Rules for Fixed Monitoring Stations

As Shown in Figure 5, Rule 1 (*Rule hasFixedAlertMaxValue*) sets the water quality alert maximum values according to river sections. Rule 2 (*Rule hasFixedAlert*) determines whether an alert of a fixed monitoring station is active. The sensor descriptions, observation data descriptions, alert descriptions, asset descriptions, and dynamic data are all semantically related in Rule 2. The key benefit of the knowledge base of this inference approach is that it is possible to directly infer whether or not an observation data reading from a sensor equipped at a fixed monitoring station should trigger an alert. If the reading is greater than the allowable maximum, the alarm would be triggered. The allowable maximum depends on the type of water quality indicators, which are described by the observational parameter in the OPO models.

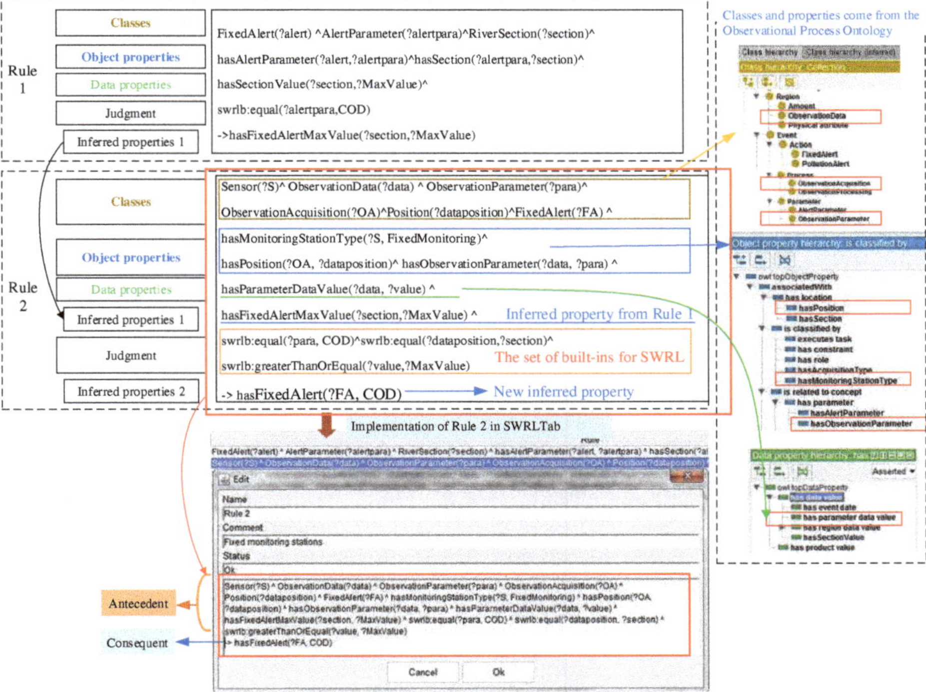

Figure 5. The definition and implementation of rules for fixed monitoring stations for water quality monitoring and warning. The COD element is used as an example, so Rule 1 could obtain the allowable maximum value according to the value of the river monitoring section, which corresponds to the data in Table 1. Suppose that the value of section reaches 4, then the maximum value of COD would be 30 mg/L. Then, the value of observation data would be greater than 30, and Rule 2 would be triggered.

3.2.2. The rules for Monitoring Stations of Polluting Enterprises

Rule 3, Rule 4, Rule 5, and Rule 6 describe the pollution alert for monitoring stations of the polluting enterprises, which can infer the pollution type, the process output, and determine if an alert is triggered. Examples of the four rules are illustrated in Figure 6.

Figure 6. Rules for monitoring stations of the polluting enterprises. The rules in Figure 6 differ from the rules in Figure 5 for the fixed monitoring stations in that the alert for the polluting enterprises outlet require the prediction of pollutants (Rule 3) and the calculation of the pollution diffusion concentration (Rule 4). The warning is a judgment on the value of the observation products generated after diffusion. The position and sampling time of the observation data can be obtained according to Rules 3 and 4, which facilitate the traceability of pollution sources for the discovery service.

When the data value of the observation data acquired by a sensor is greater than the maximum acceptable value, the relationship of 'hasPollutionType' between the observation data and the observation parameter is inferred and built. According to the type of observation data and the situation of the watershed, the rule sets the maximum acceptable value. In addition, the position of related sensor can be acquired. This capability is provided by Rule 3 (*Rule hasPollutionType*).

If the data has the relationship of 'hasPollutionType', which is activated by Rule 3, then Rule 4 (*Rule DataProcessOutput*) is required for data processing. In order to present the processing method and the processed results, it is useful to infer the link between the data method and the observation products. This is accomplished using Rule 4.

The Qingyi River is a small river with a relatively slow flow velocity, and the discharged pollutants can be uniformly mixed in the entire cross section in a short time. The one-dimensional water quality model is more suitable to describe the pollution spread for the Qingyi River basin. The model corresponds to '*opo:DataProcessMethod*' class in the ontology and rules. Therefore, the equations of process method for COD and NH_3N are described as follows:

$$C_x = C_0 \exp\left(-\frac{KL}{v \times 86400}\right), \tag{1}$$

where C_0 (mg/L) is the concentration of pollutants from the monitoring station, K represents the pollutant degradation coefficient, and it is a constant, L (m) is the longitudinal distance along the stream, and V (m/s) is the velocity.

Rule 5 (*Rule hasAlertMaxValue*) has the same function as Rule 1, which can acquire the allowable maximum. The Rule 6 (*Rule isActivePollutionAlert*) relates an alert to a Boolean literal, which is different from Rule 2 (*Rule hasFixedAlert*). The Rule 6 works with the output results from Rule 4 and Rule 5. This can occur if the reading is greater than the maximum.

4. Deployment of the Proposed Approach and Discussion

The OPO and rules are used to deploy the semantic-enhanced discovery system. The data of observational process is mapped to the OPO for polluter tracing through logical inference rules. In addition, according to the proposed approach, the three aspects of modeling methods, rules, and discovery service are discussed.

4.1. Demonstration Using a Case Scenario

The semantic-enhanced discovery system for water quality monitoring that was designed and developed is shown in Figure 7. The function of water quality discovery is to provide a quick operation of user query requests. Users can obtain observation data from the station type of sensors, the station list, and the time-required. Then, data processing of the observation data can be performed by setting the diffusion distance and flow velocity. The service also includes other functional modules, such as the data management functions, map layers, and other GIS functions.

Once there is a pollutant concentration that exceeds the limitation value, the system will automatically alert with the experiential and mechanistic processing. Then, when it is judged that there is alert information according the rules in Section 3.2, the warning content will be displayed on the interface. Figure 7 shows the query of a fixed monitoring station and the alert results for water pollution. As can be seen in the query results, the semantic reasoning and querying can arrive at the expected observation product and judge whether to publish an alert.

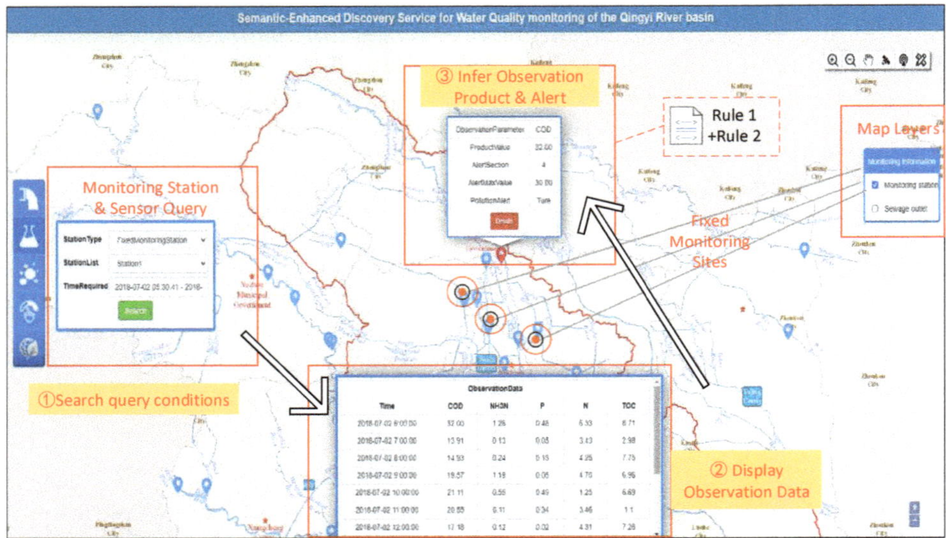

Figure 7. Semantic-enhanced discovery interface of the software system. The discovery service includes monitoring station information, observation data, the observation products, and warning based on Rule 1 and Rule 2.

4.2. Discussion

The observational process ontology (OPO) proposed in this study is able to describe the observational process for water quality monitoring and support the observational resource discovery (see Section 3.1). Water quality early warning information can be deduced through the definition of the rules, which make conditional judgments of classes and attributes from the OPO (see Section 3.2). The semantic-enhanced discovery service for alerting for water pollution uses the proposed ontology and rules (see Section 4.1). Therefore, this section makes a comparison using the three aspects of modeling methods, rules, and discovery services.

4.2.1. Comparison with Other Related Modeling Methods

(1) Comparison with non-semantic models

Over the past decade, there has been a drastic increase in the use of decision support systems. A decision support system has been developed to address water resource problems, including water allocation, water use management and water supply management. Many researchers have been working on the development of various models and decision support tools to address the complexity in the management of water resources by integrating utilization models and tools for water environmental decision support [23–26]. The models used in these decision support systems are defined as non-semantic models, which lack the observational process description.

Among these models, the typical model is WaterML 2.0 [27], which is being used in the representation of water observation data and water quality monitoring. The WaterML standard describes a time series that results from direct observations and processed data for encoding water observations data for exchange, such as forecasts and derived results. The WaterML 2.0 defines five primary components that describe water observations: time series, observation specializations, procedures, observation metadata, and the location description. The observation resources in the above model are described separately, which cannot meet the requirements for expression of the water quality monitoring process comprehensively and dynamically. The proposed ontology based on the DOLCE ontology has comprehensive descriptions of the observational process, including observation acquisition, observation data, observation processing, observation product, and other factors, as shown in Figure 8. In addition, the proposed ontology makes it possible to obtain more observational resources collaboratively.

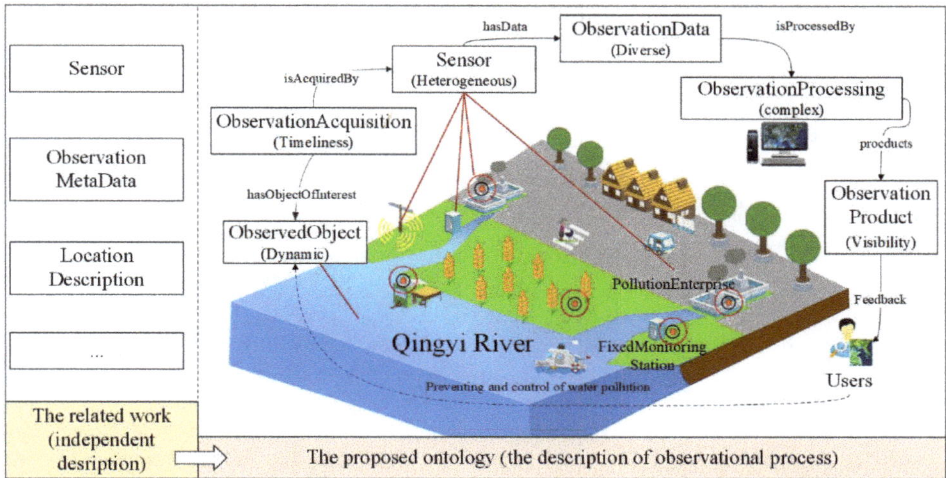

Figure 8. Characteristics of the Observational Process Ontology. The description of the ontological structure in Section 3.1 corresponds to each stage of the observational process in the real world. The proposed method can not only visualize the results to users, but also help users trace information regarding pollution sources.

(2) Comparison with other ontologies

The difference between the proposed ontology and other ontologies is the key concepts. For example, Elag and Goodall [28] proposed a water resources component (WRC) ontology for water resource model components, which describes core concepts (symbol, equation, and mathematical classification) and relationships using the OWL. An ontology [29] based on the SSN was proposed to describe the time series and assess the reliability of a hydrological sensor network. Liu et al. [30] presented a WaterML and SSN-based ontology framework that represented information from sensors, observation values, and timeseries for a distributed water information system. In summary, most related studies have been based on the SSN ontology, and the following is a comparison of the SSN ontology with the proposed ontology.

The sensor and sensor network ontology (SSN Ontology), developed by W3C [6,7], aims to improve sensing applications. It contains a comprehensive representation of information. A comparison between the SSN ontology and the proposed ontology (OPO) is shown in Table 2. Some of the differences are as follows: (1) the SSN has more comprehensive concepts of sensor metadata than the OPO, while the OPO describes more detailed observational processes. (2) The SSN and OPO can be used for in-situ sensors, and the OPO also defines descriptions related to water quality monitoring and pollution alerting.

Table 2. Comparison between the SSN and OPO ontologies.

Features	Observational Process Ontology (OPO)	Semantic Sensor Network Ontology (SSN)
Target	A semantic description of the observational process for better discovery	Development of ontologies for describing sensors
Key concepts	Sensors, observation acquisition, observation data, observation processing, observation product	Sensors, system, actuations, observation
Range of subject matter	Moderately broad	Comprehensive
Adoption	Water quality monitoring and pollution alerting	Numbers of examples, such as weather stations
Best features	Supports the description of observational processes	Interoperability and broader applicability
Weakest features	Insufficient description of sensor metadata	Lack of observational process descriptions, especially for water quality monitoring

In addition, there are some ontology modeling methods that have been applied to water resource management, but their application fields are slightly different from the proposed method. For example, Klien et al. presented [31] an application in disaster management that examined to what extent ontology-based service discovery can solve these semantic heterogeneity problems for storm hazards. An ontology engineering method [32] was applied for the description of the Water–Energy–Food nexus domain for enhancing the compatibility of qualitative descriptions logically or objectively. The ontology includes the concepts and relationships of 'precipitation' and 'fish'. Future research in the proposed ontology will add a description of aquatic ecology based on the observational process for more efficient water resource management.

4.2.2. Comparison with Pollution Alert Rules

With the rapid development of semantics into the sensor web, more rules are being defined for water pollution early-warning. As well as promoting semantic interoperability, inference represents a key advantage of ontological modeling. Earlier studies of water pollution alert rules focused on the inference of one type from a monitoring station. For example, Shu et al. [33] defined an ontology for the water data transfer format and validated data with respect to constraints through a query technique using OWL and SWRL. Additionally, the key factors of the rules would be modified according the river situation. The proposed rules in this paper describe the pollution type, the position, the sampling time, the observation product, and the pollution alert. However, Howell et al. [34] described the rules of water alert that affected the entity of the water utility decision support by leveraging the semantic web to address the heterogeneity of web resources.

It is significant to define suitable factors of water quality monitoring into the rules. However, the selection of pollution alert factors in the rules is relatively simple. For example, the reasoning rules of an emergency response system for water pollution accidents was proposed by Meng et al. [35]. This study used the GB3838 standard: "The standard value of the Category 3 functional water is the allowable maximum", to verify the usability of ontology models based on reasoning. The maximum value of the COD was 20, and the maximum valued of NH_3N was 1.0, according to the Chinese regulation GB3838.

However, the mode of setting a uniform value is not applicable in the case of the Qingying River Basin, and this article sets a maximum value for each river section. Figure 9 shows the influencing factors of the allowable maximum. According to the location of the river sections, the maximum alert value of the pollutants was selected for the proposed rules. In addition, this study defined the pollution alert process for two types of water quality monitoring stations. The results of the experiment indicated that the OPO with rules could be a very useful methodology for quality assurance with complex data processing.

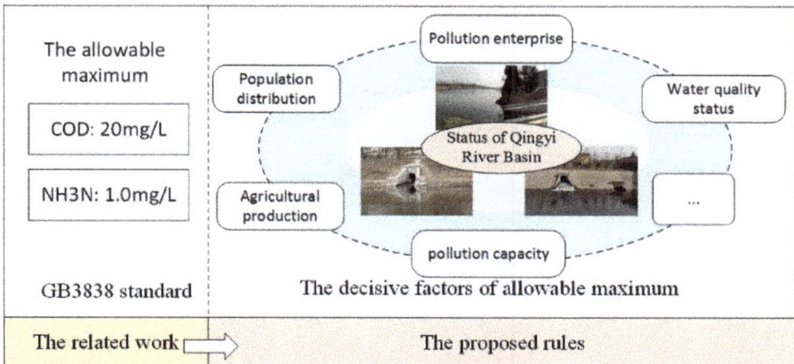

Figure 9. A comparison of the maximum selection methods. The setting of the allowable maximum value refers to Figure 1 and Table 1.

4.2.3. Comparison to Query Results Based on the OPO

Related research has focused on the visualization of data by using GIS analysis. For example, Xia et al. [36] presented a generic framework and the development steps of a decision tool prototype that used a geographic information systems (GIS)-based decision support system for the diagnosis of river health. Zhang et al. [37] proposed the development and application of an integrated environmental decision support system for water pollution control. The system included relations and links between various environmental simulation models, model integration, visualization, and real-time simulation methods. Horsburgh et al. [1] described the observations data model (ODM) Python tools allow users to query and export, visualize, and perform quality control post processing on time series of environmental observation data. Compared with the above systems, the proposed system in this study lacks professional data analysis, but it is characterized by enhancing the semantics of the observational process for tracing pollution sources.

In addition, compared with other semantic query services (for example, a given SPARQL query was proposed in [38]), the proposed system provides a richer visual representation. A query for a pollution enterprise was conducted that used "Data Acquired Start Time = 2019-09-02T5:27:51" and "Data Acquired End Time = 2019-09-02T20:00:00". Figure 10 shows the result of the search request. The observation data were targeted from the constructed ontology library. Using the observation data, the observation processing module calculated the pollutant diffusion. Then, the rules of pollution alert were judged based on the calculated results (observation products). In the discovery service, the observation information representation assisted in (1) browsing the enriched observational metadata properties of all the retrieved results in a table, (2) viewing the observation processing related to the data, and (3) viewing the observation products information and pollution alert. The realization of water polluter tracing is attributed the proposed ontology and rules that describe the observational process. Users can directly obtain the polluting sources that caused the water pollution, as shown in Figure 10b. The proposed method is generally applicable for other applications, such as precision agriculture, air pollution monitoring, and surface monitoring, which requires damage traceability.

A pollutant source tracing example is shown in Figure 10 to realize and verify the observation data processing. It obtained the suspicious source of pollution, where the pollution came from the nearest upstream sewage outlet. In the semantic-enhanced discovery service, all aspects of the observational process are displayed on the operation interface. For example, Figure 10 shows that the aspects that include the observation products, observation data, and related sensors. These aspects of observational process are then mapped to the OPO when executing the reasoning for pollutant source tracing.

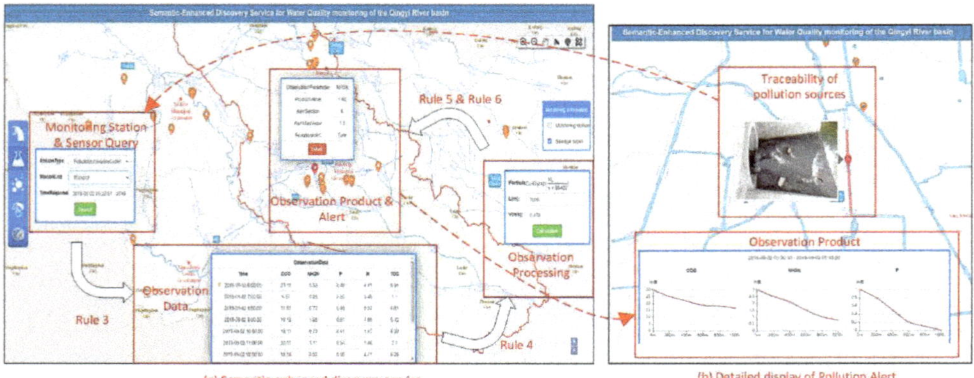

Figure 10. Query results of the sewage outlet alert. The discovery service included the monitoring station information, observation data, observation processing, observation products, and pollution alert using four rules which are defined in Section 3.2.2.

5. Conclusions

In this study, an ontology for water quality monitoring that is used for water pollution alerts was proposed. This ontology was developed by constructing the subclasses of the DOLCE ontology. This work provided semantic-enhanced pollution alert inference rules for the integration of heterogeneous water quality sensors and the mining of deep knowledge from observational resources. The water pollution early-warning and water pollution-processed models were able to be used for water pollution alerts using the rules. The proposed OPO and rules can also be developed for other application domains.

In addition, more classes, object properties, and data properties can be used in the discovery service to adapt the ontology to other purposes. The feasibility and usability of the proposed ontology were demonstrated using semantic discovery. The proposed ontology and reasoning rules allow the integration of query results in the system of observation data, observation product, sensors, and model of water quality warning. The proposed system primarily used data specifications from the Qingyi River to illustrate and describe details regarding water quality monitoring and pollution alerts for pollutant tracing. This paper proposes the adoption a semantic-driven combination of observational processes for water pollution accidents.

The following research directions will be studied more and refined in the future. More instances and rules will be described for acquiring more observational resources and for the mining of more useful knowledge. The system still needs to be able to use reasoning from external semantic sensor web services.

Author Contributions: X.W. proposed the research ideas and methods, wrote the manuscript, and created the figures. H.W. and N.C. designed the experiments and contributed to the discussions. X.H. and Z.T. are responsible for data collection, manuscript draft revision. All authors have read and agreed to the published version of the manuscript.

Funding: This work was supported by Open Research Fund of State Key Laboratory of Information Engineering in Surveying, Mapping and Remote Sensing, Wuhan University (Grant No. 18102).

Acknowledgments: We thank LetPub (www.letpub.com) for its linguistic assistance during the preparation of this manuscript.

Conflicts of Interest: The authors declare no conflict of interest.

References

1. Horsburgh, J.S.; Reeder, S.L.; Jones, A.S.; Meline, J. Open source software for visualization and quality control of continuous hydrologic and water quality sensor data. *Environ. Modell. Softw.* **2015**, *70*, 32–44. [CrossRef]
2. Rasyid, M.U.H.A.; Sayfudin, A.; Basofi, A.; Sudarsono, A. Development of semantic sensor web for monitoring environment conditions. In Proceedings of the 2016 International Seminar on Intelligent Technology and Its Applications (ISITIA), Lombok, Indonesia, 28–30 July 2016; pp. 607–612. [CrossRef]
3. Umar, U.; Al Rasyid, M.U.H.; Sukaridhoto, S. Distributed Database Semantic Integration of Wireless Sensor Network to Access the Environmental Monitoring System. *Int. J. Eng. Technol. Innov.* **2018**, *8*, 157–172.
4. Nativi, S.; Craglia, M.; Pearlman, J. Earth Science Infrastructures Interoperability: The Brokering Approach. *IEEE J-STARS* **2013**, *6*, 1118–1129. [CrossRef]
5. Sheth, A.P.; Henson, C.A.; Sahoo, S.S. Semantic Sensor Web. *IEEE Internet Comput.* **2008**, *12*, 78–83. [CrossRef]
6. Compton, M.; Barnaghi, P.; Bermudez, L.; García-Castro, R.; Corcho, O.; Cox, S.; Graybeal, J.; Hauswirth, M.; Henson, C.; Herzog, A.; et al. The SSN ontology of the W3C semantic sensor network incubator group. *J. Web Semant.* **2012**, *17*, 25–32. [CrossRef]
7. Haller, A.; Janowicz, K.; Simon Cox, C.; Le Phuoc, D.; Taylor, K.; Lefrançois, M. Semantic Sensor Network Ontology. Available online: https://www.w3.org/TR/2017/REC-vocab-ssn-20171019/ (accessed on 30 December 2019).
8. Chen, Z.; Gangopadhyay, A.; Holden, S.H.; Karabatis, G.; McGuire, M.P. Semantic integration of government data for water quality management. *Gov. Inf. Q.* **2007**, *24*, 716–735. [CrossRef]
9. Regueiro, M.A.; Viqueira, J.R.R.; Stasch, C.; Taboada, J.A. Semantic mediation of observation datasets through Sensor Observation Services. *Future Gener. Comput. Syst.* **2017**, *67*, 47–56. [CrossRef]
10. Calbimonte, J.P.; Jeung, H.; Corcho, O.; Aberer, K. Enabling Query Technologies for the Semantic Sensor Web. *Int. J. Semant. Web Inf. Syst.* **2012**, *8*, 43–63. [CrossRef]
11. Ristoski, P.; Paulheim, H. Semantic Web in data mining and knowledge discovery: A comprehensive survey. *J. Web Semant.* **2016**, *36*, 1–22. [CrossRef]
12. W3C OWL Working Group. OWL 2 Web Ontology Language Document Overview (Second Edition). Available online: https://www.w3.org/TR/owl2-overview/#ack (accessed on 30 December 2019).
13. Horrocks, I.; Patel-Schneider, P.F.; Boley, H.; Tabet, S.; Grosof, B.; Dean, M. SWRL: A Semantic Web Rule Language Combining OWL and RuleML. Available online: https://www.w3.org/Submission/2004/SUBM-SWRL-20040521/ (accessed on 30 December 2019).
14. Wang, C.; Chen, N.; Wang, W.; Chen, Z. A Hydrological Sensor Web Ontology Based on the SSN Ontology: A Case Study for a Flood. *ISPRS Int. J. Geo-Inf.* **2018**, *7*, 2. [CrossRef]
15. Xiaomin, Z.; Jianjun, Y.; Xiaoci, H.; Shaoli, C. An Ontology-based Knowledge Modelling Approach for River Water Quality Monitoring and Assessment. *Procedia Comput. Sci.* **2016**, *96*, 344–353. [CrossRef]
16. Dong, J.; Wang, G.; Yan, H.; Xu, J.; Zhang, X. A survey of smart water quality monitoring system. *Environ. Sci. Pollut. Res.* **2015**, *22*, 4893–4906. [CrossRef] [PubMed]
17. Huang, X.; Jianjun, Y.; Shaoli, C.; Xiaomin, Z. A Wireless Sensor Network-Based Approach with Decision Support for Monitoring Lake Water Quality. *Sensors* **2015**, *15*, 29273–29296. [CrossRef] [PubMed]
18. Mihaela, O. A knowledge modelling framework for intelligent environmental decision support systems and its application to some environmental problems. *Environ. Modell. Softw.* **2018**, *110*, 72–94. [CrossRef]
19. Zheng, D. Research of Hair Product Industry Polluted Status and New Neutralization Process. Master's Thesis, Henan Normal University, Xinxiang, China, 2017.
20. Borgo, S.A.M.C. Foundational Choices in DOLCE. In *Handbook on Ontologies*; Staab, S., Studer, R., Eds.; Springer: Berlin/Heidelberg, Germany, 2009; pp. 361–381. [CrossRef]
21. Musen, M.A. The Protege Project: A Look Back and a Look Forward. *AI Matters* **2015**, *1*, 4–12. [CrossRef]
22. Shao, D.; Nong, X.; Tan, X.; Chen, S.; Xu, B.; Hu, N. Daily Water Quality Forecast of the South-To-North Water Diversion Project of China Based on the Cuckoo Search-Back Propagation Neural Network. *Water* **2018**, *10*, 1471. [CrossRef]
23. Ge, Y.; Li, X.; Huang, C.; Nan, Z. A Decision Support System for irrigation water allocation along the middle reaches of the Heihe River Basin, Northwest China. *Environ. Modell. Softw.* **2013**, *47*, 182–192. [CrossRef]

24. Weng, S.Q.; Huang, G.H.; Li, Y.P. An integrated scenario-based multi-criteria decision support system for water resources management and planning—A case study in the Haihe River Basin. *Expert Syst. Appl.* **2010**, *37*, 8242–8254. [CrossRef]
25. Kort, I.A.T.D.; Booij, M.J. Decision making under uncertainty in a decision support system for the Red River. *Environ. Modell. Softw.* **2007**, *22*, 128–136. [CrossRef]
26. Salewicz, K.A.; Nakayama, M. Development of a web-based decision support system (DSS) for managing large international rivers. *Global Environ. Chang.* **2004**, *14*, 25–37. [CrossRef]
27. Taylor, P. OGC WaterML 2.0: Part 1-Timeseries. In *Open Geospatial Consortium: OGC 10-126r4*; OGC: Wayland, MA, USA, 2014; p. 137.
28. Elag, M.; Goodall, J.L. An ontology for component-based models of water resouvrce systems. *Water Resour. Res.* **2013**, *49*, 5077–5091. [CrossRef]
29. Dutta, R.; Morshed, A. Performance Evaluation of South Esk Hydrological Sensor Web: Unsupervised Machine Learning and Semantic Linked Data Approach. *IEEE Sens. J.* **2013**, *13*, 3806–3815. [CrossRef]
30. Liu, Q.; Bai, Q.; Kloppers, C.; Fitch, P.; Bai, Q.; Taylor, K.; Fox, P.; Zednik, S.; Ding, L.; Terhorst, A.; et al. An ontology-based knowledge management framework for a distributed water information system. *J. Hydroinform.* **2013**, *15*, 1169–1188. [CrossRef]
31. Klien, E.; Lutz, M.; Kuhn, W. Ontology-based discovery of geographic information services—An application in disaster management. *Comput. Environ. Urban Syst.* **2006**, *30*, 102–123. [CrossRef]
32. Endo, A.; Kumazawa, T.; Kimura, M.; Yamada, M.; Kato, T.; Kozaki, K. Describing and Visualizing a Water-Energy-Food Nexus System. *Water* **2018**, *10*, 1245. [CrossRef]
33. Shu, Y.; Liu, Q.; Taylor, K. Semantic validation of environmental observations data. *Environ. Modell. Softw.* **2016**, *79*, 10–21. [CrossRef]
34. Howell, S.; Rezgui, Y.; Beach, T. Water utility decision support through the semantic web of things. *Environ. Modell. Softw.* **2018**, *102*, 94–114. [CrossRef]
35. Meng, X.; Chao, X.; Xinxia, L.; Junming, B.; Wenhan, Z.; Hao, C.; Zhuo, C. An Ontology-Underpinned Emergency Response System for Water Pollution Accidents. *Sustainability* **2018**, *10*, 546. [CrossRef]
36. Xia, J.; Lihuai, L.; Junqiang, L.; Laounia, N. Development of a GIS-Based Decision Support System for Diagnosis of River System Health and Restoration. *Water* **2014**, *6*, 3136–3151. [CrossRef]
37. Zhang, S.; Li, Y.; Zhang, T.; Peng, Y. An integrated environmental decision support system for water pollution control based on TMDL—A case study in the Beiyun River watershed. *J. Environ. Manag.* **2015**, *156*, 31–40. [CrossRef]
38. Si, H.; Qi, Y.; Zheng, M.; Ren, Y.; Yu, L. Structured peer-to-peer-based publication and sharing of ontologies to automatically process SPARQL query on a semantic sensor network. *Int. J. Distrib. Sens. Netw.* **2018**, *14*, 1–13. [CrossRef]

Article

Land Cover and Water Quality Patterns in an Urban River: A Case Study of River Medlock, Greater Manchester, UK

Cecilia Medupin [1,*], Rosalind Bark [2] and Kofi Owusu [1]

[1] Department of Earth & Environmental Sciences, University of Manchester, Manchester M13 9PL, UK; Kofi.Owusu@manchester.ac.uk

[2] School of Environmental Sciences, University of East Anglia, Norwich NR4 7UA, UK; R.Bark@uea.ac.uk

* Correspondence: cecilia.medupin@manchester.ac.uk

Received: 31 January 2020; Accepted: 13 March 2020; Published: 17 March 2020

Abstract: Urban river catchments face multiple water quality challenges that threaten the biodiversity of riverine habitats and the flow of ecosystem services. We examined two water quality challenges, runoff from increasingly impervious land covers and effluent from combined sewer overflows within a temperate zone river catchment in Greater Manchester, North-West UK. Sub-catchment areas of the River Medlock were delineated from digital elevation models using a Geographical Information System. By combining flow accumulation and high-resolution land cover data within each sub-catchment and water quality measurements at five sampling points along the river, we identified which land cover(s) are key drivers of water quality. Impervious land covers increased downstream and were associated with higher runoff and poorer water quality. Of the impervious covers, transportation networks have the highest runoff ratios and therefore the greatest potential to convey contaminants to the river. We suggest more integrated management of imperviousness to address water quality, flood risk and, urban wellbeing could be achieved with greater catchment partnership working.

Keywords: water quality monitoring; water quality status; sources and pathways; land cover; digital elevation model; urban river; ArcGIS

---

1. Introduction

Urban land covers such as infrastructure, urban greenspace and woodland, residential, and industry and commercial covers can modify water quality either by enhancing or reducing runoff [1] and in turn contaminant levels entering freshwaters from surrounding terrestrial ecosystems [2,3]. Changes in land cover from human activities or natural drivers, can both affect the quality of the river catchment and thereby compromise the ecosystem services they provide [2,3]. With increased pressures from population growth and urban sprawl, the challenges to water quality are imminent [4,5]. Contemporary, management strategies include restoring ecohydrological processes [6] to renaturate, remediate, and revitalize urban rivers.

Urbanized areas are often characterized by rapid development, modified or canalized rivers, and a high proportion of impervious land [4,7–12]. Urban land cover, particularly, impervious surface covers which prevent infiltration of water to the soil, include transport systems (roads, parking lots), roof tops and residential development are a key indicator of the environmental impacts of urbanisation [3]. Point source pollution collected from impervious urban areas contributes to river flows, changes river hydrology, alters physicochemical processes [13] and degrades assemblages of benthic macroinvertebrates [4,12]. During high flows, combined sewer overflows (CSOs) which collect runoff from impervious land covers are designed to overflow and discharge directly, without treatment, into the receiving water body. CSO discharges can degrade water quality including physicochemical and microbial indicators [14–17].

In the UK, CSOs have long been recognized as one of the major causes of urban river pollution [18–20]. They are also a major contributor to coastal pollution and the UK's Environment Agency (EA) estimates that 30% of coastal pollution in the North West of England can be attributed to CSOs through a combination of breaches and cumulative impact (see, https://www.unitedutilities.com/services/wastewater-services/bathing-waters/what-are-combined-sewer-overflows/). It is therefore critical to understand the risks associated with CSOs particularly with changing climate and greater future flood risk [21–23] and there are challenges in retrofitting CSOs to increase their capacity [23]. Other point source pollution includes industrial effluent, effluent from wastewater treatment works (WwTW), and drainage from agricultural activities.

The application of a Geographic Information System (GIS) to environmental data analyses in river sub-catchments has been explored by researchers [9,10,24–27]. Studies show that information on land cover patterns can improve our understanding of water quality within catchments [28] and the management of pollution sources [27]. Through ArcGIS tools, topographical characteristics, e.g., slope, flow length, drainage area or catchment area [29,30] that drive surface water flow, can be analyzed through the digital elevation model (DEM) data [31]. DEM data is also used for modelling the urban environment [32] and to delineate catchments for spatial analysis [33–35]. Such information is critical for urban river quality management. Additionally, partnership working can be decisive in urban settings, where partnership groups can coordinate across: Threats, policy domains, e.g., land and water management, urban re/development including integration with urban design, i.e., green infrastructure and sustainable urban drainage systems [36,37]. Partnership groups can also coordinate restoration of the physical structure of the riverbed and the riparian corridor [38].

In this study, our objectives are to (a) analyze river water quality variables at the sub-catchment scale (b) investigate the relationship between the land cover and water quality variables, and (c) examine the influence of CSO discharge at our study sites.

2. Materials and Methods

2.1. Study Area: River Medlock

The River Medlock catchment, Figure 1 (a. catchment areas and b. Orthophotomap) has an annual average rainfall of approximately 1025 mm and an average flow rate of 0.883 m^3/s (based on the mean monthly flows for the period of record, CEH, 2018). The river rises in the hills of Greater Manchester (National Grid Reference: SD 95308 05431) in North-West England and flows for 22 km through a steep-sided wooded area before it passes through a gauging station (SJ 85781 97858) downstream of the urbanized Manchester City Centre. The region and river has a long industrial history that includes legacy pollution from industrial effluent. Added to this, is a history of inadequately treated sewage [39–42] and CSO spill events from localized storm events [20,43] and accidental spills (for example, in 2017 a sewage tank leak led to a spill of 21,700 cubic meters of raw sewage, killing fish, see, https://www.bbc.co.uk/news/uk-england-manchester-41160909).

Within the river catchment, there are fifty CSOs, including twenty-nine within the surveyed reach (United Utilities' personal communication, 2013) as well as numerous surface water drains. The surveyed reach also has a continuously operational WwTW at Failsworth (NGR: SJ 89674 99800). With regard to the CSO discharge points, the main WwTW, and site access, we identified five sampling points (Figure 1a). Sampling site 1 (S1) (NGR: SD 94183 02262) and S2 (NGR: SD 93489 01798) are located upstream of the WwTW and sites S3 (NGR: SJ 89272 99554), S4 (NGR: SJ 86052 98382) and S5 (NGR: SJ 85781 97858) are all located downstream of it. The River Medlock catchment is drained at the gauging station (GS) (NGR: SJ 84900 97518). The number of CSOs which discharge into the sampling sites varies, with seven at S1 and S2, sixteen at S3, five at S4 and one at S5 (United Utilities, personal communication [44]). Characteristics of the five sampling sites are shown on Table 1. Overflow volume data for each sample site was obtained from the water company, United Utilities.

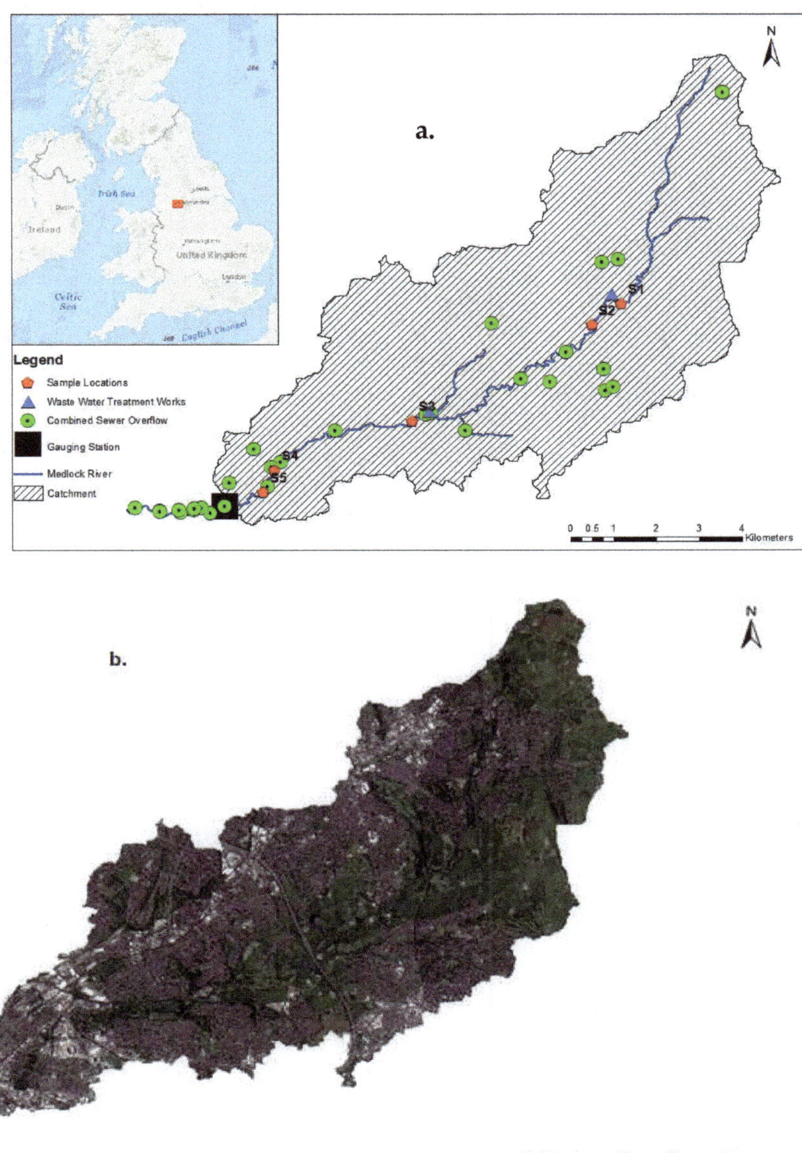

Figure 1. River Medlock catchment area and the study sites. Insert is the location of the river in Greater Manchester, UK (**a**); Orthophotomap of the river catchment (**b**).

Table 1. Catchment characteristics including the sub-catchment sampling sites S1–S5 and GS obtained from the digital elevation model (DEM).

Sub-Catchment	Grid Reference	Latitude	Longitude	Distance (km from Source)	Average Slope (%)	Sample Point Catchment Area (km^2)	Elevation (Mean Sea Level)
S1	SD 94183 02295	53.5173	−2.0892	6.6	10.72	13.65	182
S2	SD 93489 01798	53.5128	−2.0996	8.5	9.45	20.55	159
S3	SJ 89272 99554	53.4925	−2.1631	13	7.56	43.98	114
S4	SJ 86052 98382	53.4819	−2.2116	16.1	7.24	50.16	93
S5	SJ 85781 97858	53.4772	−2.2157	17.4	7.21	50.85	90
GS	SJ 84800 97500	53.4740	−2.2305	18	6.57	65.94	89

2.2. Water Quality Sampling and Measurement

The Area Ratio (AR) method [45] was used to estimate river discharge (m^3/s) at each sampling site. The principal water quality measurements were pH, temperature (°C), conductivity (µS/cm), biochemical oxygen demand (BOD, mg/L), dissolved oxygen (DO, % saturation), suspended solids (SS, mg/L), nitrate-N (NO_3-N, mg/L), ammonia-N, phosphate-P (PO_4-P, mg/L), ammonia-N (NH_3-N), total phosphorus (TP, mg/L), total organic matter (TOM, mg/L), and discharge (m^3/s).

A pre-calibrated hand-held multiparameter water quality meter (YSi 556 Multi probe system YSI Incorporated, Yellow Springs, Ohio, USA) was used to take measurements of pH, DO, temperature and conductivity. In a 14-month field campaign between March 2013 to April 2014, monthly spot samples at five sampling sites (S1 to S5) were obtained for the measurement of BOD, SS, TOM, and nutrients. Over the sampling period, a total of 735 samples were collected. These include 70 samples per variable for the analysis of nitrate-N (NO_3-N mg/L), phosphate-P (PO_4-P), ammonia-N(NH_3-N), BOD, SS, conductivity, temperature, pH, DO; TOM (65 samples), and 40 samples for TP. Samples were processed within 24 hours of the sample collection using a SEAL Auto Analyzer 3 High Resolution instrument (SEAL Analytical Ltd, Southampton, UK). Detection limits for phosphate measured as P is 0.004 mg/L and nitrate (measured as N) is 0.05 mg/L. Ammonia-N (NH_3-N) concentration (mg/L) was analyzed by spectrophotometry using the Hanna low range reagent kit HI-93700-01 (Hanna Instruments Ltd, Leighton Buzzard, Bedfordshire). The limit of detection for ammonia-N (NH_3-N) measured as N is 0.01 mg/L. All chemical variables were analyzed according to the [46].

2.3. Classification of Land Cover Patterns

The process used for mapping and determining the spatial distribution of pervious and impervious land cover for the Medlock River catchment was as follows: an orthophotomap of Greater Manchester was generated from Google (Google Maps, 2020. *Greater Manchester*. Google Maps (online) Available through: University of Manchester (https://www.manchester.ac.uk/) (Accessed 21 February 2020)) satellite imagery with a resolution of 15 cm (Figure 2) was georeferenced (Figure 2a) and the detailed image was classified in ArcGIS 10.4.1 (Figure 2b). The land cover was vectorized (Figure 2c) and further classified according to the CORINE nomenclature [47] into six categories: Continuous Urban Fabric—Transport Networks, Continuous Urban Fabric—Urban Green, Discontinuous Urban Fabric—Residential, Industrial/Commercial, Inland Waters, and Woodland—Broadleaf. The percentage total area of each category was then computed for use in further analyses. The number of pixels (m) recorded for each land cover class was processed for each site and classified by means of their spectral responses.

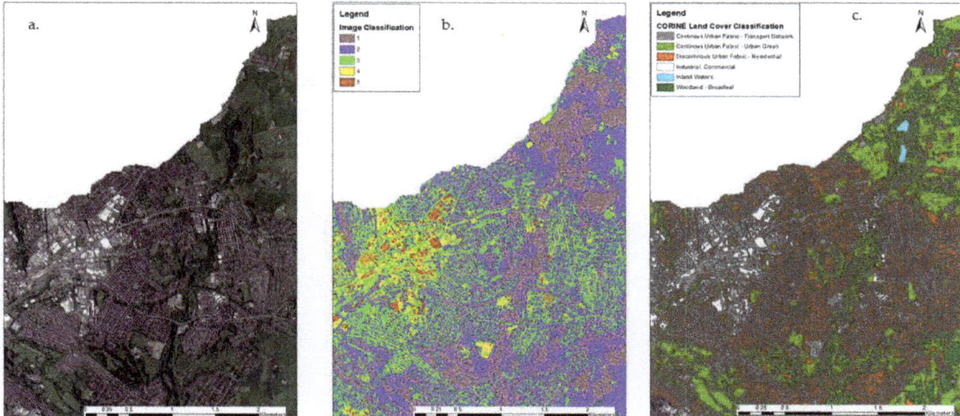

Figure 2. Georeferenced map of the River Medlock (**a**); Detailed orthophotomap classified in ArcGIS 10.4.1 (**b**); Vectorized land cover (**c**).

The land cover map for each pour point (drainage outlet) of the Medlock catchment was determined in relation to catchment area, slope (%) (Figure A1, Appendix A), elevation (mMSL) (Table 1) at a distance of 1 km. Slope was calculated using the ArcGIS Spatial Analyst tool with a DEM (elevation raster) as input. Slope can be measured in percent rise or degrees (with for example 90 degrees representing the slope of a cliff):

$$Slope\ (\%) = \frac{rise}{run} \times 100 \tag{1}$$

$$Slope(^{\circ}) = \theta \tag{2}$$

where

$$Slope\ (^{\circ}) = \frac{rise}{run}\ tan\theta \tag{3}$$

2.4. Drainage Network and Sub-Catchment Delineation

A high-resolution (12.5 m) terrain-corrected DEM was obtained from the ALOS PALSAR (https://earthdata.nasa.gov/eosdis/daacs/asf) dataset and ArcGIS 10.4.1 was used to delineate the drainage networks at the five sample locations. Depression of the DEM data was done with the *"fill"* tool which ensures there is uniformity in the DEM by filling any holes (or areas of missing data). The *"flow direction"* was determined with the filled DEM as input. The flow direction grid serves as an input for computing the flow accumulation grid (Figure A2, Appendix A). The *"flow accumulation" grid* was used to determine the river network to a specific point on the DEM. ArcGIS's Watershed tool was used to delineate the catchment of the River Medlock (Figure A3, Appendix A). The contributing area to the flow accumulation for each cell ranged from 0 to 423,074 m^2. In order to visualize the range of flow accumulation at the sample sites, a threshold of 100 m^2 was computed. *Pour points and catchment delineation*: The pour point is the lowest point (outlet) in the sub-catchment (and the lowest point at the sampling sites) through which water flows under gravity (it is the stream gauge) and delineation takes place upstream of the stream gauge. The point delineation, batch watershed delineation and sub-watershed delineation tools were used in the hydrological analysis-expanding module (Arc Hydro Tools) to divide each sub-catchment either in a diagonal or adjacent pattern. This approach specifies flow directions in the catchment by assigning pixels or cells to one of its eight neighbours, in the flow direction of steepest descent, this is identified as the pour point [31]. This extraction influences the size of the sub-catchment area, location, gradient and range. Thus, the larger the catchment area, the lower the pour point.

2.5. Data Analysis

Using the SPSS Statistics software package version 22.0 Pearson correlation analysis was carried for the physicochemical variables obtained at the sampling points. A stepwise multiple linear regression was applied in order to allow backward selection or removal of variables. The dependent variables were BOD, NO_3-N, PO_4-P, NH_3-N, SS, conductivity, discharge, pH, temperature, TOM, and DO. Correlation between land cover, i.e., compositional attribute (%) and water quality variables were determined. The CORINE land cover nomenclature [47] provides information based on the highest contributing (%) land covers at the spatial scales. The runoff coefficient was determined based on the method of Hvitved-Jacobsen (2010) cross referenced by [5]. That is, the runoff coefficient is found by dividing river discharge at each site by the average precipitation intensity (the annual precipitation (1025 mm)/hr/Area) and drainage area.

All results were compared with the European Union's Water Framework Directive (EU WFD) standards [48] except for SS, which were compared with the EU's Freshwater Fisheries Directive. Nitrate standard was determined from the European Commission's Nitrates Directive.

3. Results

3.1. Sub-Catchment Characteristics of the River Medlock

Maximum elevation in the catchment is 430 m, other characteristics including average slope, catchment area and minimum elevation are shown in Table 1. The size of the catchment area at the GS was calculated at 65.9 km^2, which is larger than the catchment area reported by the Centre for Ecology & Hydrology (CEH, 57.5 km^2). The discrepancy is that the study catchment area takes into consideration other aspects of the sub-catchments not included in the CEH maps.

3.2. Land Cover Analysis at the Catchment Level

The distribution of land covers extracted from CORINE for the sampling locations and the GS are shown on Figure 3a–f. These land covers include transport network, residential, industrial, and commercial (now classified as impervious cover); inland water, urban green and woodland (broadleaf). Overall, the percentage of impervious covers increases from S1 to the GS and it is the upland catchment that has the highest percentage of non-impervious land covers.

Table A1 provides information on the land covers by sub-catchment and Table A2 provides more detail on the impervious land covers. The proportion of impervious surfaces is 27% at S1 increasing to 45% at GS. For the catchment as a whole, impervious cover has an area of 98.57 km^2, i.e., 40% of the total area (Table A1). From Table A1 and Figure 4, the dominant impervious cover in the catchment is the "residential" cover; this is highest at S1, likely reflecting larger property sizes. Residential cover contributes more than 50% at each of the locations. The next dominant cover is the "transport network" which constitutes more than 30% of the impervious cover while the "industrial and commercial" areas contribute less than 10%. The proportion of different types of impervious cover varies between the sites with a generally increasing/decreasing proportion of transport and industrial and commercial/residential moving downstream. Although the runoff coefficient determined from impervious covers of residential and the industrial/commercial per site were insignificant, the runoff coefficient from the transport networks ranged from 1.02 to 4.53. There was a step-change in runoff between sub-catchments S2 and S3 and another large increase between sub-catchment S3 and S4.

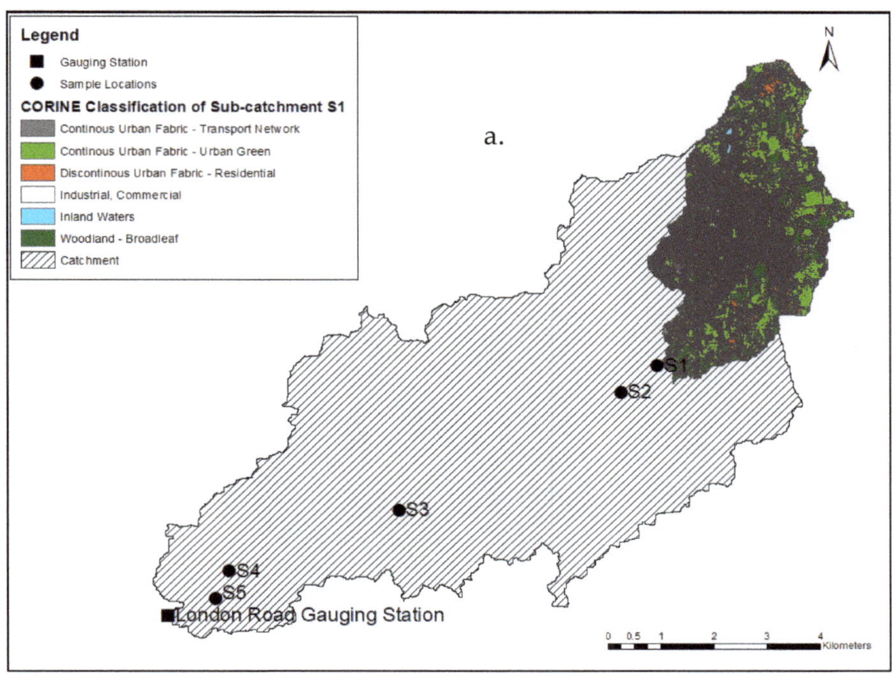

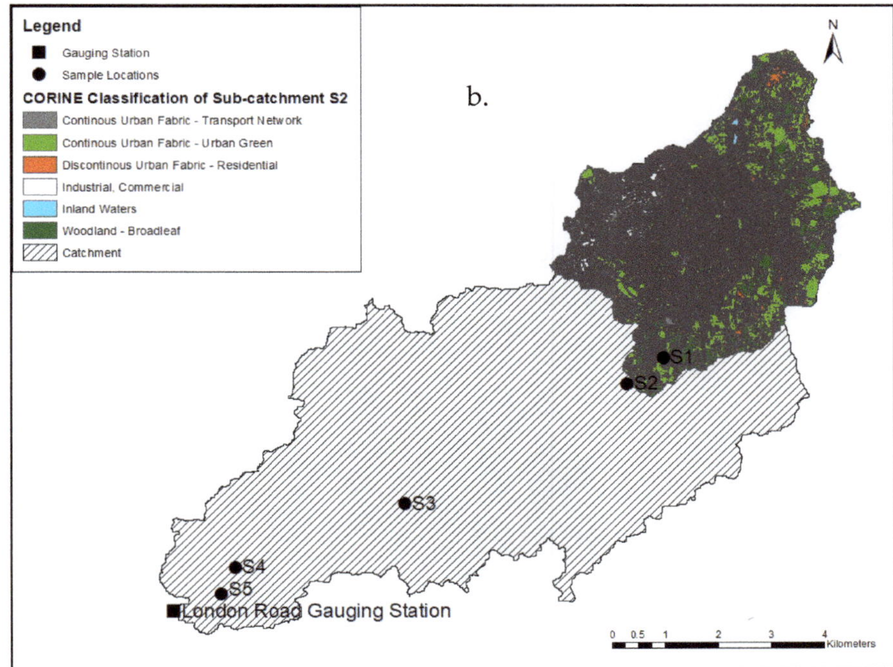

Figure 3. *Cont.*

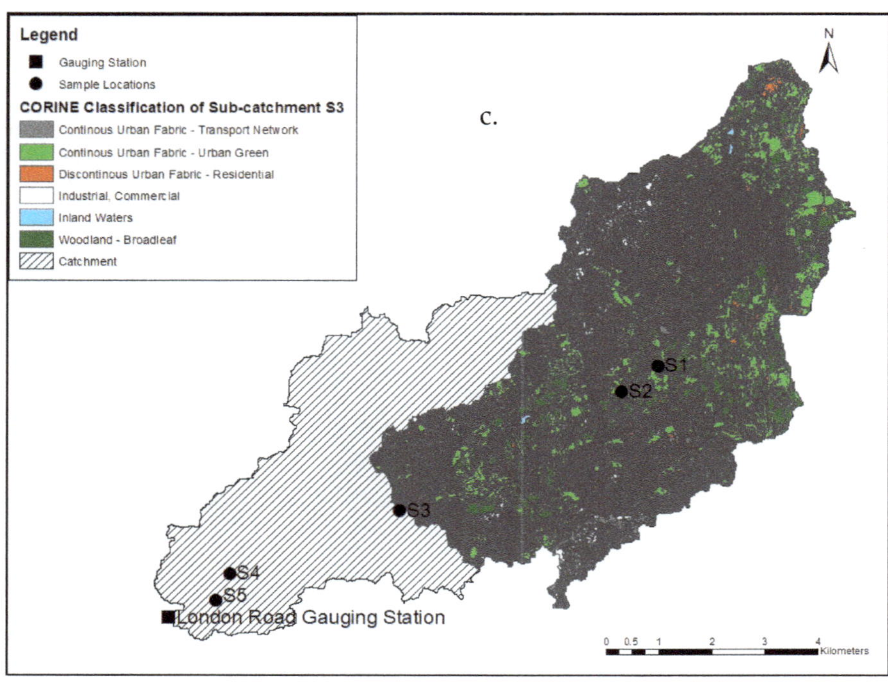

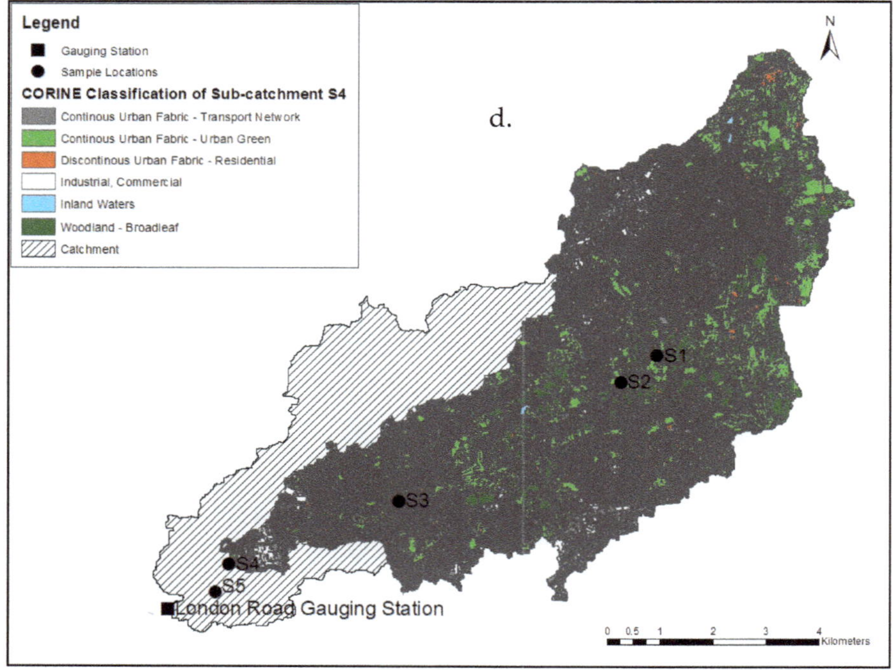

Figure 3. *Cont.*

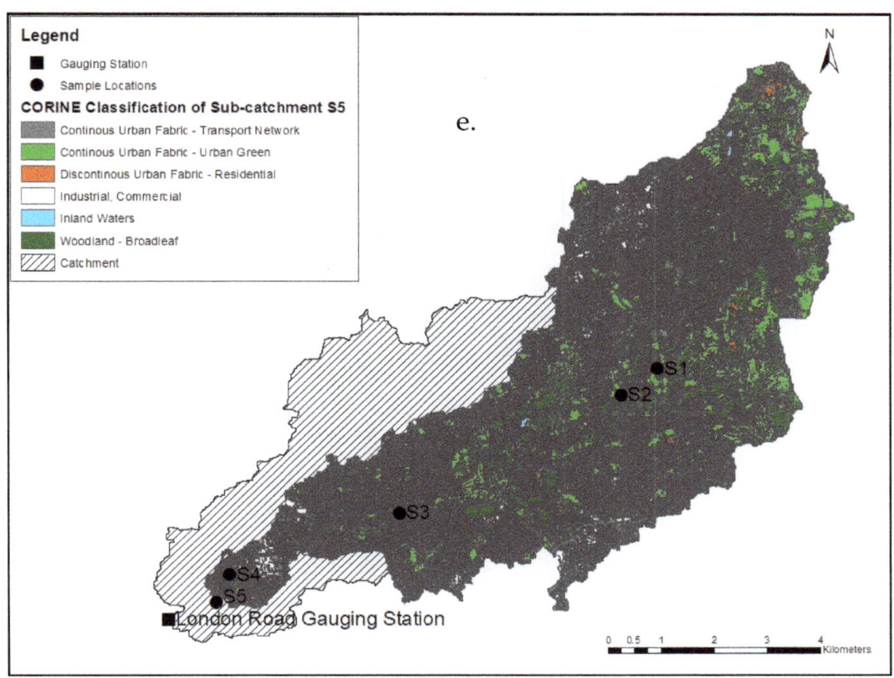

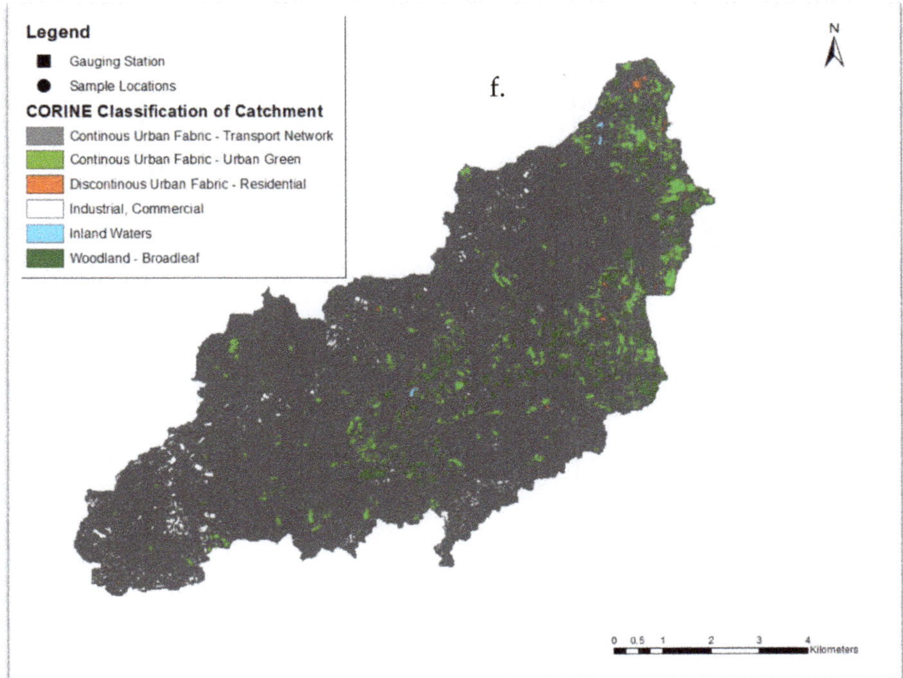

Figure 3. CORINE land cover patterns at sub-catchments S1 to S5 (**a**–**e**) and the Gauging Station (**f**).

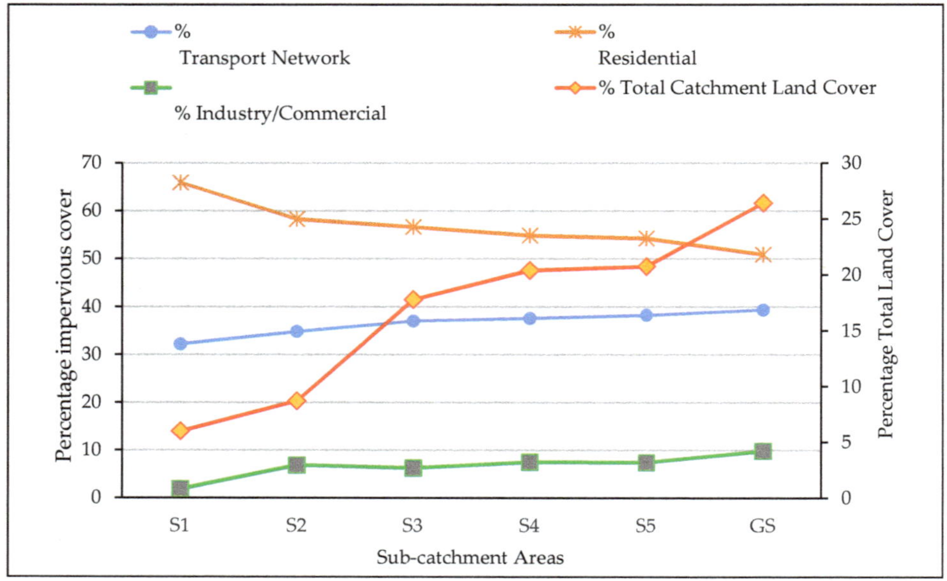

Figure 4. The percent total land cover verses impervious cover for transport networks, industry/commercial and residential. Percentage total land cover found to increase downstream of the catchment.

3.3. Land Cover Patterns and Water Quality Variables

Table 2 presents the significant correlation coefficients between monthly water quality variables and the percentage land cover. Conductivity correlates negatively with the woodland (broadleaf) and urban green covers but is positively correlated with the impervious covers. Discharge and nutrients correlated positively with the impervious covers and negatively with the urban green covers. SS concentration and TOM had positive correlations with residential cover and urban green cover respectively. DO had positive correlation with Urban Green and woodland (broadleaf cover) and had a negative correlation with the transport network and industry/commercial covers.

Table 2. Significant correlations between water quality variables and land cover, Number of sampling sites = 5, r = correlation coefficient.

Land Cover and Water Quality Variables	r	*p*
Transport Network		
DO mg/L	−0.937	0.019 *
Conductivity	0.991	0.001 *
TOM mg/L	−0.885	0.046 *
Discharge m^3/s	0.964	0.008 **
NO_3-N mg/L	0.904	0.035 *
Urban Green		
DO mg/L	0.934	0.020 *
Conductivity	−0.991	0.001 *
TOM mg/L	0.904	0.035 *
Discharge m^3/s	−0.962	0.009 **
PO_4-P mg/L	−0.881	0.049 *
NO_3-N mg/L	−0.913	0.031 *

Table 2. *Cont.*

Land Cover and Water Quality Variables	r	*p*
Residential		
Conductivity	0.993	0.001 **
Suspended solids mg/L	0.904	0.035 *
Discharge m^3/s	1.000	0.000 **
PO_4-P mg/L	0.908	0.033 *
NO_3-N mg/L	0.959	0.001 *
Industrial/Commercial		
DO mg/L	−0.897	0.039 *
Conductivity	0.928	0.023 *
Discharge m^3/s	0.915	0.029 *
Woodland (broadleaf)		
DO mg/L	0.899	0.038 *
Conductivity μS/cm	−0.999	0.000 **

** Correlation is significant at the 1% and at the * 5% level (2-tailed).

3.4. Water Quality Parameters and Standards by Sub-Catchment

Table 3 reports the physicochemical water quality status for the five sampling locations. Note for EU WFD requirements failure in one category results in the failure to meet "good ecological status". Variables are either classified as "High" which indicates excellent condition, "Very Good"; "Good"; "Moderate"; "Poor" and "Bad". The levels of pH, temperature, DO and, SS were high indicating excellent water quality while PO_4-P, BOD, and NH_3-N ranged from poor to very good. For NO_3-N, the concentrations were very good. Increasing concentration of conductivity, nutrients, BOD and river discharge were noted at the downstream sections of the river at sites S3, S4, and S5.

Table 3. Water quality measurements by sample site and EU Directives. The median values for all variables were determined except for SS where the mean values were provided to align with classification. Number of samples per site per variable S1 = 14, S2 to S5 = 15 each. All variables compared with EU Water Framework Directive (EU WFD) except suspended solids (mg/L) compared with the * Freshwater Fisheries Directives and ** Nitrates Directive and Drinking Water Directive applied to NO_3-N (mg/L).

Variables	Sub-Catchment Areas	Median	% Percentile	Standards EU WFD, * Freshwater Fisheries Directive ** Nitrates Directive and Drinking Water Directive	Interpretation
Temperature (°C) 2% and 98% percentile	S1	9.8	5–15.3	≤20	High
	S2	9.5	5.1–14.8		
	S3	10.4	5.4–17.7		
	S4	10.2	4.9–17.3		
	S5	10.3	4.8–18.5		
pH 5% and 10th percentile	S1	7.858	7.40–7.46	≤6–9	High
	S2	7.79	7.50–7.61		
	S3	8.005	7.74–7.77		
	S4	8.04	7.80–7.81		
	S5	8.06	7.83–7.86		
DO (% sat) 10%–90th percentile	S1	103.7	97.94–115.5	≥70%	High
	S2	101	85.8–111.6		
	S3	101.5	94.81–117.2		
	S4	100.9	93.82–113.7		
	S5	99.5	82.04–113.9		
*** Suspended solids ($mg L^{-1}$) Freshwater Fisheries Directive**	S1	4.183		<25	High
	S2	6.029			
	S3	11.59			
	S4	15.03			
	S5	12.05			

Table 3. *Cont.*

Variables	Sub-Catchment Areas	Median	% Percentile	Standards EU WFD, * Freshwater Fisheries Directive ** Nitrates Directive and Drinking Water Directive	Interpretation
NH_3-N (mg L^{-1}) 1%–99th percentile	S1	0.35	0–2.48	0.04	Poor
	S2	0.4	0–2.18		
	S3	0.59	0–2.50		
	S4	0.4	0.01–2.18		
	S5	0.43	0–2.55		
BOD(mg L^{-1}) 1%–99th percentile	S1	1.52	0–13.68	≤14	Good to Moderate
	S2	1.98	0–13.1	≤14	Good to Moderate
	S3	2	0.04–14.92	≤19	Good to Moderate
	S4	2.6	0.55–16.13	≤19	Good to Moderate
	S5	2.49	0–5.1	≤19	Very Good
TOM (mg L^{-1}) 1%–99th percentile	S1	0.01	0–28.13		
	S2	0.01	0–25.93		
	S3	0.84	0–18.81		
	S4	0.33	0–14.04		
	S5	0.67	0–17.75		
** NO_3-N (mg L^{-1}) (5%–95%) Nitrate Directive	S1	9.72	0.14–4.98	50	Very Good
	S2	9.60	0.14–4.46	50	Very Good
	S3	10.7	0.59–11.68	50	Very Good
	S4	10.3	0.611–9.60	50	Very Good
	S5	10.34	0.523–9.23	50	Very Good
PO_4-P (mg L^{-1}) (5%–95%)	S1	0.023	0–0.58	0.215–1.098	Very Good
	S2	0.027	0–1.01	0.215–1.099	Very Good
	S3	0.462	0.03–1.61	0.215–1.100	Moderately Poor
	S4	0.444	0.02–1.25	0.215–1.101	Moderately Poor
	S5	0.447	0.02–1.25	0.215–1.102	Moderately Poor

Table 4 shows the results of the correlation analysis between the water quality variables. A strong correlation between concentrations of (mg/L) NH_3-N and BOD and between NO_3-N and PO_4-P were recorded at all sample sites. A positive correlation between PO_4-P concentration and SS (mg/L) was also recorded at S4. The flux of TP calculated at the WwTW, the upstream S2 and downstream S3 sites, demonstrates that the WwTW contributes TP to the river. The flux at S2 was 0.43 kg TP ha^{-1} yr^{-1}, 0.94 kg TP ha^{-1} yr^{-1} at the WwTW and 0.92 kg TP ha^{-1} yr^{-1} at S3.

Table 4. Significant correlations between analyzed physicochemical variables by sample point. N = number of samples S1= 22; S3–S5 = 23 each.

Sites/Parameters	*p*	r
S1		
BOD (mg/L) /NH_3-N (mg/L)	0	0.729 ***
Conductivity (μS/cm)/pH	0.016	−0.517 **
NO_3-N (mg/L)/PO_4-P (mg/L)	0.001	0.652 ***
Q (m^3/s)/NO_3-N (mg/L)	0.013	0.509 **
S2		
BOD (mg/L)/NH_3-N (mg/L)	0.001	0.639 ***
Q(m^3/s)/NO_3-N (kg/d)	0.019	0.485 **
S3		
BOD (mg/L)/NH_3-N (mg/L)	0.018	0.501 **
NO_3-N (mg/L)/PO_4-P (mg/L)	0	0.797 ***
S4		
BOD (mg/L)/NH_3-N (mg/L)	0.009	0.535 ***
NO_3-N (mg/L)/PO_4-P (mg/L)	0	0.797 ***
PO_4-P (mg/L)/SS (mg/L)	0.02	0.481 **
S5		
NO_3-N (mg/L)/PO_4-P (mg/L)	0	0.838 ***

*** Correlation is significant at 1% level and ** at the 5% level (2-tailed).

3.5. CSOs: Volume Discharged

Figure 5 shows the volume of CSO versus river discharge across the sample sites. The highest overflowed volume was recorded at S3. At this site, the overflow volume released from 16 operational CSOs was estimated at 932,520 m^3. Seven operational CSOs discharge into S1 and S2 with an estimated overflowed volume of 735,617 m^3. Overflow volumes were significantly lower at S4 and S5 at 29,902 m^3 and 1005 m^3, respectively. In 2017, United Utilities categorized CSOs releasing wastewater into sites S1 to S3 as high risk due to the combination of increasing pollution incidents and higher spill volume and frequencies (Utilities, personal communication, 2017). In response to water quality non-compliance at S3, United Utilities is working with the Irwell Catchment Partnership to investigate grey and green infrastructure options to improve water quality.

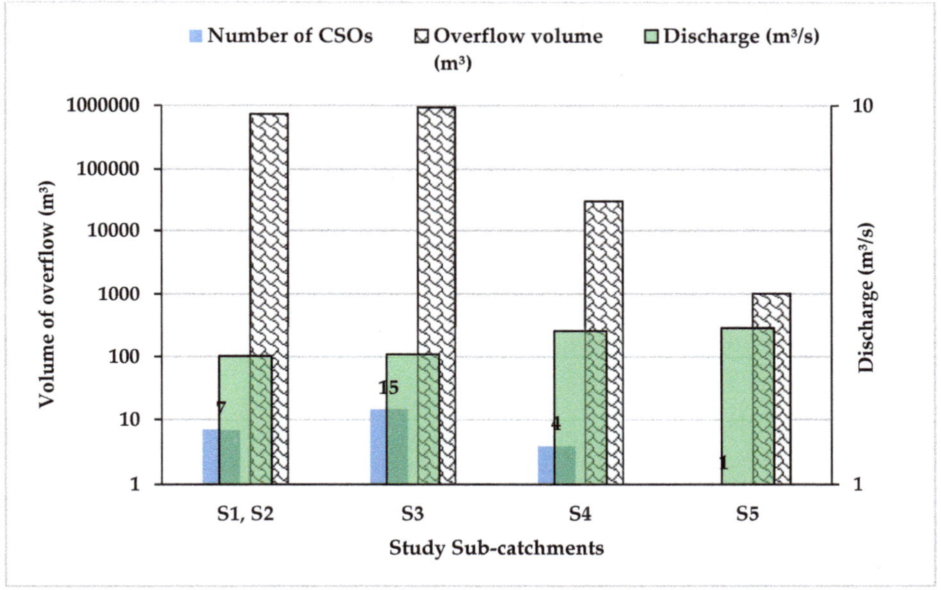

Figure 5. Volume (Log$_{10}$) (m^3) of wastewater and discharge (Log$_{10}$) (m^3/s) released from combined sewer overflow (CSO) for each sample site.

4. Discussion

4.1. Land Cover and Water Quality

Previous studies have recognized impervious covers as predictors of water quality degradation [1,3,5,12]. Generally, water quality degrades when the percentage of impervious cover is more than 10% [3]. The total area of impervious cover (i.e., transport network, residential, industrial and commercial) in our catchment is 98.57 km^2 or 40% of the total catchment land cover. Impervious covers by sub-catchment vary between 37%–57% and increases downstream. The predominant impervious land cover is residential buildings which is typical in most urban areas [3]. This complicates the water quality picture as residential areas are sources of SS, P and N pollution and contribute to the overall proportion of impervious surfaces downstream. The industrial legacy in this river catchment [39,40], has been replaced by mixed use impervious land cover (Table A2) that is correlated with high levels of surface runoff, conductivity, and NO_3 contaminants and lower DO levels (Figure 4).

Large expanses of impervious surfaces in built-up areas generate higher volumes of overland flow that transport nutrients, SS and other contaminants to artificial drainage networks and rivers [20,23].

We found, impervious covers, in particular transport networks (Table A2) increase runoff, principally from S3 downstream. Following precipitation events, nutrient transfer from impervious areas occurs rapidly [49] transporting contaminants into CSOs and the river, affecting water quality [49]. Table A2 and Figure 3 show that the higher the impervious cover, the lower the river quality (Table 3). Specifically, discharge, conductivity, SS, and nutrients are higher and DO lower at the downstream study sites.

Overall, our study catchment has not met the EU WFD's good ecological standard due to high PO_4-P concentration, which is attributed to discharge from the WwTW and intermittent discharges from CSOs (Figure 5). It is likely difficult to alter the volume of discharge from the WwTW, whereas, the degree of imperviousness in each CSO catchment also determines the conveyance of contaminants into the river [50]. Therefore, the results from the sub-catchment land cover assessment could inform targeted interventions to regulate runoff and improve water quality especially in the highly impervious built up areas, including residential areas. While there are efforts to improve CSO infrastructure and to reduce effluent discharge from point sources, the EA's Catchment Planning System predicts some improvement as shown in Table A3 (Appendix A) (see https://environment.data.gov.uk/catchment-planning/WaterBody/GB112069061152). However, increased PO_4-P concentration in the urban section of the river is forecast to remain Poor and therefore, the implementation of Best Management Practices could be deployed to the river catchment on a case-by-case basis.

Woodland cover can reduce nutrient discharge to rivers, prevent storm water runoff, reduce soil erosion and absorb pollutants. In the study area, the urban green and woodland (Broadleaf) land covers correlated positively with DO and negatively with the nutrients, discharge and conductivity. Yet, in many urban areas, including Greater Manchester, urban woodland has been crowded out by other development. Caution nevertheless should be exercised in the rush to plant trees in catchments for flood risk management or to tackle climate change, as the type of woodland matters, i.e., for this catchment broadleaf woodland might be preferred [2,51].

4.2. Spatial Scales and DEM

Even in a small catchment, where the distance between S2 and S3 is 4.5 km and from S3 and S4 is 3 km, spatial land cover patterns can affect water quality. By selecting a threshold of 100 m^2 (Section 2.4) for the delineation of the sub-catchments, the land cover and water quality variables and relationships at the riparian level was revealed. An increased buffer zone, i.e., more than 100 m^2 would weaken the influence of the land covers on water quality [25]. River monitoring at different riparian and habitat scales can reveal more information to inform management strategies [26,49].

The variation in catchment size observed in our analysis was attributed to the application of a higher resolution DEM (12.5 m), i.e., 65.9 km^2 as compared to 57.5 km^2 determined by the Centre for Ecology and Hydrology UK Integrated Hydrological Digital Terrain Model (IHDTM) on a 50 m grid interval. High resolution DEMs are essential when modelling urban environments, i.e., to predict floods [52,53].

This study demonstrates that ArcGIS and DEM provide useful tools needed to understand the impact of land cover on water quality indicators [54]. Impervious surface covers dominate the river catchment, including transport networks, residential, industrial, and commercial. The WwTW was also shown to be a major contributor to the nutrient loads and the correlation between nutrients, discharge and land covers reflects hydrogeomorphological processes that operate at varying spatial scales [55,56] and measurement scales. Nevertheless, the results of our study may be weakened by the small number of sampling sites. The process of regular sampling and monitoring of urban waterbodies requires permission and access that could be furthered by a more partnership approach to river management.

5. Conclusions

Overall, this study demonstrates that ArcGIS and DEM provide useful tools needed to understand the impact of land cover types and related diffuse source pollutants at the sub-catchment scale on urban water quality. The mix of impervious land covers in each CSO "catchment" determines the composition of overflowed effluent and is a means to separate out this load from the pollutant load discharged from a WwTW. Imperviousness emerged as a potentially useful measure to classify and model river catchments to assess water quality and biodiversity and for flood regulation. Land cover provides an additional target for interventions in river catchments to improve water quality and flow regulation. Interventions could include community and town planning strategies aimed to protect river habitats, through education on garden fertilizer management for households, the development of spatially targeted urban greenspaces, SuDS regulations, or more directly through imperviousness tax on new urban and infrastructural development. The intersection of water quality, urban habitat protection, greenspace for wellbeing, and flood risk management might provide added impetus for integrative management strategies between town planners, regulators, water companies, developers, local community groups and academic researchers.

Author Contributions: The topic was conceptualized by C.M., water quality measurement and analysis by C.M. The development of the maps and the use of the ArcGIS software was analyzed and prepared by K.O.; contributions on different versions of the manuscript discussions on background, policy through partnerships was discussed by R.B. The manuscript was written and data interpreted by C.M. and contributions and editing by R.B. and K.O. All authors have read and agreed to the published version of the manuscript.

Funding: R.B. was funded by the European Union's Horizon 2020 research and innovation program under the Marie Skłodowska-Curie grant agreement No 659449.

Acknowledgments: Thanks to Andrew Green and Keith White who gave helpful comments to improve the manuscript.

Conflicts of Interest: The authors declare no conflict of interest.

Appendix A

Table A1. Area of the land cover types extracted from the study sub-catchments.

Sites	Continuous Urban Fabric–Transport Network	Continuous Urban Fabric–Urban Green	Discontinuous Urban Fabric–Residential	Industrial, Commercial	Inland Waters	Woodland–Broadleaf	Total area (m^2) of Land Cover per Sub-Catchment	Impervious Cover %	Non-impervious Cover %
S1	1,302,096.98	7,093,284.49	2,661,520.56	76,954.12	22,569.13	3,753,243.20	14,909,668.48	27.10	72.90
S2	2,501,224.47	9,545,920.15	4,184,219.37	491,579.65	22,569.13	4,974,877.57	21,720,390.34	33.04	66.96
S3	6,153,087.00	18,284,202.32	9,394,692.58	1,039,360.66	43,686.01	9,485,514.80	44,400,543.37	37.36	62.64
S4	7,600,552.87	20,420,978.49	11,075,106.06	1,521,639.61	43,686.01	10,239,830.79	50,901,793.83	39.68	60.32
S5	7,960,135.54	20,648,100.75	11,283,991.01	1,542,921.98	43,686.01	10,311,191.26	51,790,026.55	40.14	59.86
GS	11,718,343.53	24,857,706.93	15,144,208.77	2,919,623.12	43,686.01	11,277,371.05	65,960,939.41	45.15	54.85
			Total				249,683,361.98		

Table A2. Area of the main impervious covers and average runoff coefficient (c); Average runoff coefficient only significant for transport networks. Average precipitation intensity (mm/hr/ha) (I).

Sites	Continuous Urban Fabric–Transport Network (Area-(m^2); %)	Discontinuous Urban Fabric–Residential (Area-(m^2); %)	Industrial, Commercial (Area-(m^2); %)	Sum of the Impervious cover (Area-(m^2); %)	Discharge (m^3/s) = Q	I = Average Precipitation Intensity = mm/h/(m^3/s) or ha precipitation = mm) Transp Netw	I = Average Precipitation Intensity = mm/hr/ha (precipitation = mm) Res	I == Average Precipitation Intensity = mm/hr/ha (precipitation = mm) IC Commercial	c–Transp Network = Q/I × A (Transp Networks)
S1	1,302,096.98 32.23	2,661,520.56 65.87	76,954.12 1.9	4,040,571.66	0.15	8.98×10^{-8}	4.40×10^{-8}	1.52×10^{-6}	1.28
S2	2,501,224.47 34.85	4,184,219.37 58.3	491,579.65 6.85	7,177,023.49	0.23	8.99×10^{-8}	2.80×10^{-8}	2.38×10^{-7}	1.02
S3	6,153,087.00 37.1	9,394,692.58 56.64	1,039,360.66 6.27	16,587,140.24	0.43	1.90×10^{-8}	1.25×10^{-8}	1.13×10^{-7}	3.67
S4	7,600,552.87 37.63	11,075,106.06 54.83	1,521,639.61 7.53	20,197,298.54	0.53	1.54×10^{-8}	1.06×10^{-8}	7.69×10^{-8}	4.53
S5	7,960,135.54 38.29	11,283,991.01 54.28	1,542,921.98 7.42	20,787,048.53	0.53	1.47×10^{-8}	1.04×10^{-8}	7.58×10^{-8}	4.53
GS	11,718,343.53 39.35	15,144,208.77 50.85	2,919,623.12 9.80	29,782,175.42 98,571,257.88	0.53	9.99×10^{-9}	7.73×10^{-9}	4.01×10^{-8}	4.53

Table A3. EA CPS (2019) Chemical and ecological classification for 2016 and 2027 by reach, where Up = upstream, Down = Downstream. P = Poor, M = Moderate, sG = Supports Good and G = Good. The broad classification of UK waterbodies covers the ecological (biological quality, hydromorphological, physico-chemical, supporting) and chemical (priority substances, hazard substances) statuses. Each element of the classification informs the overall classification status of the waterbody.

Classification Item	2016		2027	
	Up	**Down**	**Up**	**Down**
Ecological (overall)	**M**	**M**	**G**	**M**
Biological quality elements	P	P	M	M
Hydromorphological supporting elements	sG	sG	sG	sG
Physico-chemical quality elements	M	M	G	M
Supporting elements (surface water)	M	M	G	G
Chemical (overall)	G	G	G	G

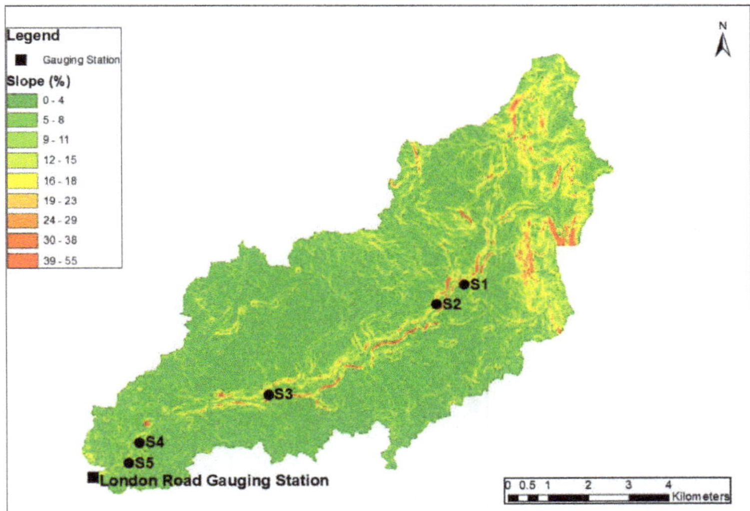

Figure A1. The percent slope map for the River Medlock catchment area.

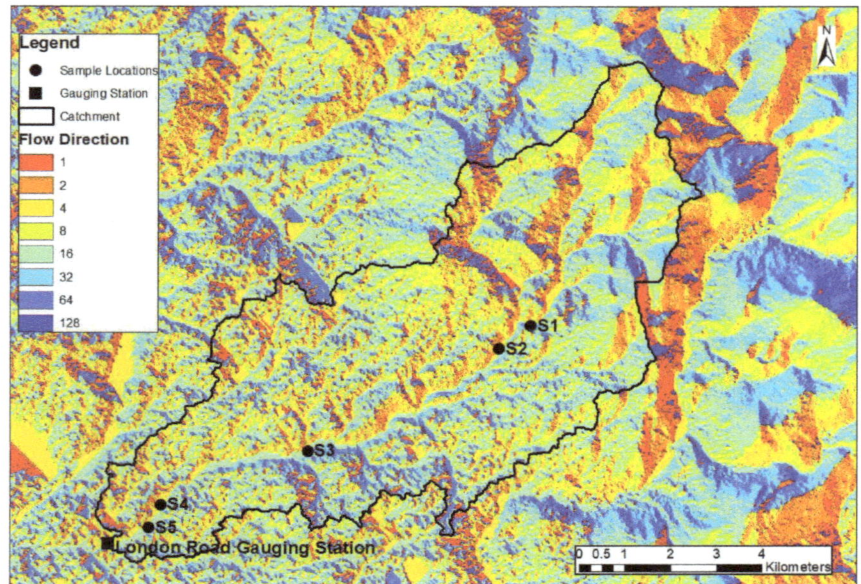

Figure A2. Flow direction map of the River Medlock catchment. Insert is the D8 legend showing flow direction values increase by a factor of 2 in a clockwise direction of a compass.

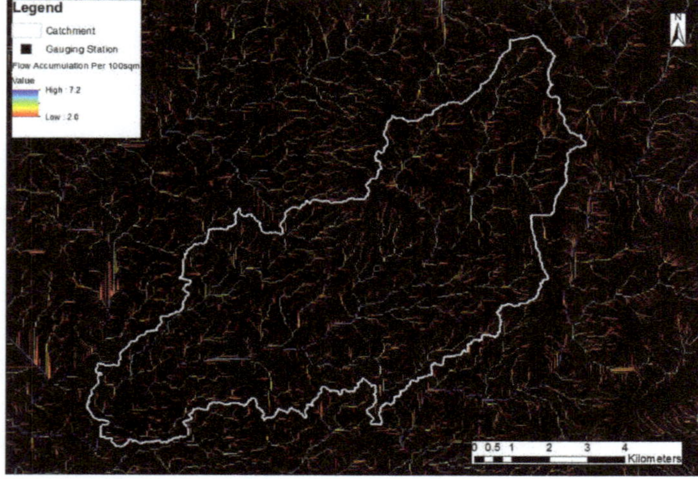

Figure A3. Flow accumulation map of the River Medlock catchment indicates the drainage areas of the sample points (S1–S5) and the gauging station. The flow accumulations were computed over a contributing area of 100 m^2.

References

1. Beck, S.M.; McHale, M.R.; Hess, G.R. Beyond Impervious: Urban Land-Cover Pattern Variation and Implications for Watershed Management. *Environ. Manag.* **2016**, *58*, 15–30. [CrossRef] [PubMed]
2. Nisbet, T.; Silgram, M.; Shah, N.; Morrow, K.; Broadmeadow, S. *Woodland for Water: Woodland Measures for Meeting Water Framework Directive Objectives*; FR(GB-JH) PDF/June11/0099; Forest Research: Surrey, UK, 2011.

3. Arnold, C.L.; Gibbons, C.J. Impervious Surface Coverage: The Emergence of a Key Environmental Indicator. *J. Am. Plan. Assoc.* **1996**, *62*, 243–258. [CrossRef]
4. Walsh, C.J.; Roy, A.H.; Feminella, J.W.; Cottingham, P.D.; Groffman, P.M.; Morgan, R.P. The urban stream syndrome: Current knowledge and the search for a cure. *J. N. Am. Benthol. Soc.* **2005**, *24*, 706. [CrossRef]
5. Lepeška, T. The Impact of Impervious Surfaces on Ecohydrology and Health in Urban Ecosystems of Banská Bystrica (Slovakia). *Soil Water Res.* **2016**, *11*, 29–36. [CrossRef]
6. Zalewski, M. Ecohydrology—The use of ecological and hydrologîcal processes for sustainable management of water resources. *J. des Sci. Hydrol.* **2003**, *47*, 823–832. [CrossRef]
7. Chadwick, M.A.; Dobberfuhl, D.R.; Benke, A.C.; Huryn, A.D.; Suberkropp, K.; Thiele, J.E. Urbanization Affects Stream Ecosystem Function by Altering Hydrology, Chemistry, and Biotic Richness. *Ecol. Appl.* **2010**, *16*, 1796–1807. [CrossRef]
8. Chin, A. Urban transformation of river landscapes in a global context. *Geomorphology* **2006**, *79*, 460–487. [CrossRef]
9. Ding, J.; Jiang, Y.; Liu, Q.; Hou, Z.; Liao, J.; Fu, L.; Peng, Q. Influences of the land use pattern on water quality in low-order streams of the Dongjiang River basin, China: A multi-scale analysis. *Sci. Total Environ.* **2016**, *551–552*, 205–216. [CrossRef]
10. Huang, J.; Zhan, J.; Yan, H.; Wu, F.; Deng, X. Evaluation of the impacts of land use on water quality: A case study in the Chaohu lake basin. *Sci. World J.* **2013**, *2013*, 7. [CrossRef]
11. Paul, M.J.; Meyer, J.L. Streams in the urban landscape. *Annu. Rev. Ecol. Syst.* **2001**, *32*, 333–365. [CrossRef]
12. Walsh, C.J.; Fletcher, T.D.; Vietz, G.J. Variability in stream ecosystem response to urbanization: Unraveling the influences of physiography and urban land and water management. *Prog. Phys. Geogr.* **2016**, *40*, 714–731. [CrossRef]
13. Cuffney, T.F.; Mcmahon, G.; Kashuba, R.; May, J.T.; Waite, I.R. Responses of Benthic Macroinvertebrates to Urbanization in Nine Metropolitan Areas. *Ecol. Appl.* **2010**, *20*, 1384–1401. [CrossRef] [PubMed]
14. Passerat, J.; Ouattara, K.N.; Mouchel, J.-M.; Rocher, V.; Servais, P. Impact of an intense combined sewer overflow event on the microbiological water quality of the Seine River. *Water Res.* **2011**, *45*, 893–903. [CrossRef] [PubMed]
15. Madoux-Humery, A.S.; Dorner, S.; Sauvé, S.; Aboulfadl, K.; Galarneau, M.; Servais, P.; Prévost, M. The effects of combined sewer overflow events on riverine sources of drinking water. *Water Res.* **2016**, *92*, 218–227. [CrossRef] [PubMed]
16. Wang, J. Combined Sewer Overflows (CSOs) Impact on Water Quality and Environmental Ecosystem in the Harlem River. *J. Environ. Prot. (Irvine Calif.)* **2014**, *11*, 1373–1389. [CrossRef]
17. Xu, Z.; Wu, J.; Li, H.; Chen, Y.; Xu, J.; Xiong, L.; Zhang, J. Characterizing heavy metals in combined sewer overflows and its influence on microbial diversity. *Sci. Total Environ.* **2018**, *625*, 1272–1282. [CrossRef]
18. APEM, A. *Manchester Ship Canal Strategic Review of Fish*; APEM: Heaton Mersey, UK, 2007; p. 139.
19. Myerscough, P.E.; Digman, C.J. Combined Sewer Overflows-Do they have a Future? In Proceedings of the 11th International Conference on Urban Drainage, Edinburgh, Scotland, 31 August–5 September 2008; pp. 1–10.
20. Rees, A.; White, K.N. Impact of Combined Sewer Overflows on the Water-Quality of an Urban Watercourse. *Regul. Rivers-Res. Manag.* **1993**, *8*, 83–94. [CrossRef]
21. Committee on Climate Change. *UK Climate Change Risk Assessment 2017. Synthesis Report for the Next Five Years*; Committee on Climate Change: London, UK, 2016.
22. National Flood Resilience Review. Available online: https://www.gov.uk/government/publications/national-flood-resilience-review (accessed on 16 March 2020).
23. Thorne, C. Geographies of UK flooding in 2013/4. *Geogr. J.* **2014**, *180*, 297–309. [CrossRef]
24. Cheng, P.; Meng, F.; Wang, Y.; Zhang, L.; Yang, Q.; Jiang, M. The impacts of land use patterns on water quality in a trans-boundary river basin in northeast China based on eco-functional regionalization. *Int. J. Environ. Res. Public Health* **2018**, *15*, 1872. [CrossRef]
25. Dai, X.; Zhou, Y.; Ma, W.; Zhou, L. Influence of spatial variation in land-use patterns and topography on water quality of the rivers inflowing to Fuxian Lake, a large deep lake in the plateau of southwestern China. *Ecol. Eng.* **2017**, *99*, 417–428. [CrossRef]
26. Shi, P.; Zhang, Y.; Li, Z.; Li, P.; Xu, G. Influence of land use and land cover patterns on seasonal water quality at multi-spatial scales. *Catena* **2017**, *151*, 182–190. [CrossRef]

27. Xia, L.L.; Liu, R.Z.; Zao, Y.W. Correlation Analysis of Landscape Pattern and Water Quality in Baiyangdian Watershed. *Procedia Environ. Sci.* **2012**, *13*, 2188–2196. [CrossRef]
28. Potter, K.M.; Cubbage, F.W.; Blank, G.B.; Schaberg, R.H. A watershed-scale model for predicting nonpoint pollution risk in North Carolina. *Environ. Manag.* **2004**, *34*, 62–74. [CrossRef] [PubMed]
29. Buakhao, W.; Kangrang, A. DEM Resolution Impact on the Estimation of the Physical Characteristics of Watersheds by Using SWAT. *Adv. Civ. Eng.* **2016**, *2016*, 9. [CrossRef]
30. Lin, Z.; Oguchi, T. Drainage density, slope angle, and relative basin position in Japanese bare lands from high-resolution DEMs. *Geomorphology* **2004**, *63*, 159–173. [CrossRef]
31. Jenson, K.S.; Dominigue, O.J. Extracting Topographic Structure from Digital Elevation Data for Geographic Information System Analysis. *Photogramm. Eng. Remote Sens.* **1988**, *54*, 1593–1600.
32. Hawker, L.; Bates, P.; Neal, J.; Rougier, J. Perspectives on Digital Elevation Model (DEM) Simulation for Flood Modeling in the Absence of a High-Accuracy Open Access Global DEM. *Front. Earth Sci.* **2018**, *6*, 233. [CrossRef]
33. Nigel, R.; Rughooputh, S. Mapping of monthly soil erosion risk of mainland Mauritius and its aggregation with delineated basins. *Geomorphology* **2010**, *114*, 101–114. [CrossRef]
34. Wechsler, S.P. Uncertainties associated with digital elevation models for hydrologic applications: A review. *Hydrol. Earth Syst. Sci.* **2007**, *11*, 1481–1500. [CrossRef]
35. Wang, G.; Yinglan, A.; Xu, Z.; Zhang, S. The influence of land use patterns on water quality at multiple spatial scales in a river system. *Hydrol. Process.* **2014**, *28*, 5259–5272. [CrossRef]
36. Gurnell, A.; Lee, M.; Souch, C. Urban Rivers: Hydrology, Geomorphology, Ecology and Opportunities for Change. *Geogr. Compass* **2007**, *1*, 1118–1137. [CrossRef]
37. Whitford, V.; Ennos, A.R.; Handley, J.F. "City form and natural process"—Indicators for the ecological performance of urban areas and their application to Merseyside, UK. *Landsc. Urban Plan.* **2001**, *57*, 91–103. [CrossRef]
38. Nolan, P.A.; Guthrie, N. River rehabilitation in an urban environment: Examples from the Mersey Basin, North West England. *Aquat. Conserv. Mar. Freshw. Ecosyst.* **1998**, *8*, 685–700. [CrossRef]
39. Burton, L.R. The Mersey Basin: An historical assessment of water quality from an anecdotal perspective. *Sci. Total Environ.* **2003**, *314–316*, 53–66. [CrossRef]
40. Douglas, I.; Hodgson, R.; Lawson, N. Industry, environment and health through 200 years in Manchester. *Ecol. Econ.* **2002**, *41*, 235–255. [CrossRef]
41. MacKillop, F. Climatic city: Two centuries of urban planning and climate science in Manchester (UK) and its region. *Cities* **2012**, *29*, 244–251. [CrossRef]
42. Williams, A.E.; Waterfall, R.J.; White, K.N.; Hendry, K. Manchester Ship Canal and Salford Quays: Industrial legacy and ecological restoration. In *Ecological Reviews: Ecology of Industrial Pollution*; Cambridge University Press: Cambridge, UK, 2010; pp. 276–308.
43. Blackburn, H.; O'Neill, R.; Rangeley-Wilson, C. *Flushed Away: How Sewage is Still Polluting the Rivers of England and Wales*; Hydro International: Clevedon, UK, 2017; pp. 1–37.
44. Medupin, C. Spatial and temporal variation of benthic macroinvertebrate communities along an urban river in Greater Manchester, UK. *Environ. Monit. Assess.* **2020**, *192*, 84. [CrossRef]
45. Archfield, S.A.; Vogel, R.M. Map correlation method: Selection of a reference streamgage to estimate daily streamflow at ungauged catchments. *Water Resour. Res.* **2010**, *46*, 1–15. [CrossRef]
46. Environment Agency-Standing Committee of Analysts (SCA library): Methods for the Examination of Waters and Associated Material—Index of Methods for the Examination of Waters and Associated Materials 1976–2011 Blue Book 236. 2011. Available online: http://www.standingcommitteeofanalysts.co.uk (accessed on 15 March 2020).
47. Barbara, K.; György, B.; Gerard, H.; Stephan, A. *Updated CLC Illustrated Nomenclature Guidelines*; European Environment Agency: Copenhagen, Denmark, 2017; pp. 1–124.
48. WFD_UKTAG (2013) UKTAG WFD Environmental Standards River Basin Management 2015–2021. November 2013. Available online: http://www.wfduk.org/sites/default/files/Media/Environmental%20standards/UKTAG%20Environmental%20Standards%20Phase%203%20Final%20Report%2004112013.pdf (accessed on 15 March 2020).

49. Luo, K.; Hu, X.; He, Q.; Wu, Z.; Cheng, H.; Hu, Z.; Mazumder, A. Impacts of rapid urbanization on the water quality and macroinvertebrate communities of streams: A case study in Liangjiang New Area, China. *Sci. Total Environ.* **2018**, *621*, 1601–1614. [CrossRef]
50. Klein, R.D. Urbanization and Stream Quality Impairment. *Water Resour. Bull.* **1979**, *15*, 948–963. [CrossRef]
51. Gagkas, Z.; Heal, K.; Stuart, N.; Nisbet, T.R. Forests and Water Guidelines: Broadleaf Woodlands and the Protection of Freshwaters in Acid-Sensitive Catchments. In Proceedings of the BHS 9th National Hydrology Symposium, Durham, England, 11–13 September 2006; pp. 53–58.
52. Savage, J.T.S.; Bates, P.; Freer, J.; Neal, J.; Aronica, G. When does spatial resolution become spurious in probabilistic flood inundation predictions? *Hydrol. Process.* **2016**, *30*, 2014–2032. [CrossRef]
53. Fewtrell, T.J.; Bates, P.D.; Horritt, M.; Hunter, N.M. Evaluating the effect of scale in flood inundation modelling in urban environments. *Hydrol. Process.* **2008**, *22*, 5107–5118. [CrossRef]
54. Flaherty, D.J.; Drelich, J. *The Use of Grasslands to Improve Water Quality in the New York City Watershed*; Session 27—Integration of Environmental and Agricultural Policy; European Environment Agency: Copenhagen, Denmark, 1996.
55. Grabowski, R.C.; Surian, N.; Gurnell, A.M. Characterizing geomorphological change to support sustainable river restoration and management. *Wiley Interdiscip. Rev. Water* **2014**, *1*, 483–512. [CrossRef]
56. Gurnell, A.M.; Rinaldi, M.; Belletti, B.; Bizzi, S.; Blamauer, B.; Braca, G.; Buijse, A.D.; Bussettini, M.; Camenen, B.; Comiti, F.; et al. A multi-scale hierarchical framework for developing understanding of river behaviour to support river management. *Aquat. Sci.* **2016**, *78*, 1–16. [CrossRef]

Article

Assessing the Impact of Storm Drains at Road Embankments on Diffuse Particulate Phosphorus Emissions in Agricultural Catchments

Gerold Hepp * and Matthias Zessner

Institute for Water Quality and Resource Management, Technische Universität Wien, Karlsplatz 13/226, 1040 Wien, Austria; sekretariat@iwag.tuwien.ac.at

* Correspondence: ghepp@iwag.tuwien.ac.at

Received: 8 September 2019; Accepted: 12 October 2019; Published: 17 October 2019

Abstract: This study presents a simple mapping key suitable for quick and systematic assessments of the types of agricultural and civil engineering structures present in a certain agricultural catchment as well as the impact they may have on the spatial distribution of critical source areas. An application of this mapping key to three small sub-catchments of a case study catchment with an area of several hundred square kilometres (one-stage cluster sampling) in Austria clearly reveals that road embankments with subsurface drainage can exert a major influence on emissions and transport pathways of sediment-bound pollutants like particulate phosphorus (PP). Due to this, the semi-empirical, spatially distributed PhosFate model is extended to separately model PP emissions into surface waters via storm drains along road embankments. Furthermore, the overall share of road embankments with subsurface drainage on all road embankments in the case study catchment is inferred with the help of a Bayesian hierarchical model. The combination of the results of these two models shows that the share of storm drains at road embankments on total PP emissions ranges from about one fifth to one third in the investigated area.

Keywords: diffuse pollution; field mapping; storm drains; Bayesian statistics; distributed modelling; PhosFate

---

1. Introduction

Many efforts to control diffuse pollution of surface waters in agricultural catchments are based on the concept of critical source areas. This concept commonly refers to relevant source areas connected to surface waters in a way that allows significant amounts of pollutants to enter surface waters (e.g., [1–3]). While the concept of critical source areas is basically applicable to many different pollutants, it is particularly relevant in the case of particulate phosphorus (PP) and other sediment-bound emissions.

PP emissions in agricultural catchments are in fact often dominated by soil erosion processes and PP transport into surface waters by overland flow. In this context, a common approach to identify critical source areas is the use of spatially distributed models. De Vente et al. [4] compared different models developed to predict soil erosion rates, sediment yield and in case of spatially distributed models also sediment sources and sinks. They conclude that while empirical/conceptual, spatially distributed models show good model performances in comparison to data requirements, the application of this type of models is only appropriate in cases where the considered erosion (e.g., sheet, rill and ephemeral gully erosion) and sediment transport processes (e.g., overland flow, soil creep and debris flow) effectively dominate the catchment of interest.

This implies that in order to successfully apply such a model to a certain catchment, a priori knowledge on its dominant erosion and sediment transport processes as well as pathways is required.

Jetten et al. [5] even advocate including observed patterns in modelling studies, since this can greatly improve model results. Such knowledge, however, is only scarcely available.

A number of studies report that agricultural and civil engineering structures (e.g., roads, ditches and culverts) may exert a decisive influence on erosion and sediment transport within catchments. Roads are found to "accelerate water flows and sediment transport" [6] and are responsible for altering flow routing patterns as well as slope lengths [7,8]. Particularly in combination with roadside ditches and culverts, they can directly connect sections of catchment areas to stream networks, which otherwise would only be indirectly connected or not connected at all [9–14].

Moussa et al. [15] furthermore point out that inter-field ditch networks increase the portions of catchment areas directly contributing to the discharge of surface waters. In short, linear agricultural and civil engineering structures hold a large potential to boost direct pollutant emissions into surface waters even from remote and at first glance disconnected fields.

This a priori knowledge is very valuable when it comes to applying a model to a certain catchment. In case of catchments where roads play a major role, they are capable of extending the natural stream network considerably. Or to say it with the words of Hösl et al. [12]: "Ditches may occur near almost every road or track to drain road systems."

Despite all these observations, linear structures are rarely considered in models and modelling studies [16]. A notable exception is the WaTEM/SEDEM model [8,17,18], which optionally takes a roads data layer as additional input. Instead of integrating linear structures into models, there is, however, also the alternative of digital elevation model (DEM) pre-processing. The RIDEM model [19,20] takes this approach. It utilises cross-sectional templates of different configurations of roads and roadside ditches to enforce their effect on flow directions. Although DEM pre-processing has the advantage of not necessarily being dependent on a certain model, it does not seem to be widely applied in the context of linear structures.

While the role of roads and ditches in altering overland flow patterns is generally rather well-known, still vague is the role of another common agricultural and civil engineering structure: storm drains [21]. This point-shaped structure is similarly capable of directly connecting remote fields to surface waters. The main difference to ditches lies in its subsurface transport pathway via sewers. In direct comparison with ditches, this means there is less retention, given that roadside ditches are often at least partly vegetated with grass and occasionally cleaned.

Several applications of the semi-empirical, spatially distributed PhosFate model [22,23] in the Austrian federal state of Upper Austria [24,25] suggest that storm drains may have a significant influence on sediment emissions into surface waters. Verstraeten et al. [26] state that the effect of riparian vegetated buffer strips on catchment scale may be rather low, since a considerable amount of sediment bypasses them through sewers, among others. Another study found that artificial zones like farm courtyards are responsible for about half of the fields connected to the stream network [27]. While they do not decidedly mention storm drains or sewers, it can be reasoned, that they are a major cause of this. In addition, Djodjic and Villa [28] express that surface water inlets are common in Sweden and due attention should be paid to them.

As a result of a field mapping campaign in a catchment in north-western Switzerland, Bug [29] reports that storm drains occur mainly at asphalt roads leading across a hillside and in areas with flow convergence. However, for another catchment in the German federal state of Lower Saxony, he depicts a dense network of ditches but no storm drains at all. This indicates that there exist different traditions in implementing agricultural and civil engineering measures.

Since detailed datasets on the existence of ditches and storm drains are rarely available but a prerequisite for modelling studies aiming at the identification of critical source areas, there is a need to map them. Although several studies have carried out related field mapping campaigns in the past (e.g., [10,12,29]), no ready to use mapping key exists for this task. This study therefore intends to develop, present and discuss a simple mapping key suitable for practitioners interested in a quick and systematic assessment of the types of agricultural and civil engineering structures present in a certain catchment as well as the impact they may have on the spatial distribution of critical source areas.

A further objective is to estimate the share of PP emissions entering surface waters via storm drains and sewers at road embankments. The relevance of these structures is assessed for a case study catchment with an area of several hundred square kilometres. This is seen as essential information for policy makers facing the task to meet water quality targets in catchments of similar size.

2. Material and Methods

2.1. Case Study Catchment

This study was performed on the catchment of the river Pram in the Austrian federal state of Upper Austria, which is located in the northern part of the country. Its area is approximately 380 km^2 in size and it belongs to the geologic formation Molasse basin with small parts in the north and northeast belonging to the crystalline Bohemian Massif. The prevalent soil texture is silt loam (nearly 50% of the soil surface) followed by other types of loam. Only 5% of the soil surface are dominated by clay. Soil textures dominated by sand are not present in the catchment area.

Annual precipitation ranges from roughly 900 mm in the north with an elevation of about 300 m a. s. l. to around 1200 mm in the south with an elevation of about 800 m a. s. l. Agricultural land (approx. 45% arable land and 25% grassland) is prevailing in the predominantly hilly terrain. Forests account for about 20% and built-up areas for almost 10% of the catchment area. The most important crops are winter grains and maize covering approximately 40 and 30% of the arable land respectively. Additional catchment properties are listed in Table 1. Diffuse PP emissions are totally dominated by erosion from, especially, agricultural land. Other types of sources are negligible [30]. Altogether, anthropogenic influence can be considered as high in this catchment.

Table 1. Additional properties of the case study catchment.

	Min.	1st Qu.	Median	Mean	3rd Qu.	Max.
Field size in ha	0.01	0.36	0.90	1.60	2.02	28.16
Slope [a] in %	0.00	4.59	8.41	10.22	13.50	241.23
Discharge [b] in $m^3\,s^{-1}$	0.94	1.87	2.68	4.58	4.23	117.00

[a] Based on a DEM with 10 × 10 m resolution. [b] Period 2008 to 2013 of the gauge 204867, Pramerdorf/Pram close to the outlet with a catchment area of approx. 340 km^2 [31].

2.2. Mapping Key

The mapping key was developed based on the hydrological connectivity definitions related to landscape features and hillslope scale as collected by Ali and Roy ([32], Table 1). Among these definitions, the one from Croke et al. [10] distinguishes "two types of connectivity: direct connectivity via new channels or gullies, and diffuse connectivity as surface runoff reaches the stream network via overland flow pathways."

For this mapping key, the direct connectivity type was extended to include not only channels and gullies but also ditches, culverts, storm drains and sewers. Direct connectivity was furthermore split into known and unknown direct connectivity in order to differentiate between areas connected to an existing (e.g., governmental mapped) stream network dataset and areas connected to the actual stream network via features unknown to such a dataset (cf. [33]). The term diffuse connectivity was kept as it is.

A major issue with all connectivity mapping methods is that they merely record "snapshots" [34]. Within the concept of hydrological connectivity one can distinguish static/structural and dynamic/functional elements of connectivity [35,36]. A simplified view on these elements comprehends space as structure and space-time as function [37]. Omitting the functional element leads to a purely structural approach, which is solely capable of pointing out potential connectivity [34].

De Vente et al. [4] note that empirical/conceptual, spatially distributed models are frequently built on the RUSLE [38]. Since this erosion model as well as its precursor USLE [39,40] estimate long term yearly erosion rates under average climatic conditions only, which can likewise be considered a potential (erosion potential), they match comparatively well with purely structural connectivity approaches.

Ultimately, the development of this mapping key aimed at aiding practitioners in preparing modelling studies with empirical/conceptual, spatially distributed models for water quality management at catchment scale. It therefore was based on a purely structural approach focusing on discrete mapping units like fields or (parts of) zero-order catchments. With the term zero-order catchments we refer to the catchment areas of all zones in a landscape exhibiting an overland-channel transition (Figure 1).

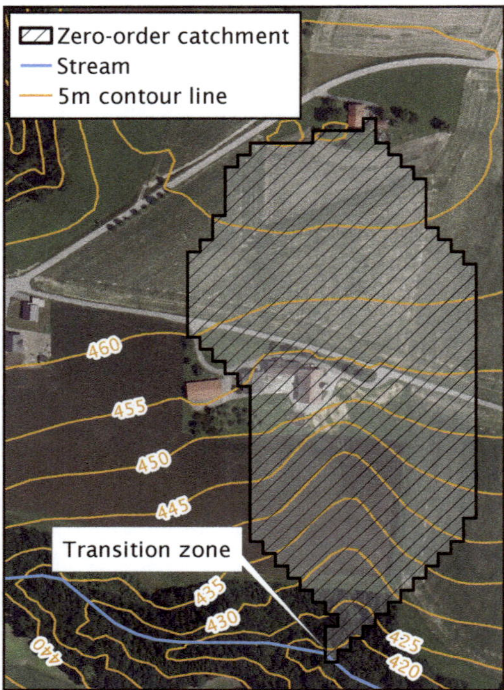

Figure 1. Example of a zero-order catchment (catchment area belonging to a single overland-channel transition zone) based on a DEM with 10 × 10 m resolution and the D8 single flow direction algorithm [41].

For each mapping unit, this mapping key fundamentally records if it is (i) connected or (ii) not connected in a natural way and/or if an (additional) artificial connection is (i) present, (ii) not present or (iii) present in a downstream mapping unit. It is thus required to map from downstream to upstream. Both of these types of connectivity are further supported by explanatory details, which include the cause of a mapping unit being connected or not in a natural way and, in case one is present, the type of the artificial connection.

Additionally, this mapping key allows for mapping artificial point-shaped and linear structures as well as natural features like channels missing in existing datasets. This can help with retracing flow pathways in case of questions once the mapping has been completed. The full ready to use and basically adaptable mapping key tables can be found in Appendix A.

Since connectivity is linked, among others, to discharge frequency, especially the categorisation of mapping units regarding connected in a natural way or not is somewhat arbitrary and in danger of subjectivity of the person(s) carrying out a mapping campaign. This issue can be addressed by defining a threshold obstacle height. Mosimann et al. [33] acknowledge obstacles higher than 50 cm as effective barriers for surface runoff, preventing connectivity, and so do we.

In order to efficiently apply this mapping key, a mobile geographic information system (GIS) supporting a global navigation satellite system (GNSS) with at least two datasets is required: (i) an existing stream network and (ii) borders of the desired mapping units as well as the ability to alter them. Additional digital orthophotos help furthermore with orientation and locating certain landscape features.

2.3. Application of the Mapping Key

For the application of the mapping key, agricultural fields were chosen as mapping units. The statistical population of the case study catchment is therefore the total amount of its fields. Survey sampling was carried out with the help of one-stage cluster sampling. Cluster sampling is particularly useful when the statistical population under consideration can be grouped geographically. In the present case, it was grouped into small sub-catchments, which allowed for minimising travel times.

Generally speaking, one-stage cluster sampling refers to the random selection of a subset of all clusters and the systematic assessment of all elements within the selected clusters. On the contrary, two-stage cluster sampling refers to not only the random selection of a subset of all clusters but also to the random selection of a subset of the elements to be assessed within the selected clusters. In this context, three small sub-catchments/clusters (Figure 2) were randomly selected from the total of 154 small sub-catchments/clusters of the case study catchment. The one-stage cluster sampling was then carried out by systematically mapping all of their fields.

For this task, a tablet computer running Microsoft Windows 10 with Esri ArcPad version 10.2.1 employing an external Global Positioning System (GPS) receiver was used. The field borders could be taken from a (geo-)database related to the Integrated Administration and Control System (IACS) of the Common Agricultural Policy (CAP) of the European Union (EU) [42]. A high-quality governmental mapped stream network was available from the Federal Office of Metrology and Surveying [43] and digital orthophotos were contributed by the State Government of Upper Austria. Moreover, for referencing, a picture geodatabase was created containing overview pictures and pictures showing the relevant details of each field.

About 650 ha of agricultural land corresponding to 323 fields were mapped in total (Table 2). These are approximately 2.5% of the overall agricultural land within the case study catchment. Each mapped cluster has its own characteristics: cluster A is elongated with tendentiously steep slopes and roads mainly on the ridges, cluster B has a tree-like stream network, tendentiously gentle slopes and roads mainly on the ridges as well and cluster C has roads primarily leading across hillsides and partially cased streams. The fairly smaller average field size of cluster C may be due to its roads leading across hillsides, which split more fields apart. While cluster A's diffuse PP emissions reaching surface waters are with roughly 3.4 kg ha^{-1} rather high due to its steep slopes, those of cluster B and C are lower and amount to roughly 1.4 and 1.1 kg ha^{-1} respectively.

Table 2. Mapped agricultural land and number of fields per catchment and agricultural land use type.

		Cluster A	Cluster B	Cluster C	Total
Mapped agricultural land in ha	Arable land	128	206	149	483
	Grassland	23	69	76	168
	Total	151	275	225	651
No. of fields	Arable land	37	67	73	177
	Grassland	21	49	76	146
	Total	58	116	149	323
Average field size in ha	Arable land	3.5	3.1	2.0	2.7
	Grassland	1.1	1.4	1.0	1.2
	Total	2.6	2.4	1.5	2.0

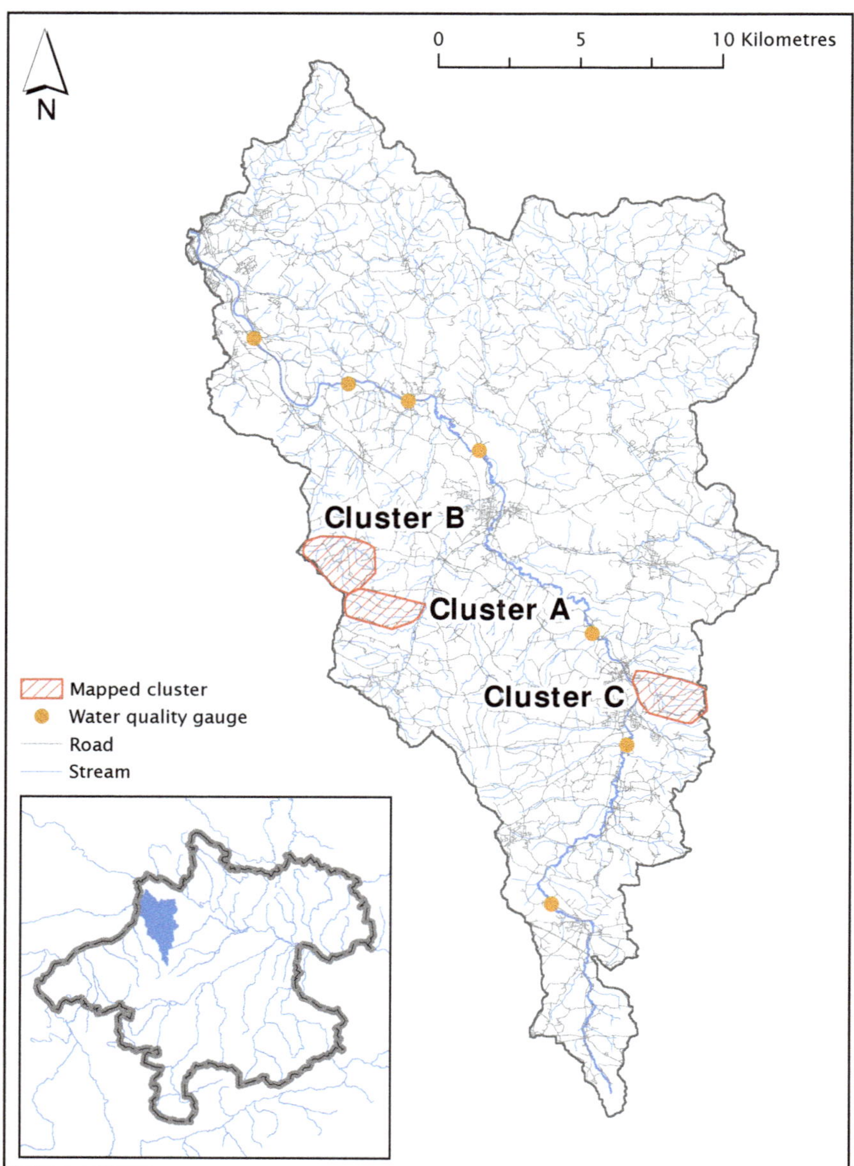

Figure 2. The case study catchment with the three randomly selected and systematically mapped clusters as well as the seven water quality gauges along the river Pram used for calibrating PhosFate.

2.4. *Modelling the Impact of Storm Drains at Road Embankments on PP Transport*

2.4.1. The PhosFate Model

PhosFate is a semi-empirical, spatially distributed phosphorus emission and transport model created for the identification of critical source areas in a management context at catchment scale [22,23]. It is based on raster GIS data and utilises the (R)USLE [38–40] incorporating a single flow algorithm version of the slope length factor of Desmet and Govers [44]. PP emissions are then calculated from the erosion and PP content of each raster cell considering a local enrichment ratio [24].

The PP transport part of PhosFate consists of the computation of the PP retention via an exponential function of the cell residence time and a mass balance equation considering the inflowing PP load, the local PP emission and the local as well as the transfer PP retention. Computing the cell residence time requires the calculation of the hydraulic radius among others. This variable in turn involves model parameters related to discharge frequency [24]. So again, a potential (transport potential) is calculated. PP transport as calculated by PhosFate thus reflects maximum potential PP transport for a chosen discharge frequency.

2.4.2. Storm Drains Model Extension

The current version of PhosFate does not consider any linear structures potentially influencing PP transport. As a result of the field mapping campaign (see Section 3.1), it was deemed necessary to take storm drains at varying distances along road embankments often in combination with roadside ditches (Figure 3) into account. An algorithm incorporating known locations of storm water infrastructure is the (W)ASI algorithm [45,46]. At catchment scale, however, particularly the locations of rural storm drains are hardly known.

Figure 3. Roadside ditch with multiple storm drains at varying distances. The left picture was taken upstream of the right.

It was therefore decided to regard every raster cell bordering and flowing in the direction of a road as a storm drain with its outlet at the nearest stream cell (cf., [9]). Since the flow lengths in roadside ditches between the actual locations of surface runoff leaving a field and entering a storm drain remain unknown choosing such an approach, the retention potential in roadside ditches is likewise unknown. Retention in roadside ditches was thus considered globally by means of a transfer coefficient. This transfer coefficient represents the share of PP emissions reaching a storm drain, which is further routed to a stream cell. In this way, it emulates retention in roadside ditches.

In order not to account retention in roadside ditches to the bottom cells of fields bordering roads, it was necessary to introduce an additional raster layer taking care of this. The purpose of this layer is to store the amount of PP emissions further routed to a stream cell, which allows for calculating the amount of PP retained in roadside ditches among others.

2.4.3. Modelling Case Study

The storm drains model extension was tested at a spatial resolution of 10×10 m on several scenarios within the case study catchment: two road and three transport coefficient scenarios. Both types of scenarios were combined one by one adding up to six scenarios in total. The first road scenario takes all asphalt roads (termed all roads) and the second just major roads into account. Road data was taken from an up-to-date governmental reference routing dataset [47] and while the major roads scenario encompasses about 100 urban and 200 km non-urban roads, the all roads scenario encompasses about 500 urban and 850 km non-urban roads.

Compared to the natural stream network, which is about 700 km long, this means that the non-urban road network is even longer than the natural stream network. In this comparison, it has to be considered, though, that roads usually have one uphill side only, whereas streams are fed from both of their sides.

Regarding the transport coefficient scenarios, values of 0.4, 0.6 and 0.8 were chosen and kept constant during calibration. The channel deposition rate was furthermore calibrated on a fixed total in-channel PP retention at catchment outlet of approximately 20%. This value was adopted from a modelling study [25] with the lumped catchment model MONERIS [48–50]. The period of the modelling case study ranged from the year 2008 to the year 2013 and all model parameters related to discharge frequency for the calculation of the hydraulic radius were set accordingly, i. e. to a recurrence interval of six years.

As most important input data for the model served the already mentioned (geo-)database related to the IACS of the CAP of the EU [42]. It not only contains field borders of agricultural land but also detailed information on cultivated crops as well as the different factors of the USLE. Other important input data were a DEM with 10×10 m resolution utilised for flow routing as well as slope calculation, a dataset based on the digital cadastre map providing non-agricultural land use and another dataset derived from the digital soil map of Austria encompassing top soil characteristics. The State Government of Upper Austria contributed the latter three datasets. In addition, Manning's roughness coefficients were taken from Engman [51] and data on PP accumulation in top soil from Zessner et al. [30,52]. Zessner et al. [53,54] give more detailed information on the input data.

Calibration quality was assessed with the help of Nash-Sutcliffe efficiency (NSE), modified Nash-Sutcliffe efficiency (mNSE), percent bias (PBIAS) and ratio of the root mean square error to the standard deviation of measured data (RSR) [55,56] by comparing observed mean annual PP loads of seven water quality gauges along the river Pram (Figure 2) with modelled mean annual PP loads at the same locations (Table 3).

Table 3. Modelled shares of storm drains at road embankments on total PP emissions including the calibration quality of the six modelled scenarios and plausibility check of the chosen PP transfer coefficients in roadside ditches utilising the estimated PP retention per metre roadside ditch range. The closer the values of the modelled mean PP retention in roadside ditches and the mean estimated PP retention per metre roadside ditch range for a given transfer coefficient scenario are, the more plausible we consider the coefficient.

	All Roads			Major Roads		
PP transfer coefficient	0.4	0.6	0.8	0.4	0.6	0.8
	Calibration quality					
NSE	0.95	0.94	0.94	0.95	0.95	0.95
mNSE	0.83	0.83	0.82	0.84	0.84	0.84
PBIAS	−2.3	−1.6	−1.5	−2.6	−2.6	−2.6
RSR	0.22	0.22	0.22	0.21	0.21	0.21
	Plausibility check utilising the estimated PP retention per metre roadside ditch range: 3.0 to 11.3 $g\,m^{-1}$ with a mean of 7.1 $g\,m^{-1}$					
Modelled mean PP retention in roadside ditches in $g\,m^{-1}$	9.5	5.7	2.6	11.4	7.4	3.6
Share of storm drains at road embankments on total PP emissions	0.27	0.36	0.44	0.08	0.11	0.14

Water quality data was available from the surface water monitoring programme of the State Government of Upper Austria [57] for the full period of the modelling case study (2008 to 2013). It is a routine sampling programme with a sampling interval of two weeks and measures PP concentration according to EN ISO 15681-2 and EN ISO 6878 with a minimum limit of quantification (LOQ) of 0.005 mg L^{-1}. This data was then combined with stream discharge from the web GIS eHYD [31] and mean annual PP river loads were calculated according to the flow intervals method of Zoboli et al. [58].

This method facilitates the calculation of river loads for different flow conditions. Since the PP loads modelled by PhosFate represent solely rainfall induced loads stemming from erosion, they cannot be directly compared to total PP river loads. For this, the total PP river loads first have to be reduced by the amount of PP emanated during baseflow conditions.

One of the results of the extended PhosFate model is the ratio of emissions reaching surface waters via cells defined as storm drains on total emissions. In order to check the chosen transfer coefficients, which strongly affect this ratio, for plausibility, it was possible to obtain data concerning roadside ditch cleaning from the State Government of Upper Austria (personal communication). Although this data is from the year 2015 only, it integrates over a period of approximately 10 to 15 years as this is the regular interval of cleaning campaigns. It was not possible to get additional years, since data on these operations is not part of general record keeping.

In total, data on the amount of material removed from a cleaned length of about 80 km was collected. This data was then used to estimate average minimum and maximum values of PP retention per metre roadside ditch (3.0 to 11.3 g m^{-1}). Despite the consideration of different soil particle densities and material types as well as cleaning intervals to account for uncertainty, this best to worst case range has to be considered as indicative only, as it originates from little data. Nonetheless, the minimum, mean and maximum values of this estimated range could be used to check if the modelled retention rate due to the chosen transfer coefficients in roadside ditches is plausible.

2.5. *Statistical Inference of the Overall Impact of Storm Drains at Road Embankments on Diffuse PP Emissions*

While fields appear to be the "natural" mapping unit from a management point of view, they are problematic from a statistical point of view. Due to potential flow routing from one field to another, not all of them are statistically independent. This problem was solved, however, by allocating all mapped fields influenced by road embankments to delineated zero-order catchments.

Subsequently, we applied a Bayesian hierarchical model on the reallocated data. The purpose of this model was to infer the mode as well as an equal-tailed credible and highest posterior density interval (HPDI) for the share of road embankments with subsurface drainage on all road embankments in the case study catchment. In order to make a best estimate of the overall impact of storm drains at road embankments on diffuse PP emissions, the results of the Bayesian model in turn were combined with the results of the extended PhosFate model.

The Bayesian model considered all 154 clusters of the case study catchment of which the data of three was known. First, the number of zero-order catchments influenced by road embankments was drawn from a Poisson distribution with a weakly informative gamma prior for the unmapped clusters. Then the number of influenced zero-order catchments with subsurface drainage was estimated from the total number by means of a binomial distribution. Its success probability was allowed to vary on the logit-scale for each cluster. For this, a weakly informative normal prior on its mean and a weakly informative half-Cauchy prior as a conservative choice on its standard deviation [59,60] were applied. The model was written in R [61] using JAGS (Just Another Gibbs Sampler) [62].

3. Results

3.1. *Field Mapping Campaign*

Among the mapped clusters and agricultural land use types, the minimum share of fields connected in a natural way is about 50 and the maximum about 75%, whereas fields connected in an artificial way constitute approximately 10 to 30 % (Figure 4). The proportion of fields showing both attributes lies between roughly 5 and 15% and fields not connected at all exhibit a minimum share of 0 and a maximum of about 10%.

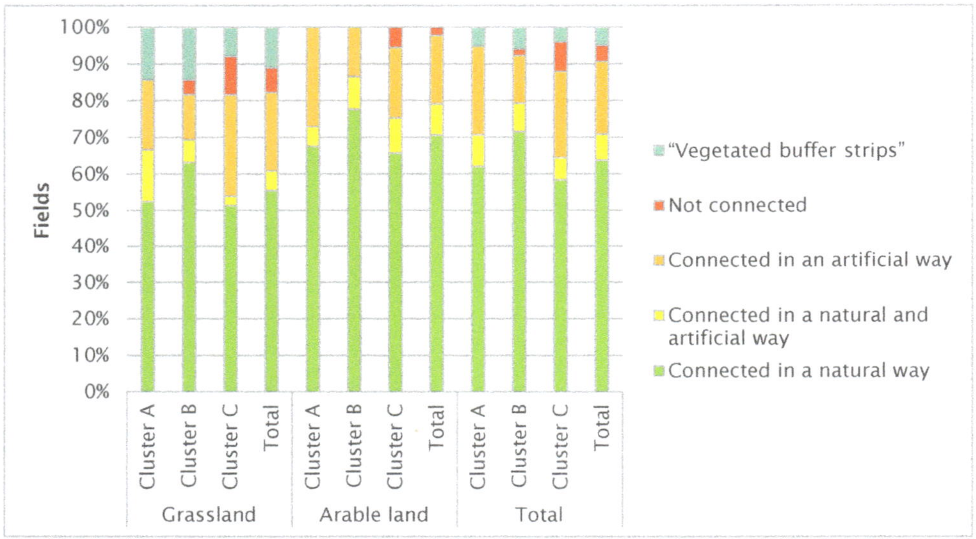

Figure 4. Proportions of the assessed connectivity types among the mapped clusters and agricultural land use types. The results indicate a rather low overall variability ("Vegetated buffer strips" can be considered as grassland fields connected in a natural way).

Furthermore, a fifth category was identified and termed "vegetated buffer strips" (VBS). These are grassland fields connected in a natural way and located between arable land and streams showing some of the properties (i.e., dimensions) of "normal" VBS. They are, however, not arable land with VBS but leftovers, which for some reason were not turned into arable land and are not subject to fertiliser restrictions. Their share varies between approximately 10 and 15% among the grassland fields and constitutes roughly 5% of all mapped fields. Moreover, a general comparison of all these results—especially when accounting the mapped "VBS" to grassland fields connected in a natural way—indicates a rather low overall variability.

Due to the assessment of the explanatory details, it was possible to analyse the causes of fields not being connected in a natural way and the different types of artificial connections present in the mapped clusters (Figure 5). Fields categorised as "VBS" were excluded from this analysis. Table 4 shows the causes and their respective shares as identified in the field mapping campaign. These data clearly reveal that road embankments are the main cause of influence on emissions and transport pathways (approx. 65%). They are followed by single storm drains unrelated to roads or roadside ditches and influences stemming from differences between the actual and the mapped stream network (e.g., as a result of cased streams or unknown channels) (both approx. 15%). The remaining almost 10% are composed of a variety of causes including sinks and settlements.

Considering all mapped fields, road embankments with subsurface drainage (i.e., storm drains at varying distances along road embankments often in combination with roadside ditches) are responsible for influencing emissions and transport pathways of almost a quarter of the fields. Given that about 45% of the mapped fields are subject to an influence, road embankments with subsurface drainage account for approximately half of all influences present in the mapped clusters. They furthermore constitute about 80% of the fields influenced by road embankments.

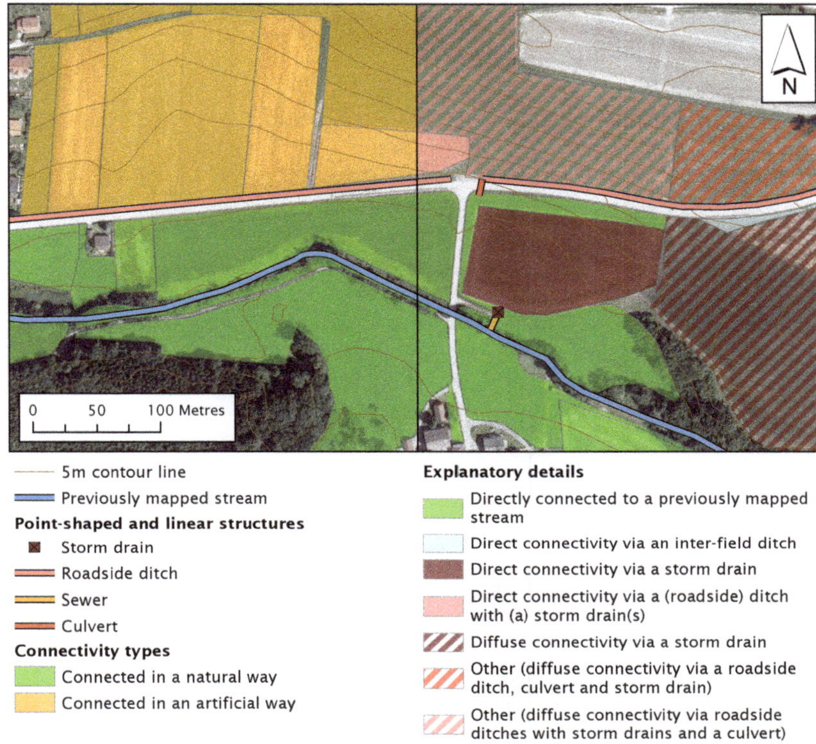

Figure 5. Exemplified visualisation of some of the results of the field mapping campaign with connectivity types shown on the left and explanatory details on the right side of the figure. The part above connectivity types on the left side of the legend is common to both sides. Please note that the top- and rightmost, bright looking field behind the north arrow does not belong to the mapped cluster.

Table 4. Causes of influence on emissions and transport pathways expressed as percentage of all influenced fields as well as all mapped fields.

Cause	Arable Land	Grassland	Total
	% of influenced fields		
Road embankments with subsurface drainage [a]	57	45	51
Road embankments with overland drainage [b]	10	5	8
Road embankments without artificial drainage	4	6	5
Single storm drains [c]	10	15	13
Stream network differences [d]	12	17	14
Other	7	12	9
	% of mapped fields		
Road embankments with subsurface drainage	24	21	23
Road embankments with overland drainage	5	2	3
Road embankments without artificial drainage	2	3	2
Single storm drains	5	7	6
Stream network differences	5	8	6
Other	3	5	4
Total	43	45	44

[a] Storm drains with or without roadside ditches. [b] Roadside ditches and/or culverts. [c] Storm drains unrelated to roads or roadside ditches. [d] Differences between the actual and the mapped stream network (e.g., due to cased streams).

3.2. Impact of Storm Drains at Road Embankments on PP Transport

The calibration quality of the six modelled scenarios (two road and three transport coefficient scenarios) indicates good model performance (cf., [56]) and is consistent across all scenarios (Table 3). Unfortunately, a validation with independent river loads could not be performed, since not all of the input data are available at the same spatial resolution in another period of time.

The plausibility check of the chosen PP transfer coefficients in roadside ditches utilising the estimated PP retention per metre roadside ditch range demonstrates that a transfer coefficient of 0.6 fits best with both road scenarios. Whereas a transfer coefficient of 0.8 scrapes at the lower limit of the estimated PP retention per metre roadside ditch range in both instances, a transfer coefficient of 0.4 only scrapes at the upper limit of the major road scenario and indicates that the best fit actually seems to be somewhere between a transfer coefficient of 0.5 and 0.6 in the other case.

A comparison of the modelled shares of storm drains at road embankments on total PP emissions between the six modelled scenarios results in a small range of 8 to 14 % for the major roads scenarios and a big range of 27 to 44 % for the scenarios considering all roads. Overall, the range encompasses a share of as little as approximately one tenth and as much as almost one half.

3.3. Best Estimate of the Overall Impact of Storm Drains at Road Embankments on Diffuse PP Emissions

Overall, the retention rate as calculated by the extended PhosFate model fits well to real world data. By comparing the values of the transfer coefficients and the resulting mean estimated PP transfer ratios in roadside ditches, it can be concluded that a transfer coefficient of 0.6 fits best and is therefore the most likely coefficient. However, neither the major roads nor the all roads scenario seems to appropriately reflect the situation in the case study catchment.

According to the results of the Bayesian model, the mode of the share of road embankments with subsurface drainage on all road embankments is 77% and the equal-tailed 95% credible interval ranges from 20 to 98%. While this is quite a big range, the 90% HPDI ranges from 54 to 100% only. Taking into account that the case study catchment appears all in all fairly homogeneous, our best estimate of the share of storm drains at road embankments on total PP emissions consequently ranges from about one fifth to one third. This estimate was obtained by combining the 90% HPDI with the PhosFate results of the all roads scenario with a transfer coefficient of 0.6.

4. Discussion

By applying the mapping key presented here, it could be shown that it is capable of collecting valuable information on agricultural and civil engineering structures potentially influencing emissions and transport pathways in agricultural catchments. It clearly revealed that the main influencing factor in the mapped clusters are road embankments with subsurface drainage and that the most important question in this case study catchment is whether a road embankment is equipped with one or more storm drains or not.

A limitation of the mapping key is that it is adequate to assess structural but not functional elements of connectivity. It thus only points out potential connectivity [34]. This, however, fits well to spatially distributed models like PhosFate, which incorporate the (R)USLE, as this equation only estimates an (erosion) potential as well. Beyond that, due to the involvement of parameters for the calculation of the hydraulic radius related to discharge frequency, PhosFate again calculates a (transport) potential only. This calculation of mere potentials can be considered a limitation of the chosen modelling approach, which otherwise shows a good ratio of model performance to data requirements (cf., [4]).

Calibrating PhosFate on several water quality gauges along the main river of the case study catchment ensured that the utilised input data reflect the spatial distribution of the relevant transport pathways and that the assumption of global calibration parameters was justified. The fact that all modelled scenarios show an equally good calibration quality further supports this.

When interpreting the channel and in particular the field deposition rate resulting from calibration, one has to bear in mind, though, that they include a temporal component. This means that apart from integrating catchment properties influencing transport (e.g., density and average width of hedges as well as reins not covered by the land use input data), they also relate the local modelled mean annual transport loads corresponding to the chosen discharge frequency of the hydraulic radius to the global observed mean annual loads at water quality gauges. In other words, they at least partly control whether a certain area contributes to the overall river load (i.e., becomes connected) or not (cf., [63]).

Another limitation of the mapping key is its use of discrete mapping units. For one thing, they are of different sizes, for another, due to potential flow routing from one unit to another, not all of them may be statistically independent. The latter can be solved by delineating zero-order catchments and using these as mapping units.

Addressing the issue of different sizes is more challenging. In fact, assuming an equal-sized sampling grid at a spatial resolution smaller than the average field size would increase the issue of dependent samples, whereas a sampling grid at a spatial resolution larger than the average field size would lead to ambiguity. This already is a problem with fields (e.g., same field on both sides of a ridge) and means that a certain property is sometimes only valid for the majority but not all of the assessed area.

Getting back to zero-order catchments, though, and assuming that the sizes of those (parts) showing a certain property and those not showing a certain property average out in a sampling cluster (i.e., small sub-catchment in this case), this issue can probably be neglected, as the weights stay the same in both cases. This would even allow for a geostatistical approach, interpolating ratios between clusters [64]. The total number of affected units and their associated areas or pollutant loads could then be estimated with the help of auxiliary data.

Ali and Roy [32] compare different spatial sampling schemes and conclude that cyclic sampling is the most adequate sampling design for geostatistical analysis. It should therefore be investigated how and with how much effort it can be successfully applied to cluster sampling within catchment areas of several hundred square kilometres and if the above assumption holds true.

So while the use of fields as mapping units worked well, there may be much precision to gain by switching to zero-order catchments in the future. This, nonetheless, has the complication of many possible statistical populations (cf., [65]), each strongly depending on the type and resolution of elevation data as well as flow routing algorithm used for their delineation. Again, future research should investigate if mapping results are sensitive to these and if it may be advantageous to only consider zero-order catchments with a certain minimum size or showing a certain shape for statistical inference.

Furthermore, another limitation of the chosen modelling approach results from using a global transport coefficient. As storm drains are usually built at locations of high overland flow concentration, it is not unlikely that their locations also exhibit high erosion and in turn high diffuse PP emissions along with steep slopes and short or even no passages through roadside ditches where retention can take place. A global transport coefficient simply considering average conditions may therefore overestimate retention under such conditions. Thus, the overall impact of storm drains at road embankments on diffuse PP emissions may be even higher. This consideration should likewise be subject of future research.

Finally, if one wanted to know the exact locations of, for example, storm drains or culverts, a conceivable solution could be to focus on a narrow strip along roads only. In this case, the question could be formulated as machine learning classification problem for which several existing model types could be utilised (e.g., logistic regression, decision trees, support vector machines and artificial neural networks).

A good start for this might be to use high resolution orthophotos in combination with high resolution airborne topographic Light Detection And Ranging (LiDAR) data as input to learn from. Another possibility would be to equip a car with appropriate sensors and collect topographic LiDAR data including pictures and videos from the roadsides by merely driving along all of them or at least along those of interest [66]. In the light of rather recent advancements in image recognition with deep neural networks (e.g., [67]) and their emerging ability to produce a measure of uncertainty even in the context of large scale image classification problems [68], such a task becomes less and less utopian. To get an idea of what is going on in a catchment area, such detail, however, is usually not required.

5. Conclusions

The influence of road embankments with subsurface drainage can be of importance in catchment areas with a strong tradition towards this drainage type. In case there is indication of this in a certain catchment, a field mapping campaign assessing the degree of their influence should be performed. This is particularly important when trying to identify critical source areas and finding suitable spots for the implementation of measures reducing pollutant emissions into surface waters (cf., [69]).

Field observations capable of collecting valuable information on agricultural and civil engineering structures potentially influencing emissions and transport pathways can already be accomplished with very simple means. Carried out with a mobile GIS application and a simple mapping key as the one presented in this study, they can easily be integrated into the preparation of modelling studies aiming at the identification of critical source areas at catchment scale. In the end, the combination and comparison of field observations with simplifying hypothesis [70] and model output [71,72] proved to be of great help in assessing the impact of agricultural and civil engineering structures in this case study catchment.

Author Contributions: Conceptualization, G.H. and M.Z.; Data curation, G.H.; Formal analysis, G.H.; Funding acquisition, M.Z.; Methodology, G.H.; Project administration, M.Z.; Supervision, M.Z.; Visualization, G.H.; Writing—original draft, G.H.; Writing—review & editing, M.Z.

Funding: This research was funded by the State Government of Upper Austria contract numbers AUWR-2012-61484/46-StU and AUWR-2015-231931/24-StU.

Acknowledgments: The authors especially thank Franz Überwimmer and Ulrike Steinmair of the State Government of Upper Austria for hosting constructive discussions of the results. Furthermore, our thank goes to Markus Langer, Andreas Lechner and Helene Trautvetter for carrying out the field mapping campaign as well as to Ottavia Zoboli and Arabel Amann for their valuable help in improving the manuscript.

Conflicts of Interest: The authors declare no conflict of interest.

Appendix A. Mapping Key Tables

Table A1. Connected or not in a natural way. Mosimann et al. [33] acknowledge obstacles higher than 50 cm as effective barriers for surface runoff, preventing connectivity, and so do we.

Code	Description
0	Not connected in a natural way
1	Connected in a natural way

Table A2. Explanatory details for mapping units not connected in a natural way. Particularly code 4 and 5 in combination with the codes of Table A3 can also be used to indicate a change in connectivity (e.g., a change from direct towards diffuse connectivity).

Code	Description
1	Road embankment
2	Levee or other linear structure
3	Depression
4	Previously mapped stream is missing
5	Previously mapped stream is cased
99	Other

Table A3. Explanatory details for mapping units connected in a natural way.

Code	Description
1	Directly connected to a previously mapped stream
2	Direct connectivity via a previously unknown channel
3	Direct connectivity via a gully
4	Diffuse connectivity
99	Other

Table A4. Artificial connection.

Code	Description
0	Not present
1	Present
2	Diffusely present in a downstream mapping unit

Table A5. Explanatory details for mapping units connected in an artificial way. Diffuse connectivity in this context means that an artificial connection is either diffusely present in a downstream mapping unit (see code 2 of Table A4) or that an artificial structure is not directly connected to a surface water (e.g., a ditch ending in a forest).

Code	Description
1	Direct connectivity via an inter-field ditch
2	Direct connectivity via a roadside ditch
3	Direct connectivity via a storm drain
4	Direct connectivity via a culvert
5	Direct connectivity via a (roadside) ditch with (a) storm drain(s)
6	Direct connectivity via a (roadside) ditch with (a) culvert(s)
7	Diffuse connectivity via an inter-field ditch
8	Diffuse connectivity via a roadside ditch
9	Diffuse connectivity via a storm drain
10	Diffuse connectivity via a culvert
11	Diffuse connectivity via a (roadside) ditch with (a) storm drain(s)
12	Diffuse connectivity via a (roadside) ditch with (a) culvert(s)
99	Other

Table A6. Presence and absence of natural features as well as artificial point-shaped and linear structures.

Code	Description
1	Channel
2	Gully
3	Inter-field ditch
4	Roadside ditch
5	Storm drain
6	Sewer
7	Culvert
8	Stream casing
9	Missing, previously mapped stream
99	Other

References

1. Heathwaite, A.L.; Quinn, P.F.; Hewett, C.J.M. Modelling and Managing Critical Source Areas of Diffuse Pollution from Agricultural Land Using Flow Connectivity Simulation. *J. Hydrol.* **2005**, *304*, 446–461. [CrossRef]
2. Pionke, H.B.; Gburek, W.J.; Sharpley, A.N. Critical Source Area Controls on Water Quality in an Agricultural Watershed Located in the Chesapeake Basin. *Ecol. Eng.* **2000**, *14*, 325–335. [CrossRef]
3. Strauss, P.; Leone, A.; Ripa, M.N.; Turpin, N.; Lescot, J.M.; Laplana, R. Using Critical Source Areas for Targeting Cost-Effective Best Management Practices to Mitigate Phosphorus and Sediment Transfer at the Watershed Scale. *Soil Use Manag.* **2007**, *23*, 144–153. [CrossRef]
4. De Vente, J.; Poesen, J.; Verstraeten, G.; Govers, G.; Vanmaercke, M.; Van Rompaey, A.; Arabkhedri, M.; Boix-Fayos, C. Predicting Soil Erosion and Sediment Yield at Regional Scales: Where Do We Stand? *Earth Sci. Rev.* **2013**, *127*, 16–29. [CrossRef]
5. Jetten, V.; Govers, G.; Hessel, R. Erosion Models: Quality of Spatial Predictions. *Hydrol. Process.* **2003**, *17*, 887–900. [CrossRef]

6. Forman, R.T.T.; Alexander, L.E. Roads and Their Major Ecological Effects. *Annu. Rev. Ecol. Syst.* **1998**, *29*, 207–231. [CrossRef]
7. Aurousseau, P.; Gascuel-Odoux, C.; Squividant, H.; Trepos, R.; Tortrat, F.; Cordier, M.O. A Plot Drainage Network as a Conceptual Tool for the Spatial Representation of Surface Flow Pathways in Agricultural Catchments. *Comput. Geosci.* **2009**, *35*, 276–288. [CrossRef]
8. Van Oost, K.; Govers, G.; Desmet, P. Evaluating the Effects of Changes in Landscape Structure on Soil Erosion by Water and Tillage. *Landsc. Ecol.* **2000**, *15*, 577–589. [CrossRef]
9. Alder, S.; Prasuhn, V.; Liniger, H.; Herweg, K.; Hurni, H.; Candinas, A.; Gujer, H.U. A High-Resolution Map of Direct and Indirect Connectivity of Erosion Risk Areas to Surface Waters in Switzerland—A Risk Assessment Tool for Planning and Policy-Making. *Land Use Policy* **2015**, *48*, 236–249. [CrossRef]
10. Croke, J.; Mockler, S.; Fogarty, P.; Takken, I. Sediment Concentration Changes in Runoff Pathways from a Forest Road Network and the Resultant Spatial Pattern of Catchment Connectivity. *Geomorphology* **2005**, *68*, 257–268. [CrossRef]
11. Doppler, T.; Camenzuli, L.; Hirzel, G.; Krauss, M.; Lück, A.; Stamm, C. Spatial Variability of Herbicide Mobilisation and Transport at Catchment Scale: Insights from a Field Experiment. *Hydrol. Earth Syst. Sci.* **2012**, *16*, 1947–1967. [CrossRef]
12. Hösl, R.; Strauss, P.; Glade, T. Man-Made Linear Flow Paths at Catchment Scale: Identification, Factors and Consequences for the Efficiency of Vegetated Filter Strips. *Landsc. Urban Plan.* **2012**, *104*, 245–252. [CrossRef]
13. La Marche, J.L.; Lettenmaier, D.P. Effects of Forest Roads on Flood Flows in the Deschutes River, Washington. *Earth Surf. Process. Landforms* **2001**, *26*, 115–134. [CrossRef]
14. Wemple, B.C.; Jones, J.A.; Grant, G.E. Channel Network Extension by Logging Roads in Two Basins, Western Cascades, Oregon. *J. Am. Water Resour. Assoc.* **1996**, *32*, 1195–1207. [CrossRef]
15. Moussa, R.; Voltz, M.; Andrieux, P. Effects of the Spatial Organization of Agricultural Management on the Hydrological Behaviour of a Farmed Catchment during Flood Events. *Hydrol. Process.* **2002**, *16*, 393–412. [CrossRef]
16. Fiener, P.; Auerswald, K.; Van Oost, K. Spatio-Temporal Patterns in Land Use and Management Affecting Surface Runoff Response of Agricultural Catchments—A Review. *Earth Sci. Rev.* **2011**, *106*, 92–104. [CrossRef]
17. Van Rompaey, A.J.J.; Verstraeten, G.; Van Oost, K.; Govers, G.; Poesen, J. Modelling Mean Annual Sediment Yield Using a Distributed Approach. *Earth Surf. Process. Landforms* **2001**, *26*, 1221–1236. [CrossRef]
18. Verstraeten, G.; Van Oost, K.; Van Rompaey, A.; Poesen, J.; Govers, G. Evaluating an Integrated Approach to Catchment Management to Reduce Soil Loss and Sediment Pollution through Modelling. *Soil Use Manag.* **2002**, *18*, 386–394. [CrossRef]
19. Duke, G.D.; Kienzle, S.W.; Johnson, D.L.; Byrne, J.M. Improving Overland Flow Routing by Incorporating Ancillary Road Data into Digital Elevation Models. *J. Spat. Hydrol.* **2003**, *3*, 23–49.
20. Duke, G.D.; Kienzle, S.W.; Johnson, D.L.; Byrne, J.M. Incorporating Ancillary Data to Refine Anthropogenically Modified Overland Flow Paths. *Hydrol. Process.* **2006**, *20*, 1827–1843. [CrossRef]
21. Gramlich, A.; Stoll, S.; Stamm, C.; Walter, T.; Prasuhn, V. Effects of Artificial Land Drainage on Hydrology, Nutrient and Pesticide Fluxes from Agricultural Fields—A Review. *Agric. Ecosyst. Environ.* **2018**, *266*, 84–99. [CrossRef]
22. Kovacs, A.S.; Honti, M.; Clement, A. Design of Best Management Practice Applications for Diffuse Phosphorus Pollution Using Interactive GIS. *Water Sci. Technol.* **2008**, *57*, 1727–1733. [CrossRef] [PubMed]
23. Kovacs, A.; Honti, M.; Zessner, M.; Eder, A.; Clement, A.; Blöschl, G. Identification of Phosphorus Emission Hotspots in Agricultural Catchments. *Sci. Total Environ.* **2012**, *433*, 74–88. [CrossRef] [PubMed]
24. Kovacs, A. Quantification of Diffuse Phosphorous Inputs into Surface Water Systems. Ph.D. Thesis, Technische Universität Wien, Vienna, Austria, 2013.
25. Zessner, M.; Hepp, G.; Kuderna, M.; Weinberger, C.; Gabriel, O.; Windhofer, G. *Konzipierung und Ausrichtung übergeordneter strategischer Maßnahmen zur Reduktion von Nährstoffeinträgen in oberösterreichische Fließgewässer*; Technical Report; Amt der Oberösterreichischen Landesregierung: Wien, Austria, 2014.
26. Verstraeten, G.; Poesen, J.; Gillijns, K.; Govers, G. The Use of Riparian Vegetated Filter Strips to Reduce River Sediment Loads: An Overestimated Control Measure? *Hydrol. Process.* **2006**, *20*, 4259–4267. [CrossRef]

27. Gascuel-Odoux, C.; Aurousseau, P.; Doray, T.; Squividant, H.; Macary, F.; Uny, D.; Grimaldi, C. Incorporating Landscape Features to Obtain an Object-Oriented Landscape Drainage Network Representing the Connectivity of Surface Flow Pathways over Rural Catchments. *Hydrol. Process.* **2011**, *25*, 3625–3636. [CrossRef]
28. Djodjic, F.; Villa, A. Distributed, High-Resolution Modelling of Critical Source Areas for Erosion and Phosphorus Losses. *AMBIO* **2015**, *44*, 241–251. [CrossRef]
29. Bug, J.F. Modellierung der linearen Erosion und des Risikos von Partikeleinträgen in Gewässer. Ph.D. Thesis, Leibniz Universität Hannover, Hannover, Germany, 2011.
30. Zessner, M.; Gabriel, O.; Kovacs, A.; Kuderna, M.; Schilling, C.; Hochedlinger, G.; Windhofer, G. *Analyse der Nährstoffströme in oberösterreichischen Einzugsgebieten nach unterschiedlichen Eintragspfaden für strategische Planungen (Nährstoffströme Oberösterreich)*; Technical Report; Amt der Oberösterreichischen Landesregierung: Wien, Austria, 2011.
31. BMLFUW. *Hydrografisches Jahrbuch von Österreich 2013. Daten und Auswertungen*; Technical Report 121; Bundesministerium für Land- und Forstwirtschaft, Umwelt und Wasserwirtschaft: Wien, Austria, 2015.
32. Ali, G.A.; Roy, A.G. Revisiting Hydrologic Sampling Strategies for an Accurate Assessment of Hydrologic Connectivity in Humid Temperate Systems. *Geogr. Compass* **2009**, *3*, 350–374. [CrossRef]
33. Mosimann, T.; Backhaus, J.; Westphal, H. *Gewässeranschluss von Ackerflächen. Ein Schlüssel für Betriebsleiter und Berater in Niedersachsen*; Landwirtschaftskammer Niedersachsen: Hannover, Germany, 2007.
34. Bracken, L.J.; Wainwright, J.; Ali, G.A.; Tetzlaff, D.; Smith, M.W.; Reaney, S.M.; Roy, A.G. Concepts of Hydrological Connectivity: Research Approaches, Pathways and Future Agendas. *Earth Sci. Rev.* **2013**, *119*, 17–34. [CrossRef]
35. Bracken, L.J.; Croke, J. The Concept of Hydrological Connectivity and Its Contribution to Understanding Runoff-Dominated Geomorphic Systems. *Hydrol. Process.* **2007**, *21*, 1749–1763. [CrossRef]
36. Turnbull, L.; Wainwright, J.; Brazier, R.E. A Conceptual Framework for Understanding Semi-Arid Land Degradation: Ecohydrological Interactions across Multiple-Space and Time Scales. *Ecohydrology* **2008**, *1*, 23–34. [CrossRef]
37. Wainwright, J.; Turnbull, L.; Ibrahim, T.G.; Lexartza-Artza, I.; Thornton, S.F.; Brazier, R.E. Linking Environmental Régimes, Space and Time: Interpretations of Structural and Functional Connectivity. *Geomorphology* **2011**, *126*, 387–404. [CrossRef]
38. Renard, K.G.; Foster, G.R.; Weesies, G.A.; McCool, D.K.; Yoder, D.C. *Predicting Soil Erosion by Water: A Guide to Conservation Planning with the Revised Universal Soil Loss Equation (RUSLE)*; Number 703 in Agriculture Handbook, U.S. Government Printing Office: Washington, DC, USA, 1997.
39. Schwertmann, U.; Vogl, W.; Kainz, M. *Bodenerosion durch Wasser. Vorhersage des Abtrags und Bewertung von Gegenmaßnahmen*, 2nd ed.; Ulmer: Stuttgart, Germany, 1987.
40. Wischmeier, W.H.; Smith, D.D. *Predicting Rainfall Erosion Losses. A Guide to Conservation Planning*; Number 537 in Agriculture Handbook; U.S. Government Printing Office: Washington, DC, USA, 1978.
41. O'Callaghan, J.F.; Mark, D.M. The Extraction of Drainage Networks from Digital Elevation Data. *Comput. Vis. Graph. Image Process.* **1984**, *28*, 323–344. [CrossRef]
42. Hofer, O.; Fahrner, W.; Pavlis-Fronaschitz, G.; Linder, S.; Gmeiner, P. *INVEKOS-Datenpool 2014 des BMLFUW. Übersicht über alle im Ordner, Invekosdaten" enthaltenen Datenbanken mit ausführlicher Tabellenbeschreibung sowie Informationen zu sonstigen verfügbaren Datenbanken*; Technical Report; Bundesministerium für Land- und Forstwirtschaft, Umwelt und Wasserwirtschaft: Wien, Austria, 2014.
43. Federal Office of Metrology and Surveying. Digitales Landschaftsmodell. Fließende Gewässer. Available online: http://www.bev.gv.at/portal/page?_pageid=713,2847785&_dad=portal&_schema=PORTAL (accessed on 31 August 2015).
44. Desmet, P.J.J.; Govers, G. A GIS Procedure for Automatically Calculating the USLE LS Factor on Topographically Complex Landscape Units. *J. Soil Water Conserv.* **1996**, *51*, 427–433.
45. Choi, Y.; Yi, H.; Park, H.D. A New Algorithm for Grid-Based Hydrologic Analysis by Incorporating Stormwater Infrastructure. *Comput. Geosci.* **2011**, *37*, 1035–1044. [CrossRef]
46. Choi, Y. A New Algorithm to Calculate Weighted Flow-Accumulation from a DEM by Considering Surface and Underground Stormwater Infrastructure. *Environ. Model. Softw.* **2012**, *30*, 81–91. [CrossRef]
47. Geoland.at. *Intermodales Verkehrsreferenzsystem Österreich (GIP.at)*; Technical Report; Geodatenverbund der Länder: Wien, Austria, 2016.

48. Behrendt, H.; Opitz, D. Retention of Nutrients in River Systems: Dependence on Specific Runoff and Hydraulic Load. *Hydrobiologia* **1999**, *410*, 111–122. [CrossRef]
49. Venohr, M.; Hirt, U.; Hofmann, J.; Opitz, D.; Gericke, A.; Wetzig, A.; Ortelbach, K.; Natho, S.; Neumann, F.; Hürdler, J. *The Model System MONERIS Version 2.14.1vba*; Leibniz-Institute of Freshwater Ecology and Inland Fisheries: Berlin, Germany, 2009.
50. Zessner, M.; Kovacs, A.; Schilling, C.; Hochedlinger, G.; Gabriel, O.; Natho, S.; Thaler, S.; Windhofer, G. Enhancement of the MONERIS Model for Application in Alpine Catchments in Austria. *Int. Rev. Hydrobiol.* **2011**, *96*, 541–560. [CrossRef]
51. Engman, E.T. Roughness Coefficients for Routing Surface Runoff. *J. Irrig. Drain. Eng.* **1986**, *112*, 39–53. [CrossRef]
52. Zessner, M.; Zoboli, O.; Hepp, G.; Kuderna, M.; Weinberger, C.; Gabriel, O. Shedding Light on Increasing Trends of Phosphorus Concentration in Upper Austrian Rivers. *Water* **2016**, *8*, 404. [CrossRef]
53. Zessner, M.; Hepp, G.; Zoboli, O.; Mollo Manonelles, O.; Kuderna, M.; Weinberger, C.; Gabriel, O. *Erstellung und Evaluierung eines Prognosetools zur Quantifizierung von Maßnahmenwirksamkeiten im Bereich der Nährstoffeinträge in oberösterreichische Oberflächengewässer*; Technical Report; Amt der Oberösterreichischen Landesregierung: Wien, Austria, 2016.
54. Zessner, M.; Hepp, G.; Kuderna, M.; Weinberger, C.; Gabriel, O. *Zustandserfassung, Nährstoffentwicklung und Quantifizierung der Maßnahmenwirksamkeiten von ÖPUL 2007 in oberösterreichischen Einzugsgebieten*; Technical Report; Amt der Oberösterreichischen Landesregierung: Wien, Austria, 2017.
55. Krause, P.; Boyle, D.P.; Bäse, F. Comparison of Different Efficiency Criteria for Hydrological Model Assessment. *Adv. Geosci.* **2005**, *5*, 89–97. [CrossRef]
56. Moriasi, D.N.; Arnold, J.G.; Van Liew, M.W.; Bingner, R.L.; Harmel, R.D.; Veith, T.L. Model Evaluation Guidelines for Systematic Quantification of Accuracy in Watershed Simulations. *Trans. ASABE* **2007**, *50*, 885–900. [CrossRef]
57. Kapfer, S. *Jahresbericht 2013—Korrektur. Zustand der Oö. Fließgewässer gem. WRRL*; Technical Report; Amt der Oberösterreichischen Landesregierung: Linz, Austria, 2014.
58. Zoboli, O.; Viglione, A.; Rechberger, H.; Zessner, M. Impact of Reduced Anthropogenic Emissions and Century Flood on the Phosphorus Stock, Concentrations and Loads in the Upper Danube. *Sci. Total Environ.* **2015**, *518–519*, 117–129. [CrossRef] [PubMed]
59. Gelman, A. Prior Distributions for Variance Parameters in Hierarchical Models (Comment on Article by Browne and Draper). *Bayesian Anal.* **2006**, *1*, 515–534. [CrossRef]
60. Gelman, A.; Jakulin, A.; Pittau, M.G.; Su, Y.S. A Weakly Informative Default Prior Distribution for Logistic and Other Regression Models. *Ann. Appl. Stat.* **2008**, *2*, 1360–1383. [CrossRef]
61. R Core Team. *R: A Language and Environment for Statistical Computing*; R Core Team: Vienna, Austria, 2016.
62. Plummer, M. JAGS: A Program for Analysis of Bayesian Graphical Models Using Gibbs Sampling. In Proceedings of the 3rd International Workshop on Distributed Statistical Computing, Vienna, Austria, 20–22 March 2003.
63. Lane, S.N.; Reaney, S.M.; Heathwaite, A.L. Representation of Landscape Hydrological Connectivity Using a Topographically Driven Surface Flow Index. *Water Resour. Res.* **2009**, *45*, W08423. [CrossRef]
64. Krivoruchko, K.; Gribov, A.; Krause, E. Multivariate Areal Interpolation for Continuous and Count Data. *Procedia Environ. Sci.* **2011**, *3*, 14–19. [CrossRef]
65. Poeppl, R.; Parsons, A. The Geomorphic Cell: A Basis for Studying Connectivity. *Earth Surf. Process. Landf.* **2018**, *43*, 1155–1159. [CrossRef]
66. Holland, D.A.; Pook, C.; Capstick, D.; Hemmings, A. The Topographic Data Deluge—Collecting and Maintaining Data in a 21st Century Mapping Agency. *Int. Arch. Photogramm. Remote Sens. Spat. Inf. Sci.* **2016**, *XLI-B4*, 727–731. [CrossRef]
67. He, K.; Zhang, X.; Ren, S.; Sun, J. Deep Residual Learning for Image Recognition. In Proceedings of the CVPR 2015: 28th IEEE Conference on Computer Vision and Pattern Recognition, Boston, MA, USA, 7–12 June 2015.
68. Heek, J.; Kalchbrenner, N. Bayesian Inference for Large Scale Image Classification. *arXiv* **2019**, arxiv:1908.03491.
69. Blackwell, M.S.A.; Hogan, D.V.; Maltby, E. The Use of Conventionally and Alternatively Located Buffer Zones for the Removal of Nitrate from Diffuse Agricultural Run-Off. *Water Sci. Technol.* **1999**, *39*, 157–164. [CrossRef]

70. Ludwig, B.; Boiffin, J.; Chadœuf, J.; Auzet, A.V. Hydrological Structure and Erosion Damage Caused by Concentrated Flow in Cultivated Catchments. *CATENA* **1995**, *25*, 227–252. [CrossRef]
71. Jetten, V.; de Roo, A.; Favis-Mortlock, D. Evaluation of Field-Scale and Catchment-Scale Soil Erosion Models. *CATENA* **1999**, *37*, 521–541. [CrossRef]
72. van Dijk, P.M.; Auzet, A.V.; Lemmel, M. Rapid Assessment of Field Erosion and Sediment Transport Pathways in Cultivated Catchments after Heavy Rainfall Events. *Earth Surf. Process. Landf.* **2005**, *30*, 169–182. [CrossRef]

Article

Impact of Combined Sewer Systems on the Quality of Urban Streams: Frequency and Duration of Elevated Micropollutant Concentrations

Ulrich Dittmer [1,*], Anna Bachmann-Machnik [1] and Marie A. Launay [2]

[1] Institute of Urban Water Management, University of Kaiserslautern, Paul-Ehrlich-Str. 14, 67663 Kaiserslautern, Germany; anna.bachmann87@outlook.de

[2] Kompetenzzentrum Spurenstoffe Baden-Württemberg, University of Stuttgart, Bandtäle 2, 70569 Stuttgart, Germany; marie.launay@koms-bw.de

* Correspondence: ulrich.dittmer@bauing.uni-kl.de

Received: 4 February 2020; Accepted: 16 March 2020; Published: 18 March 2020

Abstract: Water quality in urban streams is highly influenced by emissions from WWTP and from sewer systems particularly by overflows from combined systems. During storm events, this causes random fluctuations in discharge and pollutant concentrations over a wide range. The aim of this study is an appraisal of the environmental impact of micropollutant loads emitted from combined sewer systems. For this purpose, high-resolution time series of river concentrations were generated by combining a detailed calibrated model of a sewer system with measured discharge of a small natural river to a virtual urban catchment. This river base flow represents the remains of the natural hydrological system in the urban catchment. River concentrations downstream of the outlets are simulated based on mixing ratios of base flow, WWTP effluent, and CSO discharge. The results show that the standard method of time proportional sampling of rivers does not capture the risk of critical stress on aquatic organisms. The ratio between average and peak concentrations and the duration of elevated concentrations strongly depends on the source and the properties of the particular substance. The design of sampling campaigns and evaluation of data should consider these characteristics and account for their effects.

Keywords: modeling; CSO; urban drainage; sewer system; trace pollutants; urban runoff; concentration duration frequency curve

1. Introduction

Flow regime and water quality of urban streams are strongly influenced by surface runoff from paved areas that are connected to the receiving surface water by pipe systems. Discharge points are stormwater outlets in separate sewer systems and combined sewer overflows (CSOs) in combined systems. With an increasing percentage of paved surfaces, the hydrological response of urban catchments is becoming more rapid, peak flows are increasing, and more pollutants are transported with the runoff. These phenomena are addressed in numerous publications e.g., [1,2]. The associated damage and depletion of aquatic ecosystems has been receiving worldwide attention during the last years and is currently also known under the term "urban stream syndrome" [3].

A more recent concern is the emission of micropollutants (MP) to urban streams, also referred to as trace pollutants or emerging contaminants, and their potential impact on aquatic ecosystems. This is particularly relevant in the case of combined sewer systems where CSO spills do not only transport pollutants from urban surfaces but also wastewater constituents.

Although the relevance of the problem is pointed out by various authors, only few papers deal with this form of pollution of urban streams. Some studies are based on grab sampling in overflowing structures or receiving waters (e.g., [4–6]), others used automated samplers to create time series over spilling events [7–10] or to obtain flow weighted event mean concentrations (EMC) [8,11–13]. The specific challenges and impediments of MP sampling during wet weather are discussed by [14].

MP concentrations in CSO discharge vary widely between events and in the course of individual events [15–20]. Any sampling at the CSO structure or in the receiving urban stream only reflects a small section of a randomly fluctuating signal. Calculation of emitted loads and assessment of effects on aquatic ecosystems require continuous data over long periods [2]. This information can be gained through long-term simulations.

Various models are available to simulate the quantity and quality of urban runoff in combined and separate sewers. They are widely applied for dimensioning and management of drainage systems. Yet, calibration of the quality component is a major problem because of the multitude of random influences on pollutant transport [21]. CSO emissions generated by simulation are therefore associated with high uncertainty. This applies particularly to MP as the high analytical costs limit the size of databases for calibration.

The domain of urban drainage models is limited to the quantification of emissions. An assessment of potential ecological effects requires time series of concentrations in the urban stream. For this purpose, models have to account for the hydrology of the undeveloped part of the catchment and of the unpaved surfaces in the urban areas. Following the terminology of [22] we refer to this sub-system as the "natural system," whereas the combination of paved surfaces and underground pipe network is called "artificial system." Both sub-systems show very different hydrological responses, which makes integration into one model difficult (e.g., [23]). The problem becomes even more difficult when receiving water quality should be simulated. In [24] an integrated model of the mixed artificial-natural system that includes the wastewater treatment plant (WWTP) is presented. The model also accounts for transformation processes in the receiving water to simulate concentrations of dissolved oxygen and unionized ammonia. From the application to a case study they conclude that a detailed representation of all model components is necessary to produce satisfactory results.

The study presented in this paper circumvents the problem of model integration by overlaying the simulation results of a sewer system model with discharge measurements from a purely natural catchment. Thus, we constructed a virtual mixed artificial-natural system from two catchments that are separate in reality. Previous studies have identified the proportions of river base flow, WWTP discharge and CSO discharge as suitable proxies for the prediction of MP concentrations [10,25]. Based on this knowledge we generate long-term time series of the percentage of base flow, WWTP effluent, and CSO discharge in the receiving water. CSO discharge is further divided into varying proportions of raw wastewater, infiltration water, and urban runoff. River concentrations are calculated by multiplying each component of the discharge with the corresponding concentration derived from previous studies.

The focus of the study is not on specific results for a given case, but on general patterns and phenomena that are typical for wet weather conditions in urban streams affected by combined sewer systems. The main aims are an appraisal of the ecological impacts of micropollutant emissions from CSOs and WWTPs and a comparative assessment of the contribution of each of these pathways to potentially critical situations in the receiving urban stream. Duration and frequency of elevated MP concentrations shall be analyzed to characterize the in situ exposure of organisms quantitatively. Furthermore, conclusions on the design of sampling campaigns shall be drawn from the analysis of the concentration time series.

2. Materials and Methods

2.1. Simulation of the Artificial System

For the case study the drainage system of the city of Möhringen, Germany was used. This combined sewer system drains into the river Körsch. The total connected area is 859 ha of which 48% are paved surfaces. The main CSO structures are equipped with storage tanks, where the water is retained for subsequent treatment in the WWTP. The total storage volume is 13,095 m^3. In relation to the impervious area, this corresponds to 3.2 mm of surface runoff that can be stored. [26]

Figure 1 shows a schematic overview of the Möhringen sewer system with selected CSO tanks. The corresponding storage volumes and controlled outflows of the simulated CSO tanks are given in Table 1. The overflows of CSO tanks 1–7 are directly connected to the river in the model. The outflows of CSO tanks 8–11 are connected to the WWTP but their overflows are discharged into tributaries that flow into the River Körsch further downstream of the WWTP. The overflows of these tanks are therefore not connected to the model river. The contribution of simple CSO structures without storage to total emissions is negligible.

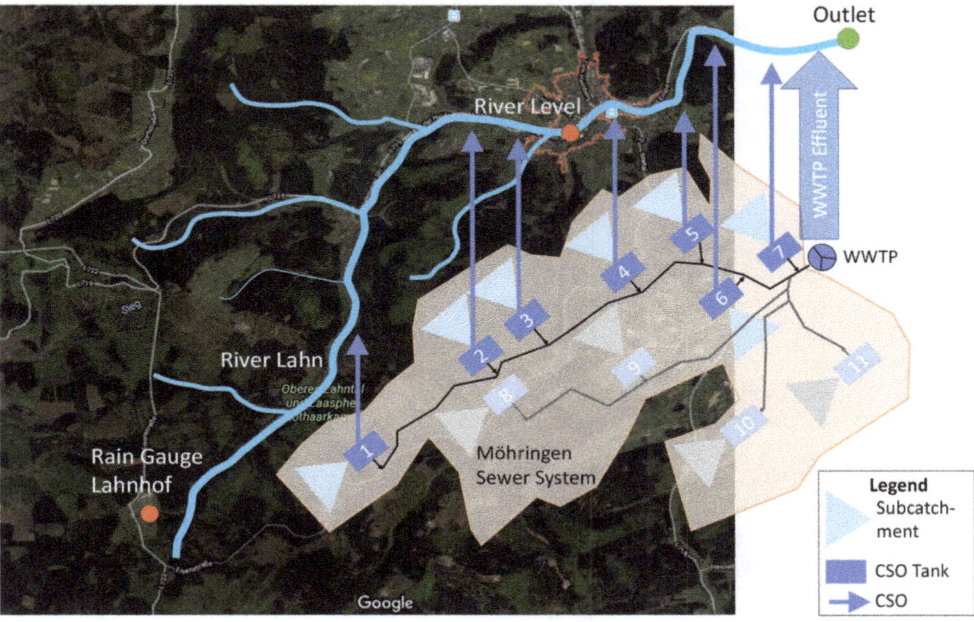

Figure 1. Schematic overview of the overlay of the Möhringen sewer system model and the natural baseflow from the river Lahn; CSO = combined sewer overflow; WWTP = wastewater treatment plant; grey sewer system structures were not considered in the model. (Source: Google Maps)

Table 1. Total connected area, storage volume and controlled outflow of main storage structures in the Möhringen sewer system. Combined sewer overflow tanks are numbered according to Figure 1.

Combined Sewer Overflow Tank No.	Total Connected Area (ha)	Storage Volume (m^3)	Controlled Outflow (L/s)
1	16	550	304
2	61	700	40
3	30	390	78
4	42	1300	140
5	127	2300	89
6	2	550	121
7	12	1750	1000
8	73	1200	53
9	228	1665	440
10	244	2290	254
11	24	400	23

The Möhringen wastewater treatment plant (WWTP) is designed for 160,000 population equivalents with a daily inflow of 22,985 m^3/d. The inflow to WWTP is limited to 1000 L/s. The daily water consumption is approx. 125 L/d. The system is located at 366 to 536 m above sea levels and the topography is characterized by mild slopes. [26]

The sewer system is represented by a lumped model. The hydrological parameters of the sub-catchments were calibrated in a previous study [26]. The representation of wastewater and infiltration water flow is based on the evaluation of monitoring data of the WWTP influent and of outflow at several CSO tanks. A detailed description of this evaluation and of the calibration process can be found in [26]. The wastewater flow from the population follows different daily patterns on weekdays and weekends. The infiltration water at the inflow of the WWTP is 100 L/s and is considered as constant over the entire year. The ratio of infiltration water and wastewater is homogeneous over the entire catchment.

The sewer system model was set up in the simulation software EPA SWMM version 5.1.007 (United States Environmental Protection Agency, Washington, DC, USA) [27]. Simulation was run under kinematic wave conditions with a flow routing time step of 30 s and a reporting time step of 15 min. Subcatchment slope was derived from catchment topography, subcatchment width was derived from subcatchment geometry. The percentage of impervious area was determined by hydraulic calibration of the sewer system model. The setup parameters of the model can be found in Table S1 in the Supplementary Material. The full model input file and further model details cannot be provided because of confidentiality reasons.

Sewer discharge consists of three components, wastewater (Q_{WW}), infiltration water (Q_{Inf}), and stormwater (Q_{SW}). Each component is marked by a virtual tracer with a constant concentration of 100 mg/L at the source. The concentrations of these tracers in the CSOs and the effluent of the WWTP thus represent the percentage of the respective component in the total flow. The resulting concentration of MP can thus be calculated by multiplying these percentages with the respective concentrations of the MP in each flow component. Transport of the virtual tracer in the sewers system is modeled in SWMM using the "tanks in series approach" that represent each conduit or storage element as a completely mixed reactor. [28].

The WWTP is implemented in the SWMM sewer system model as a cascade of six completely stirred tanks in series (storage nodes) to reproduce a realistic translation and retention of the substances passing through the WWTP (see Figure S1 in Supplementary Material). The total tank volume used for WWTP implementation is 15,500 m^3. The tanks are connected by conduits. A weir after the last storage unit ensures that the water level in the WWTP is kept constant.

Data of a rain gauge within the natural catchment of the river Lahn (station Lahnhof) is used as input for the sewer system model. The data is recorded in 1 min resolution and the period from 1 January 2004 to 31 December 2004 was used. Therefore, the measured base flow in the river and the discharge from the sewer system are hydrological responses to the same precipitation signal.

2.2. *Overlay with River Base Flow from a Natural Catchment*

Concentrations of MP in the receiving water body are the result of emission by CSO and WWTP effluent and their dilution by the base flow from the natural catchment. However, the base flow was not measured in the specific case of the River Körsch at Möhringen. This is a typical situation, as urban streams are mostly found in headwaters. These river sections are only rarely equipped with gauge stations as monitoring is mostly done on larger scales.

As a substitute for the missing data, the measured time series of a river with a natural catchment of similar size and topography (river Lahn, Germany) was used. Figure 1 shows the overlay of the natural catchment and the sewer system model. The catchment area of the River Körsch at the WWTP outlet is 23.5 km^2, while the area of the river Lahn at the gauge station Lahnhof is 25.4 km^2. The percentage of urbanized area is 37% in the real Möhringen catchment and 34% in the virtual catchment that combines the artificial system of Möhringen with the natural system monitored at the Lahnhof gauge station.

As hydrology depends on many more parameters than size and urbanization calculated river concentrations do not exactly represent the specific situation of the river Körsch. However, the virtual catchment represents a realistic scenario that can be used to assess the general level of MP concentrations in urban streams and to analyze their variability.

The river section that receives discharge from the artificial system is represented in the SWMM model as an open trapezoidal channel (bottom width = 4 m, top width = 11.6 m, maximum depth = 1.9 m, roughness = 0.25, length = 5940 m) following the natural slope of the Möhringen catchment. The channel is divided into a series of six conduits with different length. Overflows from the sewer system and the effluent from the wastewater treatment plant are directly connected to river nodes. Measured flow data from the Lahn river from the year 2004 (corresponding to the rain data as input for the sewer system simulation) is inserted directly as inflow into the first river node. The time step of the measured data is 15 min.

In analogy to the flow components in the sewer system, the base flow is assigned a virtual tracer concentration (Q_B) of 100 mg/L. The mixing of the components is simulated based on the "tank in series approach" in SWMM that represents each conduit as a completely mixed reactor. River quality is evaluated at the outlet of the virtual catchment 100 m downstream of the WWTP discharge (green dot in Figure 1).

2.3. *Pollution of the Flow Components*

The terms micropollutants or trace substances summarize a big number of substances with very different fields of application and physicochemical properties. General statements on the occurrence and behavior of micropollutants are therefore misleading. Yet, because of the unmanageable number of potentially harmful substances, each measurement or modeling must focus on a selection of substances that represent larger groups.

In a previous study, we have shown that source (wastewater or surface runoff) and elimination in WWTP are the most relevant characteristics that define the transport pathways to the receiving water during dry and wet weather [29]. For this study, we chose four substances that represent combinations of the extremes in terms of these characteristics. The idea behind this selection is that using extremes makes the influence of each characteristic on the results visible. This allows for conclusions on the behavior of other substances with less pronounced characteristics. Another criterion was a reliable database for the estimation of the concentrations.

The painkillers ibuprofen and diclofenac are chosen as indicator parameters originating from wastewater. The difference between these substances lies in their elimination in conventional municipal sewage treatment plants. The elimination of Diclofenac in WWTPs is negligible. It is assumed here that in rainy weather a reduction of the concentrations in the effluent of the WWTP plant results solely from dilution effects. Ibuprofen, on the other hand, is readily degradable and is largely eliminated in the WWTP. However, with an increased inflow into the sewage treatment plant, the elimination efficiency also decreases. It is assumed in this study that the effluent concentration is not changed by the proportion of rainwater passing through the WWTP.

Fluoranthene, from the group of polycyclic aromatic hydrocarbons (PAH) and the urban herbicide mecoprop, are mainly introduced with rainwater into the drainage systems. Fluoranthene has a strong tendency to bind to particles and is thus largely retained in the wastewater treatment plant. Mecoprop is a mainly conservative substance. As diclofenac, the concentration in the effluent of the WWTP results from the mixing of the rainwater and wastewater. In the case of fluoranthene, the higher concentrated rainwater is "diluted" with wastewater.

Table 2 summarizes the properties of the selected substances and the concentrations in the different flow components. The concentrations were derived from the results of intensive sampling in the Möhringen system [9] and data from literature [9,17,30].

Table 2. Concentrations of selected substances in the different flow components and in the outflow of the wastewater treatment plant (WWTP); RW = rainwater; WW = wastewater.

Substance	Origin	Degra-Dation	Rainwater Concentration (ng/L)	Wastewater Concentration (ng/L)	Conc. WWTP Effluent RW (ng/L)	Conc. WWTP Effluent WW (ng/L)
Fluoranthene	RW	+	400	160	2	2
Mecoprop	RW	–	500	30	500	30
Ibuprofen	WW	+	0	9000	40	40
Diclofenac	WW	–	0	1700	0	1700

Constant mean concentrations were used for the flow components wastewater (Q_{WW}), infiltration water (Q_{Inf}), and rainwater runoff (Q_{RW}). Concentrations in CSO discharge are solely the result of mixing of these components. Despite the fact that MP concentrations in rainwater runoff are not constant during rain events [31–33], knowledge on occurrence and transport of MP in wet weather flow is too scarce to justify a more detailed modeling approach. [14,34–36]

Infiltration water was assumed to be unpolluted. The elimination in the WWTP plant is described by different constant effluent concentrations for the rainwater and the wastewater fraction. These concentrations represent the effluent of a conventional biological treatment plant. The river base flow from the natural catchment is assumed to be unpolluted concerning the evaluated MP. Transformation in the river plays a minor role at the evaluated temporal and special scale and is therefore neglected.

3. Results and Discussion

3.1. Rainfall and Runoff Regime

The measured river base flow of the river Lahn and the precipitation at the rain gauge Lahnhof from January until December 2004 is shown in Figure 2. The average annual precipitation sum at the Lahnhof rain gauge is 1043 mm (2000–2010). The yearly precipitation in 2004 is 835 mm.

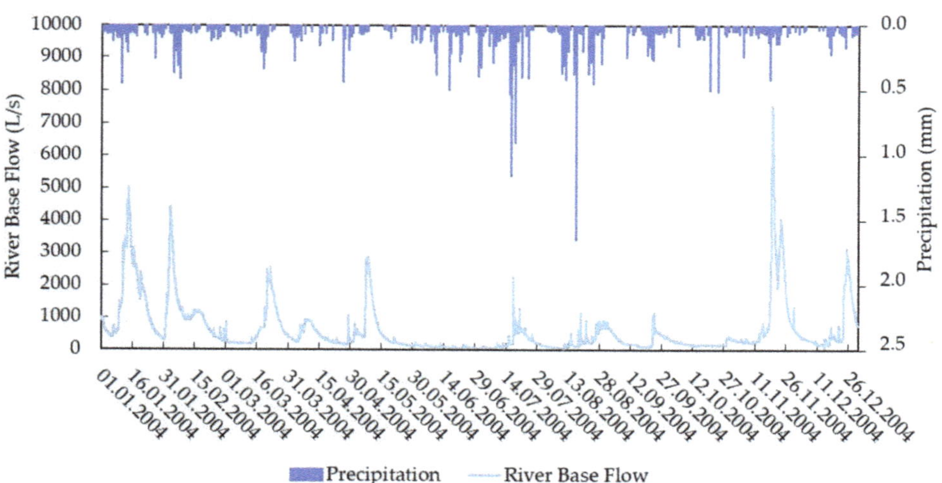

Figure 2. Hydrograph of the natural system for the year 2004.

The duration curve of the total river flow (including CSO and WWTP effluent) and the base flow from natural catchment is shown in Figure 3.

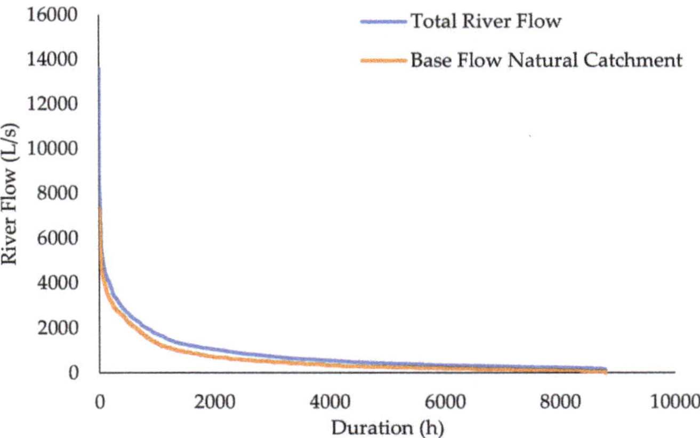

Figure 3. Duration curve of total river flow (including CSO and WWTP effluent) and base flow from natural catchment.

The average monthly base flow from the natural catchment of the river Lahn is shown in Figure 4. The discharge shows a distinct yearly pattern with low values from April to November and higher ones during winter from December to March.

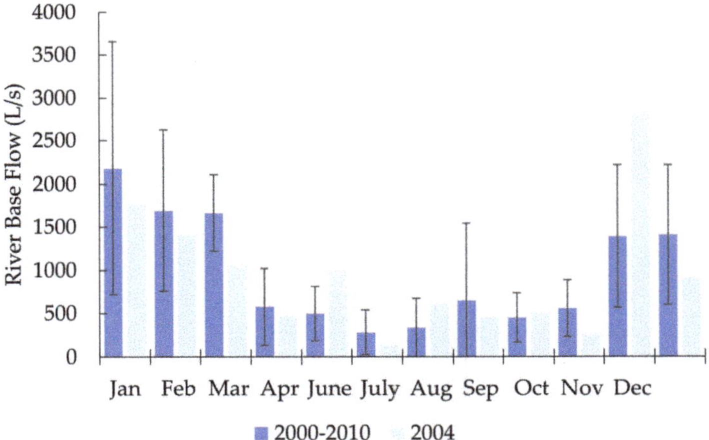

Figure 4. Monthly average base flow in the river Lahn from 2000–2010 (average value, whisker = ±standard deviation) and 2004.

3.2. Example of Mixing Ratios and Resulting Concentrations

The resulting concentrations in the river downstream of the discharge points of the CSOs and the WWTP (system outlet in Figure 1) for an exemplary overflow event of the system from 22–25 September are shown in Figure 5. The total precipitation of this event is 24 mm with an average rain intensity of 0.4 mm/h and a maximum rain intensity of 11 mm/h.

During this event, the CSO tanks in the sewer system start filling on September 22nd around 5 pm. The first overflows from the system start on 22 September at 8 pm at tank 7. The last overflow from the system ends at 5 pm on 23 September. During this rain event, a total of 2 (no. 3 and no. 7) CSO tanks discharge into the river. A total volume of 21,000 m^3 is emitted.

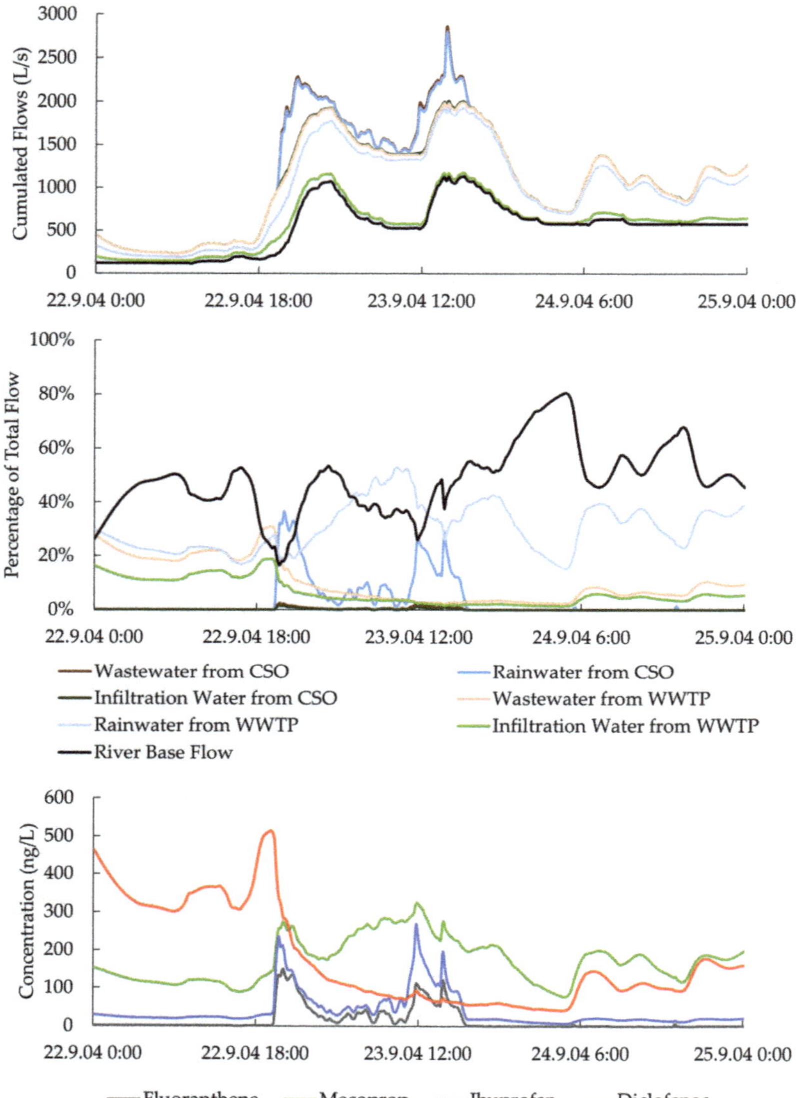

Figure 5. Simulated timeseries of flows (**top**), total flow percentage (**middle**), and corresponding micropollutant concentrations (**bottom**) for event from 22 September until 25 September, 2004; CSO = combined sewer overflow, WWTP = wastewater treatment plant.

The graph on top shows the cumulated flows of all components during this event. The proportion of infiltration water in the river during dry weather conditions is approx. 20%, the wastewater proportion is 30%. A proportion of 50% of the total river flow during dry weather is the catchment´s natural baseflow.

In the displayed section from 22–25 September a rainwater proportion, which was treated in the WWTP of around 20% is already flowing in the river, lowering the infiltration water proportion to 10% and the wastewater proportion to 20% (Figure 5, middle). Starting at 5 pm on 22 September nd the outflow of the WWTP slowly increases mainly because of the increased amount of rainwater treated in the WWTP. At 8 pm on 22 September, the first system overflows reach the observation point at the system outlet. At the same time, the baseflow from the natural catchment also increases. However, this increase is much slower than the increase of the flow from the urban system. At 8:30 pm the base flow from the natural catchment decreases to only 17% of the total flow at the observation point. The rainwater proportion treated in the WWTP is 20%, the untreated rainwater is around 30% at that time. The maximum proportion of untreated wastewater within the river during this event is 3%. In the entire year 2004, a maximum proportion of untreated wastewater within the river of 5% is reached in July during an overflow event with a very low river base flow of only 200 L/s.

On 23 September at 5:30 pm no flow proportions from CSOs are detected at the observation point anymore. After 6 pm the outflow of the WWTP decreases until the base flow from the natural catchment increases until 80%. Afterward the base flow stabilizes at around 600 L/s which is higher than the flow level before the rain event (≈140 L/s).

The concentrations of the indicator micropollutants (Figure 5, bottom) result from the multiplication of the ratios of the wastewater components in the river with the corresponding micropollutant concentration in this pathway. The river base flow and infiltration water were assumed to be unpolluted and a concentration of 0 ng/L was assigned to each micropollutant.

The concentration of diclofenac is determined only by the dilution of wastewater under dry weather conditions. As a conservative substance, no degradation takes place in the WWTP. At the beginning of the rain event the concentration of diclofenac is 300–500 ng/L in the river at the observation point corresponding to a wastewater proportion of 20–30%. Starting from 8 pm the wastewater at the observation point is increasingly diluted resulting in lower concentrations of around 50 ng/L of diclofenac in the river. The emitted load stays constant over time.

The urban herbicide mecoprop is not degraded in the WWTP. Because of the increased entry with urban runoff, the concentration profile is quantitively inverse from diclofenac. Only a small amount of mecoprop is discharged through the treatment plant during dry weather, resulting in a concentration of around 100 ng/L at the observation point. With increasing rainwater runoff, this concentration increases to over 300 ng/L. Since diclofenac and mecoprop are both not eliminated in the WWTP, it is irrelevant whether the discharge into the receiving water takes place via CSO or WWTP.

Ibuprofen and fluoranthene behave differently, both substances are effectively eliminated in the WWTP. The WWTP effluent concentration of fluoranthene is unaffected by the increased inflow to the WWTP during wet weather. In the model, it is assumed that the particle retention in secondary clarifier continues to work well also during high hydraulic loading. On the contrary, the CSO is heavily contaminated with PAHs such as fluoranthene. Short-term concentration peaks of fluoranthene are therefore observed at the system outlet. Despite a different transport path, Ibuprofen behaves similarly. Concentration peaks in the river are also directly linked to CSO events for this substance.

3.3. Annual Balance of Volume Flow and Pollutant Loads

In 2004 73% of the total flow recorded at the observation point is base flow from the natural catchment (see Figure 6). Only 1.5% are rainwater, wastewater, and infiltration water discharged via combined sewer overflows. A proportion of 10% of the total river flow is treated wastewater, 10% is rainwater treated in the wastewater treatment plant, and 6% is infiltration water. The outflow of the wastewater treatment plant makes a total of 26% of the total flow at the observation point in 2004.

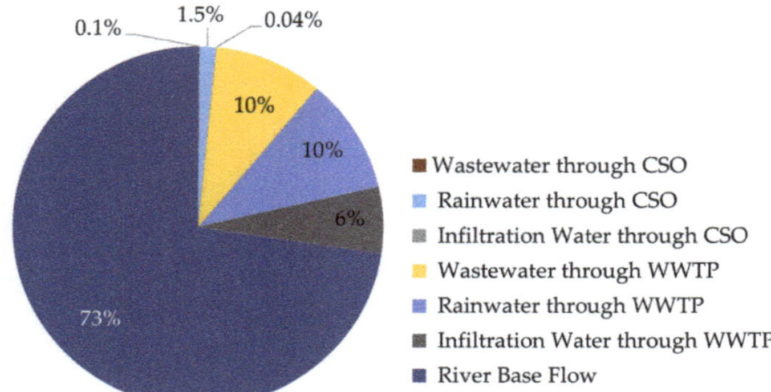

Figure 6. Contribution of different emission pathways to total emitted flow; CSO = combined sewer overflow, WWTP = wastewater treatment plant.

Multiplication of the volume of the flow components with the corresponding MP concentrations gives the annual loads shown in Figure 7. As expected, diclofenac that originates from wastewater and is not well eliminated in the WWTP is mainly emitted via the WWTP effluent. In contrast, fluoranthene that is transported mainly by surface runoff and is almost completely eliminated in the WWTP is emitted mainly via CSOs. In the case of mecoprop, the poor elimination in the WWTP is the dominating influence concerning the pathway. Although ibuprofen is only transported with the wastewater, it is emitted to one-third by CSO.

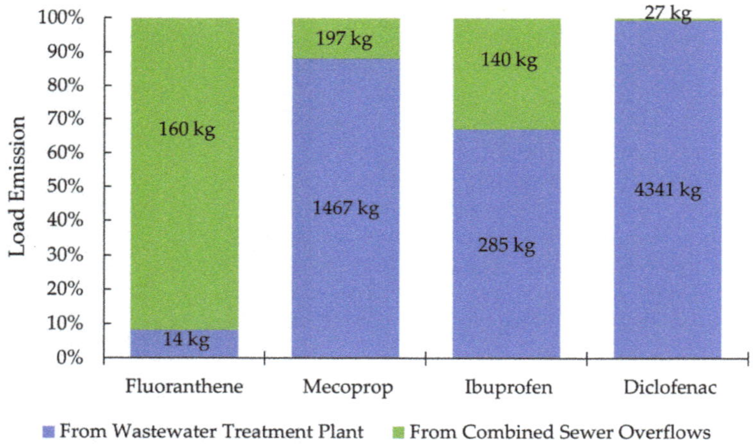

Figure 7. Annual pollutant loads emitted by CSO discharge and WWTP effluent.

The total balance of the MP emission pathways is strongly determined by MP concentrations in each of the flow components (see Table 2). The data for MP concentrations in rainwater runoff from paved areas are scarce and show wide and random variability between sites and events as well as in the course of each event. [9,17,37,38] Therefore, MP concentrations within this flow component are particularly uncertain. To assess the importance of this uncertainty, concentrations of fluoranthene and mecoprop in rainwater runoff were varied within a range derived from published data. Fluoranthene was set to 10 ng/L and 3200 ng/L as found in [17,18,38,39] and mecoprop to 1 ng/L to 6900 ng/L as found in [17,39].

The total emitted load of both substances increases linearly with increased rainwater runoff concentration. Figure 8 shows the contribution of the CSO emissions to the total emitted load depending on the concentrations of fluoranthene and mecoprop in rainwater runoff. For fluoranthene the proportion of CSO load increases with rainwater runoff concentration. For fluoranthene concentrations higher than 2000 ng/L more than 98% of the load is discharged via CSO.

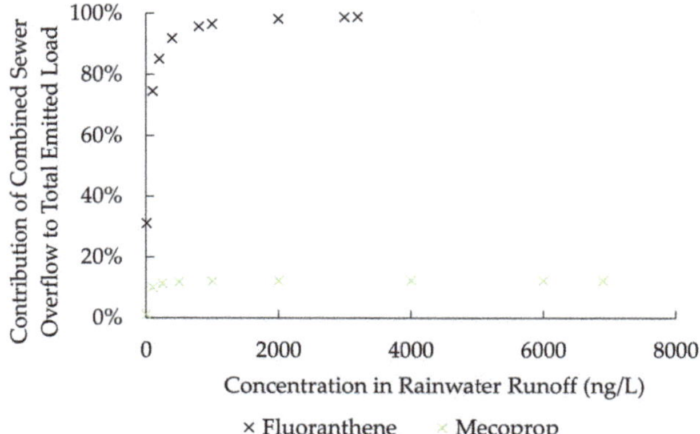

Figure 8. Influence of fluoranthene and mecoprop concentrations in rainwater runoff on the contribution of combined sewer overflow to the total emitted load.

In contrast to fluoranthene, mecoprop is not degraded in conventional WWTPs and during rain events and it is discharged from the WWTP with rainwater runoff concentration. Therefore, the contribution of CSOs does not reach more than 13% of the total emitted load. The WWTP remains the main entry path to the river for mecoprop.

3.4. *Variability of Concentrations and Mass Flow Rates*

An overview of the resulting concentrations throughout the entire observation period is shown in the concentration duration curves of Figure 9. The curve shows for which time (cumulated) a certain concentration has been reached or exceeded at the observation point in the river. The highest concentrations of diclofenac occur at the area outlet during dry weather with low base flow. As the base flow increases, the concentrations in the river decrease because of dilution. The proportion of wastewater in the river varies by up to 48% under dry weather conditions. The concentration of diclofenac in dry weather varies between 300 and 700 ng/L. Low average concentrations of 100 ng/L can only be reached when the wastewater is additionally diluted during rain events.

Ibuprofen and fluoranthene are eliminated in the WWTP. Significant concentration peaks in the river therefore occur only during CSO events. In 2004 CSOs in the system occurred during 234 h. This is clearly reflected in the concentration duration curves showing a strong peak in the first 234 h. The concentration duration curve of mecoprop is less pronounced but more concave than the curve of diclofenac. WWTP effluent contains 30 ng/L of mecoprop during dry weather. Higher values in the receiving water indicate the influence of stormwater in WWTP effluent or CSO events. The limit value of the German Surface Waters Ordinance (OGewVO 2016) for the annual average mecoprop concentration of 100 ng/L exceeds during 1500 h in 2004.

Besides monitoring river water quality, sampling may also aim to determine loads discharged from urban areas. The duration curves clearly show that time proportional sampling in an urban stream is prone to underestimate pollution related to storm events. Table 3 shows the values of volume and time proportional mean concentrations and the difference between the two approaches. This is the systematic deviation based on data in 15 min intervals over the entire year. It is not to be confounded with the uncertainty of sampling due to limited resolution in time or limited sampling periods. The results show that time proportional sampling systematically overestimates the load of diclofenac. The time proportional approach does not account for the fact, that lower concentrations are associated with large volumes. On the contrary, loads of fluoranthene are underestimated because they occur mainly during the short time of CSO events when the flow is high.

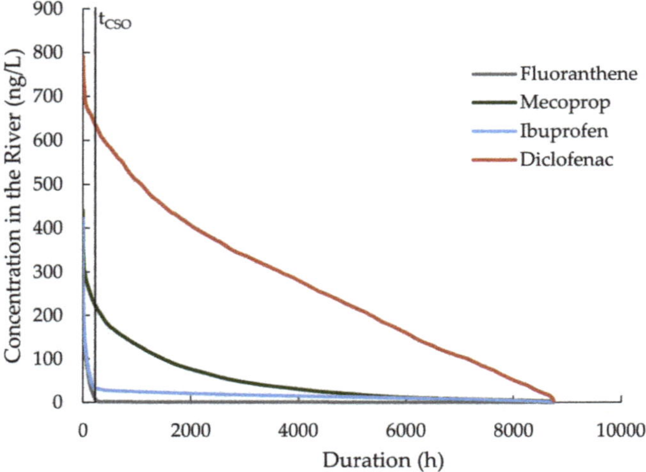

Figure 9. Concentration duration curves downstream of the WWTP for the assessed micropollutants.

Table 3. Annual mean MP concentrations using time proportional and flow proportional average (positive differences indicate an overestimation by time proportional approach).

Substance	Mean Concentration Flow Proportional (ng/L)	Mean Concentration Time Proportional (ng/L)	Difference (%)
Fluoranthene	6.5	2.6	−60%
Mecoprop	61.9	51.1	−18%
Ibuprofen	15.9	16.3	3%
Diclofenac	162.6	272.9	68%

3.5. *Impact of Combined Sewer Overflow on River Quality*

Figure 10 shows the relationship between water concentrations and total discharge at the observation point. Concentrations during CSOs are indicated with circles.

High concentrations of the drug diclofenac occur at low river flows when the wastewater is only sparsely diluted by other flow components. Concentrations in the river during overflow and without CSOs in the system are very close (780 ng/L and 790 ng/L).

Mecoprop shows a similar behavior. Since this substance, such as diclofenac, is not retained in the WWTP, the river concentration results from the dilution of the rainwater runoff from the urban areas with wastewater, infiltration water, and the river base runoff. Maximum concentrations of up to 320 ng/L are reached when no overflow from the sewer system occurs. Highest river concentrations of up to 440 ng/L occur during CSO events.

Like mecoprop, the PAHs (here: fluoranthene) are mainly transported with the rainwater runoff. Without any overflow from the system the fluoranthene concentration in the river is very low (around 10 ng/L) except for few observations up to 150 ng/L during low river flow. The reason for these low values is the high elimination rate in the WWTP for fluoranthene even during wet weather. Higher concentrations of up to 350 ng/L are only reached during CSOs. This is due to the low concentration in the effluent of the treatment plant, which is achieved even with mixed water drainage.

Without overflows from the system, the maximum river concentration of ibuprofen is 50 ng/L. This can be explained by the good removal efficiency for ibuprofen in the WWTP too. During CSO events high concentrations in the river of up to 420 ng/L are reached.

Interpreting the scatter plots one has to take into account that CSO events occur only during a very short fraction of time (2.6% in the case of Möhringen). More meaningful, but less intuitive, are cumulated frequency curves of river concentrations during dry weather, wet weather and during CSO events (see Figure S2 in Supplementary Material).

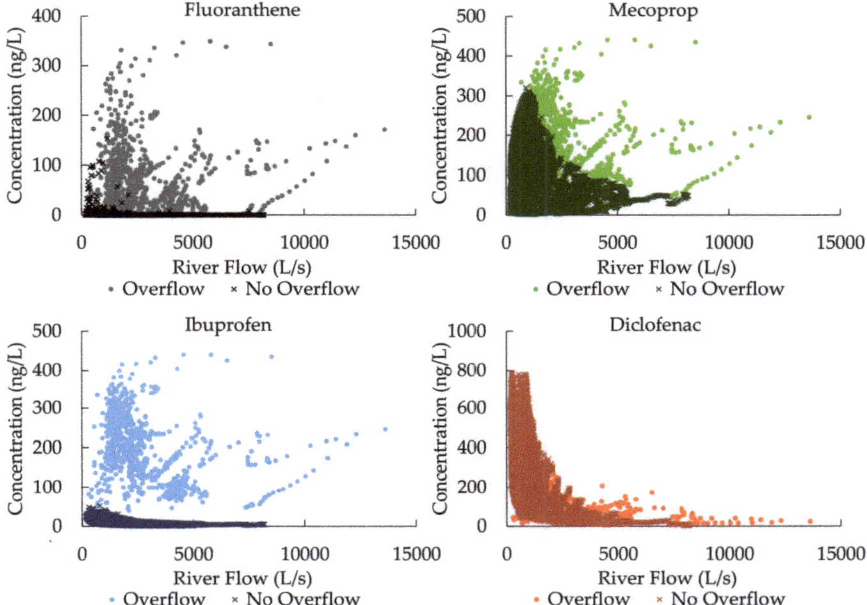

Figure 10. Micropollutant concentrations plotted over river flow downstream of the wastewater treatment plant. "Overflow" marks value during overflow events (overflow duration: 234 h, corresponding to 2.7% of total time).

3.6. Concentration-Duration-Frequency Relation

The impact of critical situations on aquatic organisms or entire ecosystems depends on the intensity of the stressor (here: concentration) and on the duration of the exposure. To assess the impact of emissions from sewers systems and WWTP concentration-duration-frequency (CDF) relations are analyzed as shown in Figure 11. This is done by processing the complete time series of the river concentration in three steps:

(1) Produce time series of the moving average over relevant durations D (here: 15 min to 48 h).
(2) Sort each time series in descending order.
(3) Select the n-th highest value for the frequency n (events/year), (here: n = 1 and n = 5).

In step 3 values should be selected from independent events to get meaningful results. The time difference between the values has to be larger than D.

Figure 11 shows the results for the system outlet (downstream of WWTP) and another reference point immediately upstream of the WWTP. The figures show concentrations that have been reached or exceeded continuously over a time D with the frequency n.

For diclofenac, we see a big difference in concentration levels upstream and downstream of the WWTP, which is expected as this substance is continuously released by the WWTP. Duration and frequency have only a minor influence. This indicates that the five highest values all occur during a similarly low base flow. In contrast to CSO spills low water extremes develop over a longer time. We therefore see no effect in the observed range of durations. The highest concentrations observed over several days continuously is around three times higher than the time averaged concentration.

For fluoranthene, we see a strong influence of the duration and lower but still significant influence of the frequency. High peaks are limited to short events. This can be explained by the fact that they are generated from CSO events. The WWTP effluent dilutes the emission and leads to lower concentrations downstream. Concentrations 100 times above the average (2.6 ng/L) occur for up to three hours in a row. In the simulated year, concentrations higher than 30 times the average were observed five times for 12 h.

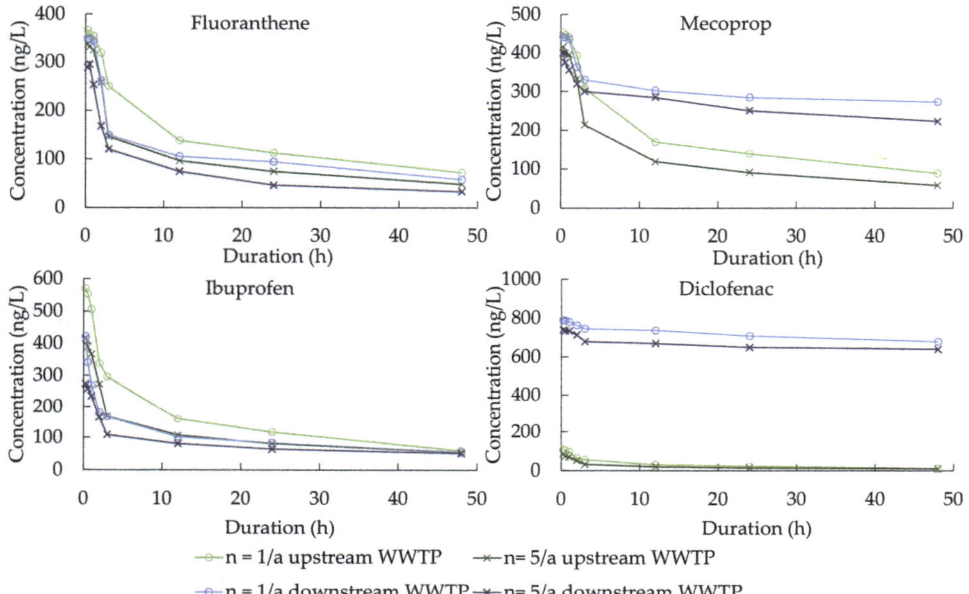

Figure 11. Concentration-duration-frequency curves in the river immediately upstream and downstream of the wastewater treatment plant (WWTP). The curves indicate the concentration that is exceeded continuously over the time D with the frequency n.

Mecoprop and ibuprofen also show a strong influence of short events (D < 3 h), presumably by CSO spills. For mecoprop, concentration downstream of the WWTP is higher than upstream for events longer than three hours as there is no relevant elimination in the WWTP. Ibuprofen shows higher concentrations upstream of the WWTP as the effluent has much lower concentrations and therefore dilutes the emissions from CSO. Concentrations that are exceeded five times in the simulated year over 12 h in a row are four to five times above the average for both substances.

The influence of the rainwater concentration of fluoranthene and mecoprop on the CDF relation downstream the WWTP is shown in Figure S3 in the Supplementary Material. Based on the ranges found in literature (see Section 3.3) rainwater runoff concentrations of fluoranthene were increased to 3200 ng/L and to 6900 ng/L for mecoprop. On the time scale of several hours up to 2 days maximum concentrations in the receiving water are directly related to heavy rain events. Discharge in the urban stream consists mainly of rainwater runoff. Therefore, the level of CDF curves changes almost linearly with the rainwater runoff concentration, while the form of the curve remains the same.

Setting the concentration of rainwater runoff to values at the lower end of the observed range decouples receiving water quality from rain events. Construction of rain event related CDF curves makes no sense under these conditions. Concentrations in the urban stream vary mainly on a larger time scale with the ratio between WWTP effluent base flow.

The comparison of potentially critical events with the average values reveals that dry weather in combination with time proportional evaluation give no indication of the acute impact on the ecosystem. However, the time scale of effects on aquatic organisms and their relation to the frequency of occurrence requires further research on the side of ecotoxicology.

4. Conclusions

The results of the simulation show how strongly the behavior of micropollutants in urban water depends on their origin and properties (here: retention in the WWTP). General statements on the behavior of all micropollutants are therefore misleading. For the conducted simulation study, indicator parameters with distinct characteristics were chosen. The results can be transferred to other substances if their specific properties are taken into account.

The precipitation characteristics directly affect the concentrations of micropollutants in urban waters. For the evaluation of the chemical status, the conditions during wet weather are most important, since most of the legally regulated substances originate from surface runoff. However, water samples are usually drawn at previously fixed time intervals or random over time. Samples of the river water during wet weather are therefore rarely examined. Water monitoring for evaluation of the waterbodies chemical status takes place not only temporally, but also spatially on a much coarser scale. Urban catchment areas are not specifically investigated. The results of the simulation suggest that even urban water bodies that may appear uncritical on a larger level, spatially (urban areas) and temporally (wet weather) massive violations of EQN concentrations may occur. It remains unclear how such stressing factors can be assessed considering the aquatic ecology.

Neglecting wet weather runoff also leads to serious misinterpretations when grab sampling is used to estimate long-term load emissions. The extrapolation underestimates loads of biocides and PAH, which are mainly transported with the wet weather runoff. Even wastewater-borne substances that are partially degraded in the wastewater treatment plants enter the receiving water bodies increasingly during wet weather.

To adequately classify the measured concentrations and the influence of the sewage system, water sampling in urban streams should generally take place more intensively during wet weather. If the proportions of all relevant flow components within the river were known at the time of sampling, the individual pathways could be extrapolated by observing or simulating flow volumes alone.

However, full monitoring of all combined sewer overflows is not feasible with reasonable effort. Alternatively, the proportions of treated and untreated wastewater in the river can be estimated by comparing concentrations of indicator substances under dry weather conditions in the inflow and outflow of the WWTP.

Mean values (time or flow weighted) of samples from urban streams do not reflect the potential impact of stress events on aquatic organisms. In discussions with ecotoxicologists, we have found CDF curves to be a suitable visualization for the communication across disciplines. However, the robustness of these curves against uncertainties in model structure and parameter settings needs to be investigated further. In future studies we want to use CDF curves to evaluate and compare different options for stormwater management and treatment.

The study assumes that the proportions of treated and untreated runoff and wastewater dominate micropollutant concentrations in urban streams. This hypothesis is based on our own previous studies and on literature. However, it needs to be tested in future studies when more sampling data are available.

Supplementary Materials: The following are available online at http://www.mdpi.com/2073-4441/12/3/850/s1, Figure S1: Propagation of inert tracer through the wastewater treatment plant (WWTP) during wet weather (WWTP inflow = 1000 L/s), Figure S2: Cumulative frequency distributions for fluoranthene, mecoprop, ibuprofen and diclofenac in the river downstream the wastewater treatment plant. The percentage on the *y*-axis corresponds to the time of the state lasts. All: total simulation period (8760 h), rainy weather: stormwater in the WWTP effluent >5% (5935 h), overflow: CSO events (234 h), Figure S3: Concentration-duration-frequency curves in the river immediately downstream of the wastewater treatment plant for fluoranthene and mecoprop with varying rainwater (RW) concentrations, Table S1: Setup parameters for the SWMM sewer system model.

Author Contributions: Conceptualization, U.D.; methodology, U.D. and A.B.-M; software, A.B.-M; validation, U.D., A.B.-M and M.A.L.; formal analysis, U.D. and A.B.-M; investigation, M.A.L.; resources, U.D.; writing—original draft preparation, U.D. and A.B.-M.; writing—review and editing, U.D. and A.B.-M.; visualization, A.B.-M.; supervision, U.D. All authors have read and agreed to the published version of the manuscript.

Funding: This research received no external funding.

Acknowledgments: The authors would like to thank Amin Ebrahim Bakshipour for his support in implementation of the wastewater treatment plant and Mehari Goitom Haile for the permission to use his calibrated sewer system model of the Möhringen catchment. Furthermore, we would like to thank the "Landesamt für Natur, Umwelt und Verbraucherschutz Nordrhein-Westfalen LANUV" (Regional Office for Nature, Environment and Consumer Protection) for the measured data used in this study as well as the Ministry of the Environment, Climate Protection, and the Energy Sector Baden-Württemberg.

Conflicts of Interest: The authors declare no conflict of interest.

References

1. Walsh, C.J.; Fletcher, T.D.; Ladson, A.R. Stream restoration in urban catchments through redesigning stormwater systems: Looking to the catchment to save the stream. *J. N. Am. Benthol. Soc.* **2005**, *24*, 690–705. [CrossRef]
2. Fletcher, T.D.; Andrieu, H.; Hamel, P. Understanding, management and modelling of urban hydrology and its consequences for receiving waters: A state of the art. *Adv. Water Resour.* **2013**, *51*, 261–279. [CrossRef]
3. Walsh, C.J.; Roy, A.H.; Feminella, J.W.; Cottingham, P.D.; Groffman, P.M.; Morgan, R.P. The urban stream syndrome: Current knowledge and the search for a cure. *J. N. Am. Benthol. Soc.* **2005**, *24*, 706–723. [CrossRef]
4. Phillips, P.; Chalmers, A.T. Wastewater effluent, combined sewer overflows, and other sources of organic compounds to Lake Champlain. *J. Am. Water Resour. Assoc.* **2009**, *45*, 45–57. [CrossRef]
5. Petrie, B.; Barden, R.; Kasprzyk-Hordern, B. A review on emerging contaminants in wastewaters and the environment: Current knowledge, understudied areas and recommendations for future monitoring. *Water Res.* **2015**, *72*, 3–27. [CrossRef]
6. Kasprzyk-Hordern, B.; Dinsdale, R.M.; Guwy, A.J. The removal of pharmaceuticals, personal care products, endocrine disruptors and illicit drugs during wastewater treatment and its impact on the quality of receiving waters. *Water Res.* **2009**, *43*, 363–380. [CrossRef]
7. Meyer, B.; Bierl, R.; Keßler, S.; Krein, A. Untersuchung von Niederschlagswassermanagementsystemen und deren stofflicher Wirkung auf den Vorfluter mittels hochaufgelöster Ereignisbeprobung. *Hydrol. Wasserbewirtsch.* **2013**, *57*, 164–173.
8. Launay, M.; Kuch, B.; Dittmer, U.; Steinmetz, H. Dynamics of selected micropollutants during various rain events in a highly urbanised catchment. In Proceedings of the 13th International Conference on Urban Drainage, ICUD, Kuching, Malaysia, 7–12 September 2014.
9. Launay, M. Organic Micropollutants in Urban Wastewater Systems during Dry and Wet Weather—Occurrence, Spatio-temporal Distribution and Emissions to Surface Waters. Ph.D. Thesis, University of Stuttgart, Stuttgart, Germany, 2017.
10. Christoffels, E.; Brunsch, A.; Wunderlich-Pfeiffer, J.; Mertens, F.M. Monitoring micropollutants in the Swist river basin. *Water Sci. Technol.* **2016**, *74*, 2280–2296. [CrossRef] [PubMed]
11. Benotti, M.J.; Brownawell, B.J. Distributions of Pharmaceuticals in an Urban Estuary during both Dry- and Wet-Weather Conditions. *Environ. Sci. Technol.* **2007**, *41*, 5795–5802. [CrossRef] [PubMed]
12. Wittmer, I.K.; Bader, H.-P.; Scheidegger, R.; Singer, H.; Lück, A.; Hanke, I.; Carlsson, C.; Stamm, C. Significance of urban and agricultural land use for biocide and pesticide dynamics in surface waters. *Water Res.* **2010**, *44*, 2850–2862. [CrossRef] [PubMed]
13. Munro, K.; Martins, C.P.B.; Loewenthal, M.; Comber, S.; Cowan, D.A.; Pereira, L.; Barron, L.P. Evaluation of combined sewer overflow impacts on short-term pharmaceutical and illicit drug occurrence in a heavily urbanised tidal river catchment (London, UK). *Sci. Total Environ.* **2019**, *657*, 1099–1111. [CrossRef] [PubMed]
14. Spahr, S.; Teixidó, M.; Sedlak, D.L.; Luthy, R.G. Hydrophilic trace organic contaminants in urban stormwater: Occurrence, toxicological relevance, and the need to enhance green stormwater infrastructure. *Environ. Sci. Water Res. Technol.* **2020**, *6*, 15–44. [CrossRef]

15. Clara, M.; Gruber, G.; Humer, F.; Hofer, T.; Kretschmer, F.; Ertl, T.; Scheffknecht, G.; Giselbrecht, G.; Windhofer, G. Spurenstoffemissionen aus Siedlungsgebieten und von Verkehrsflächen in Österreich. In Proceedings of the Abwasserkolloquium 2017, Spurenstoffe im Regen- und Mischwasserabfluss, Stuttgart, Germany, 26 October 2017; pp. 59–76.
16. Launay, M.; Dittmer, U.; Steinmetz, H. Organic micropollutants discharged by combined sewer overflows—Characterisation of pollutant sources and stormwater-related processes. *Water Res.* **2016**, *104*, 82–92. [CrossRef] [PubMed]
17. Gasperi, J.; Sebastian, C.; Ruban, V.; Delamain, M.; Percot, S.; Wiest, L.; Mirande, C.; Caupos, E.; Demare, D.; Kessoo, M.D.K.; et al. Micropollutants in urban stormwater: Occurrence, concentrations, and atmospheric contributions for a wide range of contaminants in three French catchments. *Environ. Sci. Pollut. Res.* **2014**, *21*, 5267–5281. [CrossRef]
18. Birch, H.; Mikkelsen, P.S.; Jensen, J.K.; Lutzhoft, H.-C.H. Micropollutants in stormwater runoff and combined sewer overflow in the Copenhagen area, Denmark. *Water Sci. Technol.* **2011**, *64*, 485–493. [CrossRef]
19. Bollmann, U.E.; Vollertsen, J.; Carmeliet, J.; Bester, K. Dynamics of biocide emissions from buildings in a suburban stormwater Catchment—Concentrations, mass loads and emission processes. *Water Res.* **2014**, *56*, 66–76. [CrossRef]
20. Mutzner, L.; Vermeirssen, E.L.M.; Mangold, S.; Maurer, M.; Scheidegger, A.; Singer, H.; Booij, K.; Ort, C. Passive samplers to quantify micropollutants in sewer overflows: Accumulation behaviour and field validation for short pollution events. *Water Res.* **2019**, *160*, 350–360. [CrossRef]
21. Gamerith, V. High Resolution Online Data in Sewer Quality Modelling. Ph.D. Thesis, Technische Universität Graz, Graz, Austria, 2011.
22. Braud, I.; Fletcher, T.D.; Andrieu, H. Hydrology of peri-urban catchments: Processes and modelling. *J. Hydrol.* **2013**, *485*, 1–4. [CrossRef]
23. Furusho, C.; Chancibault, K.; Andrieu, H. Adapting the coupled hydrological model ISBA-TOPMODEL to the long-term hydrological cycles of suburban rivers: Evaluation and sensitivity analysis. *J. Hydrol.* **2013**, *485*, 139–147. [CrossRef]
24. Bach, M.; Ostrowski, M. Analysis of intensively used catchments based on integrated modelling. *J. Hydrol.* **2013**, *485*, 148–161. [CrossRef]
25. Musolff, A.; Leschik, S.; Möder, M.; Strauch, G.; Reinstorf, F.; Schirmer, M. Temporal and spatial patterns of micropollutants in urban receiving waters. *Environ. Pollut.* **2009**, *157*, 3069–3077. [CrossRef]
26. Haile, M.G. Accounting for Uncertainties in Combined Sewer Overflow Modelling. Ph.D. Thesis, Oldenburg Industrieverlag GmbH, München, Germany, 2016.
27. EPA. *SWMM*; United States Environmental Protection Agency: Washington, DC, USA, 2014.
28. Rossmann, L.; Huber, W. Storm Water Management Model Reference Manual Volume III—Water Quality. US EPA, Office of Research and Development, NRMRL, Water Supply and Water Resources Division. EPA/600/R-16/093; 2016. Available online: https://nepis.epa.gov/Exe/ZyPDF.cgi/P100P2NY.PDF?Dockey=P100P2NY.PDF (accessed on 24 October 2018).
29. Launay, M.; Steinmetz, H.; Dittmer, U. Organic micropollutants discharged by combined sewer overflows (CSOs): What about inter- and intra-event variability? In Proceedings of the 14th International Conference on Urban Drainage, ICUD, Prague, Czech Republic, 10–15 September 2017.
30. Wicke, D.; Matzinger, A.; Caradot, N.; Schubert, R.-L.; Sonnenberg, H.; von Seggern, D.; Heinzmann, B.; Rouault, P. Spurenstoffemissionen im Regenwasserabfluss Berlins. In Proceedings of the Abwasserkolloquium 2017, Spurenstoffe im Regen- und Mischwasserabfluss, Stuttgart, Germany, 26 October 2017; pp. 33–42.
31. Fletcher, C.A.; Scrimshaw, M.D.; Lester, J.N. Transport of mecoprop from agricultural soils to an adjacent salt marsh. *Mar. Pollut. Bull.* **2004**, *48*, 313–320. [CrossRef] [PubMed]
32. Tu, M.-C.; Smith, P. Modeling Pollutant Buildup and Washoff Parameters for SWMM Based on Land Use in a Semiarid Urban Watershed. *Water Air Soil Pollut.* **2018**, *229*, 18. [CrossRef]
33. Siewicki, T.C. Environmental modeling and exposure assessment of sediment-associated fluoranthene in a small, urbanized, non-riverine estuary. *J. Exp. Mar. Biol. Ecol.* **1997**, *213*, 71–94. [CrossRef]
34. Musolff, A.; Leschik, S.; Reinstorf, F.; Strauch, G.; Schirmer, M. Micropollutant loads in the urban water cycle. *Environ. Sci. Technol.* **2010**, *44*, 4877–4883. [CrossRef]

35. Madoux-Humery, A.-S.; Dorner, S.; Sauvé, S.; Aboulfadl, K.; Galarneau, M.; Servais, P.; Prévost, M. Temporal variability of combined sewer overflow contaminants: Evaluation of wastewater micropollutants as tracers of fecal contamination. *Water Res.* **2013**, *47*, 4370–4382. [CrossRef] [PubMed]
36. Passerat, J.; Ouattara, N.K.; Mouchel, J.-M.; Rocher, V.; Servais, P. Impact of an intense combined sewer overflow event on the microbiological water quality of the Seine River. *Water Res.* **2011**, *45*, 893–903. [CrossRef] [PubMed]
37. Björklund, K.; Cousins, A.P.; Strömvall, A.-M.; Malmqvist, P.-A. Phthalates and nonylphenols in urban runoff: Occurrence, distribution and area emission factors. *Sci. Total Environ.* **2009**, *407*, 4665–4672. [CrossRef]
38. Kalmykova, Y.; Björklund, K.; Strömvall, A.-M.; Blom, L. Partitioning of polycyclic aromatic hydrocarbons, alkylphenols, bisphenol A and phthalates in landfill leachates and stormwater. *Water Res.* **2013**, *47*, 1317–1328. [CrossRef]
39. Wicke, D.; Matzinger, A.; Rouault, P. *Relevanz Organischer Spurenstoffe im Regenwasserabfluss Berlins—OgRe*; Abschlussbericht; Kompetenzzentrum Wasser Berlin: Berlin, Germany, 2015.

Article

Deriving a Bayesian Network to Assess the Retention Efficacy of Riparian Buffer Zones

Andreas Gericke [1,*], Hong Hanh Nguyen [1], Peter Fischer [1,2], Jochem Kail [3] and Markus Venohr [1,4]

1 Leibniz-Institute of Freshwater Ecology and Inland Fisheries, 12489 Berlin, Germany; hanh.nguyen@igb-berlin.de (H.H.N.); peter_fischer@online.de (P.F.); m.venohr@igb-berlin.de (M.V.)
2 Center for Agricultural Technology Augustenberg, 76227 Karlsruhe, Germany
3 Faculty of Biology, Department of Aquatic Ecology, University of Duisburg-Essen, 45141 Essen, Germany; jochem.kail@uni-due.de
4 Department of Geography, Humboldt-University of Berlin, 12489 Berlin, Germany
* Correspondence: gericke@igb-berlin.de

Received: 7 January 2020; Accepted: 11 February 2020; Published: 25 February 2020

Abstract: Bayesian networks (BN) have increasingly been applied in water management but not to estimate the efficacy of riparian buffer zones (RBZ). Our methodical study aims at evaluating the first BN to predict the RBZ efficacy to retain sediment and nutrients (dissolved, total, and particulate nitrogen and phosphorus) from widely available variables (width, vegetation, slope, soil texture, flow pathway, nutrient form). To evaluate the influence of parent nodes and how the number of states affects prediction errors, we used a predefined general BN structure, collected 580 published datasets from North America and Europe, and performed classification tree analyses and multiple 10-fold cross-validations of different BNs. These errors ranged from 0.31 (two output states) to 0.66 (five states). The outcome remained unchanged without the least influential nodes (flow pathway, vegetation). Lower errors were achieved when parent nodes had more than two states. The number of efficacy states influenced most strongly the prediction error as its lowest and highest states were better predicted than intermediate states. While the derived BNs could support or replace simple design guidelines, they are limited for more detailed predictions. More representative data on vegetation or additional nodes like preferential flow will probably improve the predictive power.

Keywords: model evaluation; nitrogen; nutrient retention; phosphorus; sediment

1. Introduction

The eutrophication of lakes and rivers remain a global concern despite considerable efforts in the past to reduce the input of nutrients from terrestrial environments [1]. After targeting the industrial and domestic sources, the diffuse agricultural inputs of nitrogen (N), phosphorus (P), sediments, and also pesticides gained importance [2–5]. Particularly, the legacy of long-term excess applications of fertilizers and manure became eminent [6,7]. In the European Union (EU), for instance, diffuse sources affect now more water bodies than point sources [8].

Many processes and factors control the mobilization and the diffuse transport of agricultural pollutants. Accordingly, various "best management practices" (BMP) can contribute to reduce the risk of pollutants entering water bodies. Riparian buffers zones (RBZ), also termed vegetated buffer strips, are vegetated areas along water bodies which are either too difficult to work or intentionally set aside [9]. RBZ stand among the most widely applied BMP to reduce the diffuse pollution of surface waters but also to improve their hydro-morphological and thermal conditions [9–11]. This is reflected by the rich literature on the efficacy of RBZ to retain substances from adjacent upslope areas (henceforth catchments).

The RBZ efficacy is expressed as the ratio of the amount of nutrients, pesticides, or sediment leaving the buffer towards the water body and the amount of material entering the RBZ from the catchment. It depends on RBZ properties and external factors like land management which, in turn, determine the runoff and incoming nutrients [12,13]. The large variability in RBZ efficacy reported in the literature reflects the variability of those factors.

Therefore, modelling approaches to predict RBZ efficacy in large-scale applications tend to be (too) data-demanding, e.g., process-based models like VFSMOD [14] and REMM [15], or (too) simplistic [16]. For instance, Weissteiner et al. [17] used RBZ width as single variable to estimate the efficacy of N and P retention in their model for the EU. This approach is in line with regulations and recommendations on RBZ design which specify adequate widths of RBZ as simple, easily administrable rules to reduce diffuse emissions of nutrients and sediments [18–20]. However, the use of RBZ width as the only design variable is not commonly considered sufficient [21,22].

Some studies used sub-datasets and applied multivariate regression to improve the model performance. For example, Zhang et al. [13] derived separate models for different vegetation types (grass and mixed, only trees), again only with RBZ width as the single explanatory variable for N and P retention. For sediment retention, they considered RBZ slope as the second variable which was supported by a previous study that identified RBZ width and slope to be best correlated with sediment retention [23]. In contrast, Yuan et al. [24] found only a weak relationship between sediment retention and slope. In their meta-analysis on nitrate retention, Mayer et al. [25] also tested RBZ width as the only predictor in regression models for different width classes, vegetation types, and flow pathways. The explained variability (r^2) in these subsets was partly higher than the r^2 values for the full dataset. They concluded that more factors need to be considered. In accordance, Hassanzadeh et al. [16] achieved even higher r^2 values by considering various RBZ and external variables in their regression model.

Nonetheless, such regression models have limitations which can be addressed by Bayesian networks (BN, also termed Bayesian belief networks, [26]). Firstly, BN explicitly deal with the uncertainty in data or conditions because they represent the relationships (links or edges) between system variables (nodes) probabilistically rather than deterministically [27]. Secondly, they can integrate even incomplete quantitative and qualitative information (e.g., "sandy soil") from different domains such as expert knowledge, field data, and other BN. Thirdly, BN can provide information on causes from evidence of observed effects.

Although BN have increasingly been applied for water resource management, decision-making, predictions, and assessments of ecosystem services during the last decades [26,28–30], they have only implicitly considered RBZ retention so far [31]. Therefore, the aims of this study are (i) to develop the first BN to predict RBZ efficacy to retain N and P from RBZ and external factors, (ii) to train and cross-validate the BN based on empirical data compiled from an extensive literature review, and (iii) to evaluate how node selection and the number of their states, i.e., data discretization, affect model accuracy. As our BN aims at medium- to large-scale applications, we focused on commonly available input data and grouped nutrient fractions of N and P.

2. Materials and Methods

2.1. Data Collection and Preprocessing

We compiled the measured RBZ efficacy for N, P, and sediment and information on generally available RBZ and external variables (first column in Table 1) based on 23 reviews published between 1994 and 2019 [12,13,22–25,32–48] (Supplementary Table S1). These variables were selected based on key processes of nutrient and sediment retention discussed in the literature, e.g.,

- RBZ width and nutrient form: after entering the RBZ, the velocity of surface runoff decreases due to surface roughness and soil infiltration [22,49,50]. This results in a lower transport capacity and a deposition of sediment and particulate nutrients, which is an important process for their retention [40]. Positive yet variable relationships of RBZ width and efficacy have widely been acknowledged in reviews and guidelines (e.g., [19,33,48]) with broader RBZ being needed to efficiently retain dissolved nutrients compared to sediment [46].

- RBZ vegetation, flow pathway, and nutrient form: dense aboveground vegetation increases the hydraulic roughness and reduces the transport capacity of water for particles, while dense root systems favor soil permeability, porosity, infiltration, and, thus, the sorption of dissolved nutrients (DN, DP) to soil particles as well as their uptake by plants [45,51]. Previous studies reported contrasting results in respect to the efficacy of grassed and woody RBZ [12,52,53]. Most probably, this is due the large number of processes and complex interactions among nutrients, plants, soil, and micro-organisms [54].
- Soil texture, flow pathway, and nutrient form: the larger the soil particles, the faster they settle under given flow conditions. While most of the coarse sediment can be retained even in narrow RBZ, the retention of fine sediment requires soil infiltration in RBZ which are more than 15–20 m wide in order to be effective [12,23,55]. In addition, soil composition (e.g., texture) is decisive for P sorption, hydraulic conductivity and, thus, for surface runoff, soil moisture storage, as well as residence time of nutrients in the root zone [40,47,51,56,57]. While sandy soils favor infiltration, surface flow predominates on clayey soils.
- RBZ slope and width: RBZ are expected to be less effective in reducing the transport capacity of water flow as well as in trapping sediments and nutrients in steep terrain [13,23,40] thus requiring broader buffers on steeper slopes [48]. Due to the experimental design and data availability, the influence of slope was not significant in other studies (e.g., [47,58]). RBZ slope was used in some regression models as an independent variable (e.g., [13,48]).

Table 1. List of riparian buffer zones (RBZ) and external variables used as nodes in the Bayesian networks (BN) and their discrete states. State definitions are shown as comma-separated lists. The variables (nodes) RBZ width, soil texture, RBZ slope, and RBZ efficacy were discretized using different numbers of states (given in separate lines, cf. section on BN design for details on the state definition).

Variable (Node)	Unit	States	Definition	Data per State
RBZ width	m	1, 2 1, 2, 3 1, 2, 3, 4	<10, 10–220 <5.1, <10.1, 10.1–220 <5.1, <10, <15, 15–220	315, 265 170, 214, 196 170, 145, 114, 151
RBZ vegetation	-	Grass Woody	Including e.g., giant cane With shrubs and trees	433 147
Soil texture	-	Fine, medium, coarse Fine, coarse	Clayey (fL, CL, SiC, C, SC), silty-loamy (L, SiL, Si, SCL), sandy (SL, cL, S, SSi, LS) [1] Clayey, silty to sandy	142, 330, 108 142, 408
Nutrient form	-	Particulate Mixed Dissolved	Particulate P (PP), sediment Total P (TP) and N (TN) Dissolved P (DP) [2] and N (DN) [3]	188 (17 + 171) 174 (95 + 79) 218 (87 + 131)
Flow pathway	-	Surface Subsurface	Surface runoff (Including) groundwater	480 100
RBZ slope	%	1, 2 1, 2, 3 1, 2, 3, 4	<5, 5–22 <3.1, <7.5, 7.5–22 <3, <5, <10, 10–22	262, 318 190, 191, 199 124, 138, 157, 161
RBZ efficacy	%	1, 2 1, 2, 3 1, 2, 3, 4 1, 2, 3, 4, 5	<68.3, ≥68.3 <55, <82.3, ≥82.3 <42.6, <68.3, <87.5, ≥87.5 <35.5, <60.6, <78.5, <89.8, ≥89.8	290, 290 193, 193, 194 145, 145, 145, 145 116, 116, 115, 117, 116

[1] Loam (L), clay (C), sand (S), silt (Si), fine (f), coarse (c), [2] including orthophosphate, [3] i.e., nitrate.

As far as possible, the measured RBZ efficacy values—preferably from mass (load or yield)—were extracted from the original studies to minimize compilation errors and to avoid double counting. From 139 studies, published between 1977 and 2017 (Figure 1b), we gathered 341 datasets with at least one efficacy value mainly located in North America and Western Europe (Figure 1a). Data from other continents as well as model outcomes, (constructed) wetlands, and ponds were not considered.

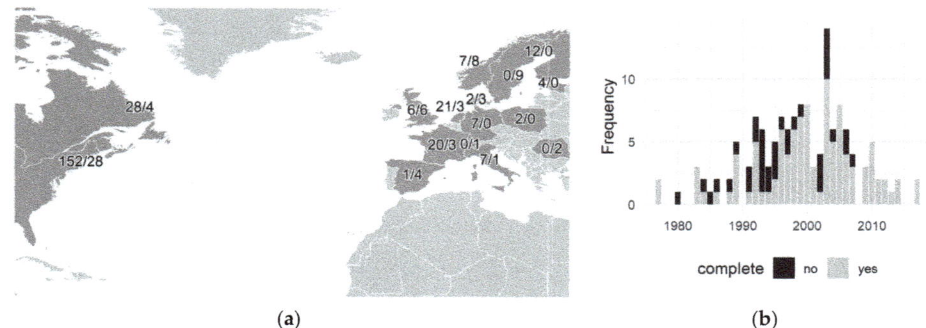

(a) (b)

Figure 1. Overview of collected data; (**a**) complete/incomplete datasets per country (n = 341, cf. Table 2), (**b**) number of studies per publication year (n = 139).

Table 2. Completeness of literature values.

Value	Complete (Count)	Incomplete (Count)
Studies [1]	104	36
Datasets	269	72
Efficacy values	580	98
Countries [2]	13	12
Variables (nodes)	All	Slope, soil texture

[1] One study with partly incomplete soil data, [2] three countries without complete data.

A total of 269 of the collected datasets were complete, i.e., with values for all variables shown in Table 1 (Table 2). The others missed RBZ slope (n = 52) and soil texture (n = 61). Slope was missing for 39 groundwater studies, i.e., 27% of all, thus shifting the dataset towards studies on surface flow. To increase the sample size, we calculated missing average values from ranges, e.g., the minimum and maximum efficacy given for RBZ of 10–15 m width was converted to an average value of 12.5 m width, and applied no quality criteria apart from data completeness.

Two thirds of the 580 efficacy values with complete data originated from plot studies and one third from field and watershed studies, with different experimental designs and time scales. Rainfall data was particularly inconsistently reported, e.g., as average values or intensities, and was therefore excluded from the data collection.

2.2. Aggregating the Nutrient Forms

We collected the efficacies for five nutrient forms and sediment. In order to reduce the number of states and increase the number of data sets for each state, we applied preliminary regression analyses to quantify the similarity of RBZ efficacy for particulates (mostly sediment) and total N and P (TN, TP) as well as the dissolved forms (DN, DP) before setting up the BN.

2.3. BN Design—Nodes and States

BN are directed acyclic graphs which link variables (called parent and child nodes) without feedback loops. The nodes and links of the BN were based on the literature listed above: the RBZ and external variables in the first column of Table 1 were considered as independent parent nodes directly influencing RBZ efficacy as the final child (output) node of the BN (Figure 2).

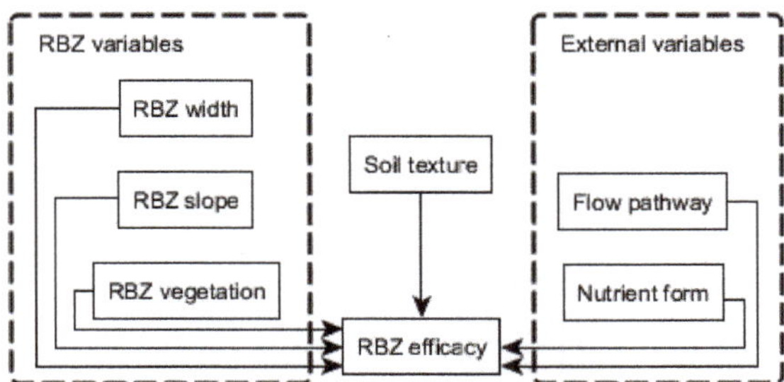

Figure 2. BN structure with nodes and links. We assumed that the reported soil textures are similar in the RBZ and their catchments.

In a BN, discrete states have to be defined for each node. They have to be exhaustive and mutually exclusive. Apart from flow pathway, the states of the nominal variables were groups of the reported values (see columns three and four in Table 1). The original nutrient forms were aggregated to the three states "particulate", "mixed", and "dissolved" based on the preliminary regression analyses. The various values of soil texture were grouped into (i) the three states "fine" (clayey soils), "medium" (loamy and silty soils), and "coarse" (sandy soils), and (ii) only two states "fine" and "coarse" (comprising loamy, silty, and sandy soils). The plant species composition, structure, and management of RBZ varied widely in the literature and were simplified to "grass" (herbaceous) and "woody" (with shrubs and trees).

In contrast to the nominal variables, the three continuous variables had to be discretized. Different discretization approaches exist but seemingly no optimal one [59,60]. Given the uneven value distribution and to acknowledge the limited data availability for parent nodes, we used the common equal-frequency approach. For RBZ width, however, we slightly adjusted the quantiles to match legal regulations. Since one aim of this study was to assess the effect of state definition on model accuracy, different numbers of quantiles were used (see Table 1).

2.4. BN Training and Evaluation—Importance of Nodes and State Definition

BNs with different number of states were trained using efficacy values with data for all parent nodes ($n = 580$) and the Maximum Likelihood parameter estimation. The training filled the conditional probability tables (CPT) which describe the probability of each state of the child node considering all possible states of the parent nodes, based on Bayes' theorem. The different state definitions of four nodes (see Table 1) resulted in 72 different BNs. A 10-fold cross-validation—i.e., 90% of the data used for training and 10% for validation—was repeated 10 times for each of these BNs to compare prediction (classification) errors [59,61], i.e., the share of wrongly predicted output states. Ten different probabilities and prediction errors were thus obtained for each BN, ranging from 0 (all predictions correct) to 1 (all predictions wrong).

In order to assess how the state definition (state number) of the parent and output nodes affected the model performance, we used the non-parametric Wilcoxon Signed Rank test to determine whether the prediction errors differed significantly for parent nodes with different numbers of states ($p < 0.05$).

We also compared the complete BNs with all parent nodes to simplified BNs where the least important nodes were removed. The node importance was evaluated in two ways. Firstly, we applied the non-parametric Kruskal–Wallis test to each parent node to assess whether the RBZ efficacy for at least one state significantly differs from the other states ($p < 0.05$). Following Moe et al. [62], we also performed a classification tree analysis for each of the 72 possible state combinations. For all parent nodes we compared their relative importance and their occurrence in the classification trees to a random binary variable. In the simplified BN, we did not consider nodes with an influence close to the random variable and with a low occurrence in the classification trees (≤33%). As the values of the random variable vary with each repetition, we averaged the outcomes of 10 analyses to avoid misclassifications. The cross-validation and state evaluation were repeated with the 72 simplified BNs. We discarded undefined output from validation data outside the training data.

2.5. Implementation

All statistical analyses were conducted with the software R [63]. For the training and cross-validation of the BNs we used the R package bnlearn [64], and rpart [65] for the classification tree analysis. To exemplarily show the influence of individual parent nodes on the output, we trained an "optimal" BN—based on the prediction errors and the CPT size—with the full dataset. For the node of interest, we determined the BN output for each possible state by setting its probability to 1 ("hard evidence") while assuming an equal probability of all states of the other parent nodes ("soft evidence"). These calculations were conducted using the R package gRain [66].

3. Results

3.1. Aggregating the Nutrient Forms

The retention of TN and TP were well correlated ($r^2 = 0.64$). Both were also more significantly correlated to sediment retention values (Figure 3, $r^2 = 0.40$–0.48, $p < 0.0001$) than the DN and DP retentions ($r^2 = 0.09, p = 0.003$). Based on these results, we considered TN and TP together as a separate "mixed" state of the node "nutrient form" and distinguished it from the "particulate" state (mainly sediment) and the "dissolved" nutrient forms, for which sedimentation is less important.

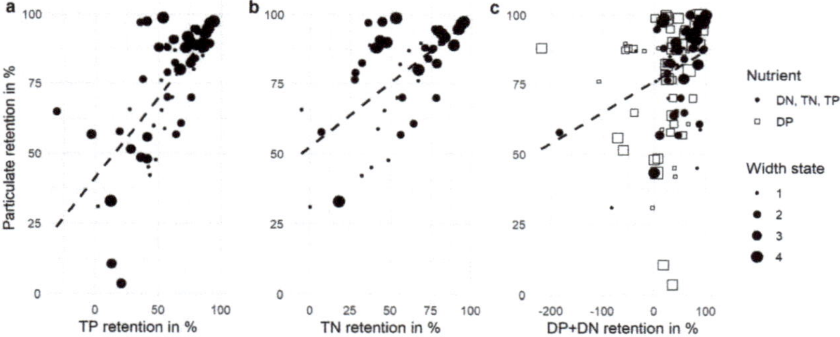

Figure 3. RBZ efficacy to retain total phosphorus (TP) (**a**), total nitrogen (TN) (**b**), as well as dissolved N and P (DP, DN) (**c**) compared to particulates as collected from the literature with (dashed) lines of best fit. Higher state numbers correspond to broader RBZ.

3.2. Variability of RBZ Efficacy and Importance of Nodes

The reported RBZ efficacy ranged from complete retention to negative values, i.e., where the RBZ acted as sources. The distribution of all values was left-skewed with a median of 68% and a mode of 87%. This variability also resulted in a strong overlap between the states of the parent nodes (Figure 4). Therefore, even wide buffers can have a rather low efficacy (Figure 4a). However, the Kruskal–Wallis test revealed significant differences in RBZ efficacy among the states of all parent nodes ($p \leq 0.0003$) except the flow pathway ($p = 0.07$) and RBZ vegetation ($p = 0.8$). For instance, RBZ are generally more effective with coarse soils than with fine and medium soils (Figure 4c) except for DP (not shown).

The classification tree analysis confirmed the strong influence of nutrient form, RBZ width, and soil texture on the RBZ efficacy (Figure 5). Despite its comparatively low importance (9%), RBZ slope occurred in 60% of the classification trees. The least important variables were RBZ vegetation and flow pathway (6%). Both variables rarely occurred in the trees, although they were more influential than a random variable (1.5%). This was especially the case for the flow pathway.

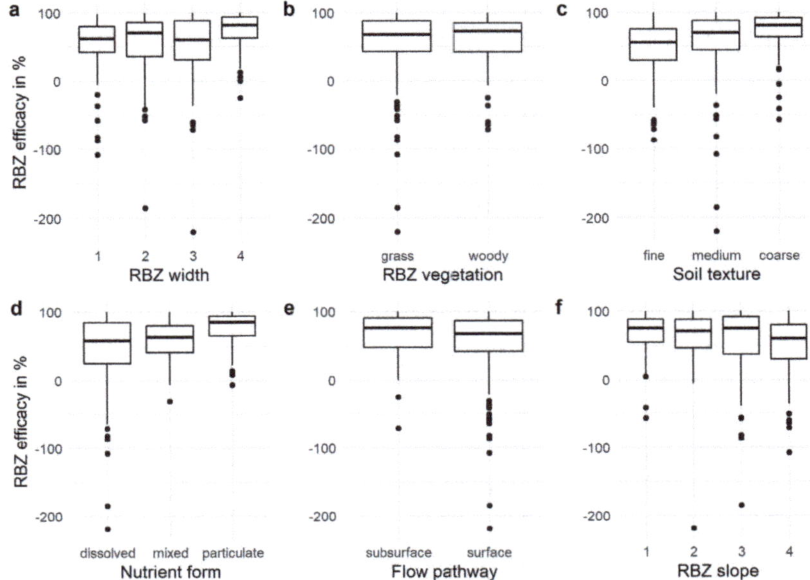

Figure 4. RBZ efficacy for the states of the parent nodes using the maximum number of states as given in Table 1. Higher state numbers correspond to broader RBZ (**a**) and steeper RBZ (**f**).

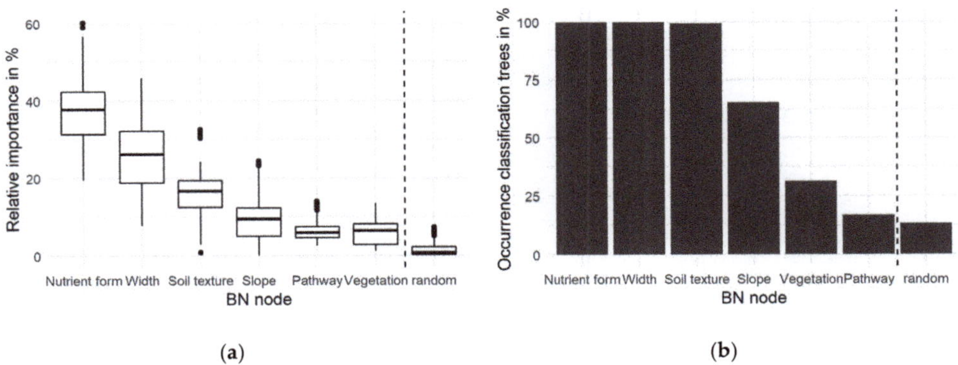

Figure 5. Relative importance (**a**) and occurrence of BN nodes (**b**) in the classification trees compared to a random binary variable. Note the changed order of (RBZ) vegetation and (flow) pathway.

3.3. Cross-Validation and Evaluation of BNs

Fewer output states resulted in lower prediction errors, with average values over all BNs ranging from 0.31 (two states) to 0.66 (five states). However, the BNs predicted the lowest and highest output states significantly better than the intermediate states (Figure 6). These errors correspond to 62% of the expected error of a random predictor (E) for the two-state output (E = 0.5) up to 82% for the five-state output (E = 0.8). The values for the minimum and maximum states equaled on average 63% of E.

The prediction errors for RBZ efficacy were significantly lower when the nutrient form was "particulate" or "mixed" rather than "dissolved", on average by 0.05 (−0.05–0.15, Figure 7). The impact of the flow pathway and the RBZ vegetation on the prediction error was negligible. The error even decreased on average by 0.01 after the BNs were simplified, i.e., by excluding the RBZ vegetation and the flow pathway. Limiting the database to the respective dominant state of these two nodes, i.e., to grassed RBZ and surface flow (n = 391), had no additional effect on the error.

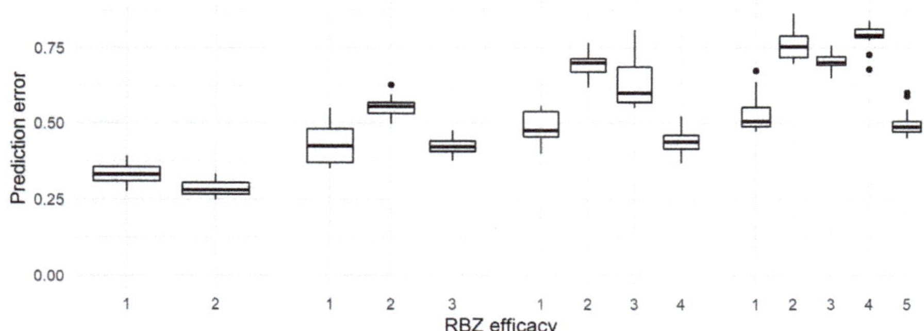

Figure 6. Cross-validation results for the different output states based on the 72 complete BNs with all nodes. Higher state numbers correspond to more effective RBZ.

The prediction errors were generally lowest when the parent nodes had three states (Table 3). However, for some BNs with all six parent nodes ("complete BN" in Table 3) choosing four states for the node "RBZ width" resulted in better predictions. Choosing two states, on the other hand, increased the predictions errors by up to 0.05 for RBZ width.

Table 3. Effect of the state number of parent nodes on prediction errors for different BNs. Mean errors and p values based on the outcomes of six BNs (RBZ width, RBZ slope) and nine BNs (soil texture). The nodes "flow pathway" and "RBZ vegetation" in complete BNs have only two states and are not shown.

BN	RBZ Efficacy Number of States	Node	Mean Prediction Error Two States	Three States	Four States	Significance (p Value) [1] 2 < 3 [2]	3 < 4
Complete	2	RBZ slope	0.32	0.26	0.31	0.031*	0.984
	3		0.51	0.45	0.46	0.016*	0.656
	4		0.59	0.55	0.56	0.016*	0.844
	5		0.67	0.65	0.64	0.016*	0.109
	2	RBZ width	0.32	0.31	0.30	0.031*	0.031*
	3		0.47	0.47	0.47	0.719	0.109
	4		0.59	0.55	0.55	0.016*	0.922
	5		0.67	0.66	0.64	0.281	0.016*
	2	Soil texture	0.32	0.30	-	0.002**	-
	3		0.48	0.47	-	0.010**	-
	4		0.58	0.55	-	0.002**	-
	5		0.66	0.65	-	0.002**	-
Simplified	2	RBZ slope	0.30	0.30	0.30	0.219	0.984
	3		0.48	0.46	0.48	0.016*	1
	4		0.55	0.54	0.56	0.500	1
	5		0.64	0.64	0.64	0.578	0.844
	2	RBZ width	0.30	0.30	0.30	0.047*	0.781
	3		0.48	0.47	0.47	0.047*	0.953
	4		0.58	0.53	0.54	0.016*	0.844
	5		0.64	0.64	0.64	0.656	0.219
	2	Soil texture	0.30	0.30	-	0.963	-
	3		0.47	0.48	-	0.898	-
	4		0.55	0.55	-	0.980	-
	5		0.64	0.64	-	0.590	-

[1] One-sided Mann–Whitney U test, symbols: $p < 0.001$ (***), < 0.01 (**), < 0.05 (*), [2] prediction error smaller with two states than with three states.

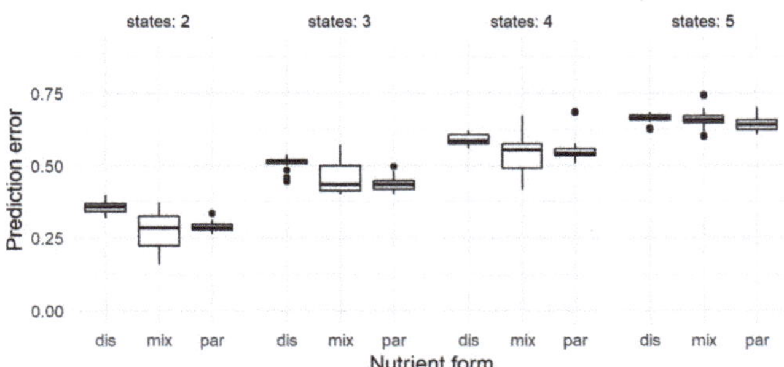

Figure 7. Cross-validation results for disaggregated (dis), particulate (par), and mixed (total, mix) nutrients and different numbers of output states.

4. Discussion

4.1. BN Cross-Validation and Evaluation

We tested the influence of nodes and states on RBZ efficacy in order to identify the minimum data requirement while ensuring no significant decrease in model performance and reliability. In this sense, the simplified BN with four out of the six considered parent nodes and with three (slope, soil texture) or four states (width) for the parent nodes as well as two to three output states can be considered as "optimal" among the tested alternatives (Figure 8A). The CPT size, i.e., the product of the state numbers, is 216 or 324 elements—which is manageable given the available input data [59]. Nonetheless, 0.3% or 0.9% of the predicted values during the cross-validation were undefined (compared to 0.2% for the smallest CPT size).

In accordance to the reviewed literature (cf. Methods), the BN reflects that a high RBZ efficacy was less probable for narrow than for broad RBZ (Figure 8B, a), for fine sediments than for coarse sediment (Figure 8B, b), and for dissolved nutrients than for particulate nutrients (Figure 8B, c). The more complex pattern for RBZ slope (Figure 8B, d) depended on the nutrient form (not shown). The higher probability of low RBZ efficacies for medium slopes occurred only for TN and TP. For particulate and dissolved nutrients, the efficacy of RBZ declined with increasing slope. Considering the dominance of studies on surface flow, this pattern can be explained by higher flow velocity and transport capacity on steeper slopes (cf. Methods). Nonetheless, it differs from findings of Zhang et al. [13], based on a smaller dataset, that sediment retention decreases with slopes steeper than 10% but increases with slopes below this threshold. At the plot scale—which dominates the collected data—flow resistance and other flow characteristics varies with slope gradient but also depends on the fragmentation of riparian vegetation [67] which is usually not available from RBZ studies. Following Zhang et al., more research on slope effects on RBZ efficacy is recommended to revise the inconsistent probabilities.

The use of these commonly available variables as nodes facilitates the envisaged application and validation of this first BN to estimate the retention efficacy of RBZ. This contributes to a better evaluation at medium to large scales of how RBZ can help to improve the water quality. The BN was significantly better in distinguishing effective (high state) from ineffective RBZ (low state) than between intermediate states (Figure 6). With two output states, the BN can for instance support or replace guidelines and design aids for effective RBZ based on single factors. For quantitative assessments, the discrete output states could be replaced by the average values of 45% and 90% efficacy in our database, or 30%, 70%, and 90% if a third state is needed. The high prediction errors hamper the use of more intermediate efficacy states at least in combination with our heterogeneous empirical dataset. Note that the BN is not intended to replace but rather complement other modelling approaches. For instance, Figure 8b illustrates that the BN can deal with uncertain evidence and provides probabilities of RBZ efficacy. Nonetheless, the high influence of RBZ width, soil texture, and slope emphasize the need of multivariate approaches in RBZ analyses.

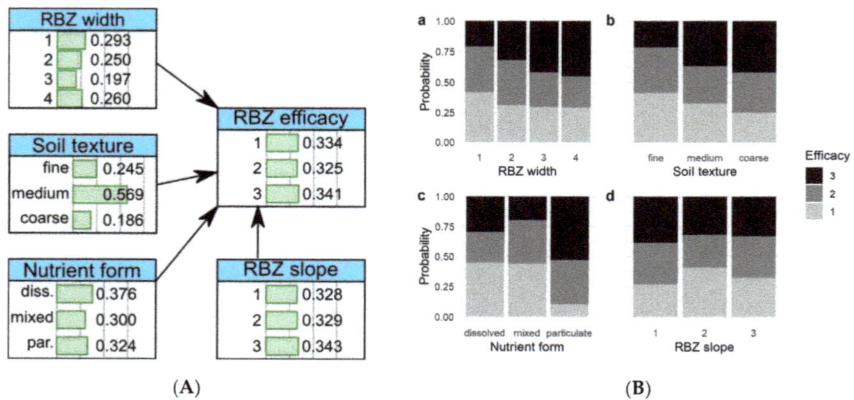

Figure 8. The "optimal" BN with three output states (**A**) influence diagram (Supplementary File S2, par. = particulate, diss. = dissolved), (**B**) probability of the output states (RBZ efficacy) given hard evidence of the parent node mentioned below the subfigures a–d (state with probability = 1 on x axes) and soft evidence (here equal probability of all states) of the other parent nodes. Higher state numbers correspond to broader RBZ (RBZ width), steeper RBZ (RBZ slope), and more efficient RBZ.

Although the loss of information is a potential limitation of discrete BN [59], defining more than two states for parent nodes was only slightly helpful to improve the model accuracy. The prediction errors as well as the variability and the significant inconsistency among the reported efficacy values may indicate the need for supplementing BN nodes. Even with more input data than in previous studies, our analyses indicate that additional parent nodes are not necessarily useful to improve predictions if the data is biased and does not cover the variability of site conditions. For instance, the low influence of RBZ vegetation is in line with the similarity of woody and grassed RBZ as observed in previous studies (cf. Methods). However, the dominant grassed RBZ were on average narrower (9 m) than the woody RBZ (22.6 m). Moreover, the common distinction between grassed and woody RBZ does not consider the influence of different plant groups and species on the nutrient retention in RBZ [68,69]. Vegetation and root density seem to be promising to refine the state definition and to improve model predictions (cf. Methods), but such data is unavailable in existing RBZ studies.

4.2. Further BN Development

The current BN can be improved in different ways. The collection of training and validation data could be limited to specific experimental designs, scales, or site conditions. This simple approach would minimize the impact of some factors not explicitly considered in the BN at the expense of restricting our intended applicability at large scales. More nodes and states, on the other hand, require enough representative training data to fill the rapidly growing CPT. Expert knowledge can reduce the data demand (e.g., [70]) but its validation can be challenging [71]. We exemplarily discuss below constraints of the available data regarding the design and application of the BN (and other empirical models).

Firstly, only less than 20% of the datasets were on subsurface flow. The underrepresentation was most striking for DP (n = 10). Although limiting the dataset to surface flow did not reduce the prediction errors, the observed low influence of the flow pathway on RBZ efficacy and the (slightly) higher errors for dissolved nutrients (Figure 7) should be scrutinized with more (DP) data. However, revising the role of flow pathway may also require considering the contrasting effects of external factors like redox conditions on the retention of P (oxidation/reduction of iron oxides) and N (denitrification) either as separate node or as additional DN and DP states of the nutrient node.

Secondly, the current BN may overestimate the efficacy of "real-world" RBZ because most literature values were measured in rather short-term studies. However, aging buffer strips may turn from sinks to sources [34,53,72] and foster eutrophication if poorly maintained [72–74]. Negative values occurred in 26 datasets and comprised mostly DP (n = 16) and DN (n = 9), followed by TN (n = 3), TP (n = 2), and PP (n = 1). Given the variety of underlying processes and the small number of cases, they could not be considered as a separate state in the output node. Apart from the remobilization of P from saturated RBZ soils as DP [34,45], P-rich colluvium in RBZ can be remobilized during heavy rainfall or flood events. Detailed maps on water-soluble P, degree of P saturation, and residual soil P in agricultural areas (e.g., [7]) might be helpful to overcome this limitation in future BN. The data availability is, however, more challenging for other processes such as the decay of plants [74,75].

Thirdly, other below-average retention efficacies e.g., for sediment retention in our data collection were explained by conservation tillage (CT) [76,77] and preferential flow [78] in the catchment of RBZ. CT, i.e., minimum- and no-till, has increasingly been used worldwide during the last decade [79], and might do so in the future [80], to reduce sediment and PP losses from agricultural land. However, the accumulation of labile P pools in top soils and the development of preferential flow paths belowground can unintendedly impair the RBZ efficacy to retain soluble, immediately bioavailable P [34,81]. No-till can also increase the water and nitrate flux entering RBZ from adjacent fields [82]. While more water flux can reduce the nitrate retention in subsurface flow [46], higher (initial) concentrations favor its retention [47].

In the few available quantitative studies on CT, no-till reduced significantly the RBZ efficacy compared to conventional till, notably except for P (Figure 9a). However, most of the values already belonged to the lowest efficacy state (RBZ efficacy < 55%). Therefore, the influence of CT on the BN output would currently be rather low. It might also not be representative for other output states. By separating the soil texture in RBZ and their catchments, the BN could consider some of the complex impacts of CT together with nutrient form.

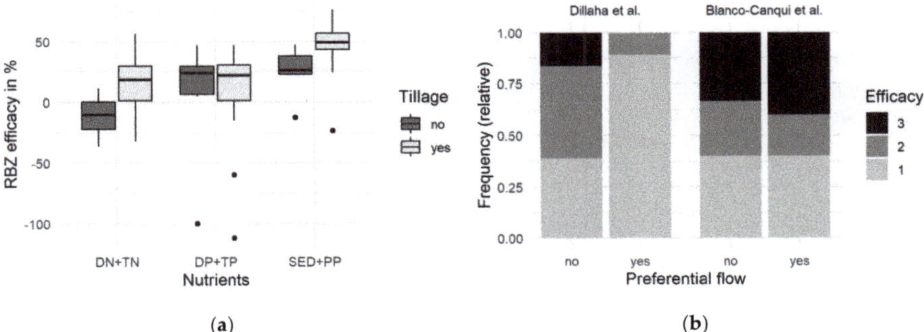

(a) (b)

Figure 9. Examples of non-representative and inconsistent effects on RBZ efficacy, (**a**) tillage (data from references [74,76,83,84], paired n = 30), (**b**) preferential flow (left: data from references [78,85], right: data from [86], without switchgrass barrier, paired n = 6).

As a side effect, these two BN nodes can generally integrate the effects of hydrology [46,87], land use, and land management [16,21]. These variables affect the concentration [47,88], timing, amount, and composition of nutrient and sediment entering RBZ and have been used in other empirical RBZ models (e.g., [16,46]).

Fourthly, most published RBZ studies were plot studies which are typically constructed with uniform slopes to favor sheet-flow conditions under which RBZ work best [48,89]. However, they neglect the tendency of water to converge and develop preferential flow paths at larger scales and within natural terrain [90–92]. According to the very few available quantitative studies, preferential flow can reduce the efficiency of RBZ [78,85,86,93] (Figure 9b, left). Nonetheless, RBZ may perform equally well depending on RBZ width and structure as well as the runoff conditions [86,91] (Figure 9b, right). According to our data collection, plot studies reported on average less effective RBZ than field-

and watershed studies which can be explained by the narrower RBZ in these plot studies. Therefore, more field studies are needed to reliably quantify how preferential flow affects the efficacy of RBZ.

In an extended BN, the current output state could be reduced based on the probability of preferential flow. The likelihood of preferential surface flow is often estimated from digital elevation models by applying area thresholds or variants thereof [94] or from linear anthropogenic features [95]. Similar slope–area relationships are known for e.g., the topographic index [96,97] and soil erosion [98] which also influence the variability and composition of nutrients.

5. Conclusions

The discrete, multivariate BNs to predict RBZ efficacy developed and evaluated in our study are flexible tools which e.g., provide probabilities instead of exact outcomes. They predict sufficiently well the RBZ efficacies collected from previous studies. The "optimal" BNs have four parent nodes (width, slope, nutrient form, and soil texture) with more than two states and up to three states for RBZ efficacy. The resulting CPTs are manageable with the available input data. RBZ vegetation and flow pathway are found to be not relevant.

The model performance is better for the highest and lowest states of RBZ efficacy compared to the intermediate states. The BNs could thus complement or replace existing design guidelines for RBZ. The expected limitations for more detailed predictions arise from missing data for other variables as discussed in the literature as well as the heterogeneous database which does not adequately represent all site conditions. This is relevant for the vegetation type which typical states "grass" and "woody" comprise different species. Additionally, most data are obtained from plot-scale studies which typically do not consider preferential flow. More field-scale studies are needed to better represent preferential flow in BNs. Future research should focus on the database to better cover the different state combinations and to provide new data to train and validate the relationships.

Supplementary Materials: The following are available online at http://www.mdpi.com/2073-4441/12/3/617/s1, Table S1: Data table, File S2: BN in Figure 8a as Hugin. net file.

Author Contributions: Conceptualization, A.G. and M.V.; methodology, A.G. and M.V.; software, A.G.; validation, A.G.; formal analysis, A.G.; investigation, A.G., P.F., and M.V.; resources, A.G. and P.F.; data curation, A.G. and M.V.; writing—original draft preparation, A.G.; writing—review and editing, A.G., H.H.N., P.F., J.K., and M.V.; visualization, A.G.; supervision, M.V.; project administration, J.K. and M.V.; funding acquisition, J.K. and M.V. All authors have read and agreed to the published version of the manuscript.

Funding: This research was funded by the German "Bundesministerium für Bildung und Forschung (Federal Ministry of Education and Research)", grant number 01LC1618A within the 2015–2016 BiodivERsA COFUND call. Peter Fischer was financed by the "Ministerium für Umwelt, Klima und Energiewirtschaft (Ministry of the Environment, Climate Protection and the Energy Sector) Baden-Württemberg" within the project "Retention of sediments, nutrients, and pesticides in riparian buffer zones".

Acknowledgments: We thank Daniela Pietras (University of Rennes) for assisting the data collection. We are thankful to three anonymous reviewers for their detailed and valuable comments.

Conflicts of Interest: The authors declare no conflict of interest. The funders had no role in the design of the study; in the collection, analyses, or interpretation of data; in the writing of the manuscript, or in the decision to publish the results.

References

1. Le Moal, M.; Gascuel-Odoux, C.; Ménesguen, A.; Souchon, Y.; Étrillard, C.; Levain, A.; Moatar, F.; Pannard, A.; Souchu, P.; Lefebvre, A.; et al. Eutrophication: A new wine in an old bottle? *Sci. Total Environ.* **2019**, *651*, 1–11. [CrossRef] [PubMed]
2. Evans, A.E.V.; Mateo-Sagasta, J.; Qadir, M.; Boelee, E.; Ippolito, A. Agricultural water pollution: Key knowledge gaps and research needs. *Curr. Opin. Environ. Sust.* **2019**, *36*, 20–27. [CrossRef]
3. FAO; IMWI. *More People, More Food, Worse Water? A Global Review of Water Pollution from Agriculture*; Mateo-Sagasta, J., Marjani Zadeh, S., Turral, H., Eds.; FAO: Rome, Italy; IMWI: Colombo, Sri Lanka, 2018; p. 207.

4. Diamantini, E.; Lutz, S.R.; Mallucci, S.; Majone, B.; Merz, R.; Bellin, A. Driver detection of water quality trends in three large European river basins. *Sci. Total Environ.* **2018**, *612*, 49–62. [CrossRef] [PubMed]
5. Schwarzenbach, R.P.; Egli, T.; Hofstetter, T.B.; von Gunten, U.; Wehrli, B. Global Water Pollution and Human Health. *Annu. Rev. Environ. Resour.* **2010**, *35*, 109–136. [CrossRef]
6. Withers, P.J.A.; Neal, C.; Jarvie, H.P.; Doody, D.G. Agriculture and Eutrophication: Where Do We Go from Here? *Sustainability* **2014**, *6*, 5853. [CrossRef]
7. Fischer, P.; Pöthig, R.; Venohr, M. The degree of phosphorus saturation of agricultural soils in Germany: Current and future risk of diffuse P loss and implications for soil P management in Europe. *Sci. Total Environ.* **2017**, *599*, 1130–1139. [CrossRef] [PubMed]
8. European Environment Agency. *European Waters: Assessment of Status and Pressures 2018*; Publications Office of the European Union: Luxembourg, Luxembourg, 2018; p. 85.
9. Stutter, M.I.; Chardon, W.J.; Kronvang, B. Riparian Buffer Strips as a Multifunctional Management Tool in Agricultural Landscapes: Introduction. *J. Environ. Qual.* **2012**, *41*, 297–303. [CrossRef]
10. Feld, C.K.; Fernandes, M.R.; Ferreira, M.T.; Hering, D.; Ormerod, S.J.; Venohr, M.; Gutiérrez-Cánovas, C. Evaluating riparian solutions to multiple stressor problems in river ecosystems-A conceptual study. *Water Res.* **2018**, *139*, 381–394. [CrossRef]
11. Mander, Ü.; Tournebize, J.; Tonderski, K.; Verhoeven, J.T.A.; Mitsch, W.J. Planning and establishment principles for constructed wetlands and riparian buffer zones in agricultural catchments. *Ecol. Eng.* **2017**, *103*, 296–300. [CrossRef]
12. Dorioz, J.-M.; Wang, D.; Poulenard, J.; Trevisan, D. The effect of grass buffer strips on phosphorus dynamics—A critical review and synthesis as a basis for application in agricultural landscapes in France. *Agric. Ecosyst. Environ.* **2006**, *117*, 4–21. [CrossRef]
13. Zhang, X.; Liu, X.; Zhang, M.; Dahlgren, R.A.; Eitzel, M. A review of vegetated buffers and a meta-analysis of their mitigation efficacy in reducing nonpoint source pollution. *J. Environ. Qual.* **2010**, *39*, 76–84. [CrossRef] [PubMed]
14. Muñoz-Carpena, R.; Parsons, J.E.; Gilliam, J.W. Modeling hydrology and sediment transport in vegetative filter strips. *J. Hydrol.* **1999**, *214*, 111–129. [CrossRef]
15. Lowrance, R.; Altier, L.S.; Williams, R.G.; Inamdar, S.P.; Sheridan, J.M.; Bosch, D.D.; Hubbard, R.K.; Thomas, D.L. REMM: The Riparian Ecosystem Management Model. *J. Soil Water Conserv.* **2000**, *55*, 27–34.
16. Hassanzadeh, Y.T.; Vidon, P.G.; Gold, A.J.; Pradhanang, S.M.; Addy Lowder, K. RZ-TRADEOFF: A New Model to Estimate Riparian Water and Air Quality Functions. *Water* **2019**, *11*, 769. [CrossRef]
17. Weissteiner, C.J.; Bouraoui, F.; Aloe, A. Reduction of nitrogen and phosphorus loads to European rivers by riparian buffer zones. *Knowl. Manag. Aquat. Ecosyst.* **2013**, *408*, 15. [CrossRef]
18. Hook, P.B. Sediment retention in rangeland riparian buffers. *J. Environ. Qual.* **2003**, *32*, 1130–1137. [CrossRef]
19. Lee, P.; Smyth, C.; Boutin, S. Quantitative review of riparian buffer width guidelines from Canada and the United States. *J. Environ. Manag.* **2004**, *70*, 165–180. [CrossRef]
20. Broadmeadow, S.; Nisbet, T.R. The effects of riparian forest management on the freshwater environment: A literature review of best management practice. *HESS* **2004**, *8*, 286–305. [CrossRef]
21. Dosskey, M.G.; Helmers, M.J.; Eisenhauer, D.E. A design aid for determining width of filter strips. *J. Soil Water Conserv.* **2008**, *63*, 232–241. [CrossRef]
22. Gumiere, S.J.; Le Bissonnais, Y.; Raclot, D.; Cheviron, B. Vegetated filter effects on sedimentological connectivity of agricultural catchments in erosion modelling: A review. *ESPL* **2011**, *36*, 3–19. [CrossRef]
23. Liu, X.; Zhang, X.; Zhang, M. Major factors influencing the efficacy of vegetated buffers on sediment trapping: A review and analysis. *J. Environ. Qual.* **2008**, *37*, 1667–1674. [CrossRef] [PubMed]
24. Yuan, Y.; Bingner, R.L.; Locke, M.A. A review of effectiveness of vegetative buffers on sediment trapping in agricultural areas. *Ecohydrology* **2009**, *2*, 321–336. [CrossRef]
25. Mayer, P.M.; Reynolds, S.K.; McCutchen, M.D.; Canfield, T.J. Meta-analysis of nitrogen removal in riparian buffers. *J. Environ. Qual.* **2007**, *36*, 1172–1180. [CrossRef] [PubMed]
26. Barton, D.N.; Kuikka, S.; Varis, O.; Uusitalo, L.; Henriksen, H.J.; Borsuk, M.; de la Hera, A.; Farmani, R.; Johnson, S.; Linnell, J.D. Bayesian networks in environmental and resource management. *Integr. Environ. Assess. Manag.* **2012**, *8*, 418–429. [CrossRef]
27. Borsuk, M.E.; Stow, C.A.; Reckhow, K.H. A Bayesian network of eutrophication models for synthesis, prediction, and uncertainty analysis. *Ecol. Model.* **2004**, *173*, 219–239. [CrossRef]

28. Kelly, R.A.; Jakeman, A.J.; Barreteau, O.; Borsuk, M.E.; ElSawah, S.; Hamilton, S.H.; Henriksen, H.J.; Kuikka, S.; Maier, H.R.; Rizzoli, A.E.; et al. Selecting among five common modelling approaches for integrated environmental assessment and management. *Environ. Model. Softw.* **2013**, *47*, 159–181. [CrossRef]
29. Marcot, B.G.; Penman, T.D. Advances in Bayesian network modelling: Integration of modelling technologies. *Environ. Model. Softw.* **2019**, *111*, 386–393. [CrossRef]
30. Phan, T.D.; Smart, J.C.R.; Capon, S.J.; Hadwen, W.L.; Sahin, O. Applications of Bayesian belief networks in water resource management: A systematic review. *Environ. Model. Softw.* **2016**, *85*, 98–111. [CrossRef]
31. McVittie, A.; Norton, L.; Martin-Ortega, J.; Siameti, I.; Glenk, K.; Aalders, I. Operationalizing an ecosystem services-based approach using Bayesian Belief Networks: An application to riparian buffer strips. *Ecol. Econ.* **2015**, *110*, 15–27. [CrossRef]
32. Castelle, A.J.; Johnson, A.W.; Conolly, C. Wetland and stream buffer size requirements-a review. *J. Environ. Qual.* **1994**, *23*, 878–882. [CrossRef]
33. Collins, A.L.; Hughes, G.; Zhang, Y.; Whitehead, J. Mitigating diffuse water pollution from agriculture: Riparian buffer strip performance with width. *CAB Rev. Perspect. Agric. Veter. Sci. Nutr. Nat. Resour.* **2009**, *4*, 15. [CrossRef]
34. Dodd, R.J.; Sharpley, A.N. Conservation practice effectiveness and adoption: Unintended consequences and implications for sustainable phosphorus management. *Nutr. Cycl. Agroecosyst.* **2016**, *104*, 373–392. [CrossRef]
35. Dosskey, M.G. Toward quantifying water pollution abatement in response to installing buffers on crop land. *Environ. Manag.* **2001**, *28*, 577–598. [CrossRef] [PubMed]
36. Dosskey, M.G.; Vidon, P.; Gurwick, N.P.; Allan, C.J.; Duval, T.P.; Lowrance, R. The Role of Riparian Vegetation in Protecting and Improving Chemical Water Quality in Streams. *JAWRA* **2010**, *46*, 261–277. [CrossRef]
37. Hickey, M.B.C.; Doran, B. A review of the efficiency of buffer strips for the maintenance and enhancement of riparian ecosystems. *Water Qual. Res. J. Can.* **2004**, *39*, 311–317. [CrossRef]
38. Hill, A.R. Nitrate removal in stream riparian zones. *J. Environ. Qual.* **1996**, *25*, 743–755. [CrossRef]
39. Hill, A.R. Groundwater nitrate removal in riparian buffer zones: A review of research progress in the past 20 years. *Biogeochemistry* **2019**, *143*, 347–369. [CrossRef]
40. Hoffmann, C.C.; Kjaergaard, C.; Uusi-Kämppä, J.; Hansen, H.C.B.; Kronvang, B. Phosphorus retention in riparian buffers: Review of their efficiency. *J. Environ. Qual.* **2009**, *38*, 1942–1955. [CrossRef]
41. Krutz, L.J.; Senseman, S.A.; Zablotowicz, R.M.; Matocha, M.A. Reducing herbicide runoff from agricultural fields with vegetative filter strips: A review. *Weed Sci.* **2005**, *53*, 353–367. [CrossRef]
42. Lacas, J.-G.; Voltz, M.; Gouy, V.; Carluer, N.; Gril, J.-J. Using grassed strips to limit pesticide transfer to surface water: A review. *Agron. Sustain. Dev.* **2005**, *25*, 253–266. [CrossRef]
43. Mayer, P.M.; Reynolds, S.K., Jr.; Canfield, T.J. *Riparian Buffer Width, Vegetative Cover, and Nitrogen Removal Effectiveness*; EPA/600/R-05/118; U.S. Environmental Protection Agency: Washington, DC, USA, 2005; p. 27.
44. Polyakov, V.; Fares, A.; Ryder, M.H. Precision riparian buffers for the control of nonpoint source pollutant loading into surface water: A review. *Environ. Rev.* **2005**, *13*, 129–144. [CrossRef]
45. Roberts, W.M.; Stutter, M.I.; Haygarth, P.M. Phosphorus retention and remobilization in vegetated buffer strips: A review. *J. Environ. Qual.* **2012**, *41*, 389–399. [CrossRef] [PubMed]
46. Sweeney, B.W.; Newbold, J.D. Streamside Forest Buffer Width Needed to Protect Stream Water Quality, Habitat, and Organisms: A Literature Review. *JAWRA* **2014**, *50*, 560–584. [CrossRef]
47. Valkama, E.; Usva, K.; Saarinen, M.; Uusi-Kämppä, J. A Meta-Analysis on Nitrogen Retention by Buffer Zones. *J. Environ. Qual.* **2018**. [CrossRef]
48. Wenger, S. *A Review of the Scientific Literature on Riparian Buffer Width, Extent and Vegetation*; Office of Public Service and Outreach, Institute of Ecology: Athens, GA, USA, 1999; p. 59.
49. Borin, M.; Vianello, M.; Morari, F.; Zanin, G. Effectiveness of buffer strips in removing pollutants in runoff from a cultivated field in North-East Italy. *Agric. Ecosyst. Environ.* **2005**, *105*, 101–114. [CrossRef]
50. Fiener, P.; Auerswald, K. Effectiveness of grassed waterways in reducing runoff and sediment delivery from agricultural watersheds. *J. Environ. Qual.* **2003**, *32*, 927–936. [CrossRef] [PubMed]
51. Pärn, J.; Pinay, G.; Mander, Ü. Indicators of nutrients transport from agricultural catchments under temperate climate: A review. *Ecol. Indic.* **2012**, *22*, 4–15. [CrossRef]

52. King, S.E.; Osmond, D.L.; Smith, J.; Burchell, M.R.; Dukes, M.; Evans, R.O.; Knies, S.; Kunickis, S. Effects of Riparian Buffer Vegetation and Width: A 12-Year Longitudinal Study. *J. Environ. Qual.* **2016**, *45*, 1243–1251. [CrossRef]
53. Osborne, L.L.; Kovacic, D.A. Riparian vegetated buffer strips in water-quality restoration and stream management. *Freshw. Biol.* **1993**, *29*, 243–258. [CrossRef]
54. Yang, F.; Yang, Y.; Li, H.; Cao, M. Removal efficiencies of vegetation-specific filter strips on nonpoint source pollutants. *Ecol. Eng.* **2015**, *82*, 145–158. [CrossRef]
55. Schauder, H.; Auerswald, K. Long-term trapping efficiency of a vegetated filter strip under agricultural use. *J. Plant Nutr. Soil Sci.* **1992**, *155*, 489–492. [CrossRef]
56. Vidon, P.; Hill, A.R. Denitrification and patterns of electron donors and acceptors in eight riparian zones with contrasting hydrogeology. *Biogeochemistry* **2004**, *71*, 259–283. [CrossRef]
57. Barling, R.D.; Moore, I.D. Role of buffer strips in management of waterway pollution: A review. *Environ. Manag.* **1994**, *18*, 543–558. [CrossRef]
58. Darch, T.; Carswell, A.; Blackwell, M.S.A.; Hawkins, J.M.B.; Haygarth, P.M.; Chadwick, D. Dissolved Phosphorus Retention in Buffer Strips: Influence of Slope and Soil Type. *J. Environ. Qual.* **2015**, *44*, 1216–1224. [CrossRef] [PubMed]
59. Chen, S.H.; Pollino, C.A. Good practice in Bayesian network modelling. *Environ. Model. Softw.* **2012**, *37*, 134–145. [CrossRef]
60. Nojavan, A.F.; Qian, S.S.; Stow, C.A. Comparative analysis of discretization methods in Bayesian networks. *Environ. Model. Softw.* **2017**, *87*, 64–71. [CrossRef]
61. Marcot, B.G. Metrics for evaluating performance and uncertainty of Bayesian network models. *Ecol. Model.* **2012**, *230*, 50–62. [CrossRef]
62. Moe, S.J.; Haande, S.; Couture, R.-M. Climate change, cyanobacteria blooms and ecological status of lakes: A Bayesian network approach. *Ecol. Model.* **2016**, *337*, 330–347. [CrossRef]
63. Team, R.C. *R: A Language and Environment for Statistical Computing*; R Foundation for Statistical Computing: Vienna, Austria, 2017.
64. Scutari, M. Learning Bayesian Networks with the bnlearn R Package. *J. Stat. Softw.* **2010**, *35*, 1–22. [CrossRef]
65. Therneau, T.; Atkinson, B.; Ripley, B. rpart: Recursive Partitioning and Regression Trees. 1–11 April 2017. Available online: https://cran.r-project.org/web/packages/rpart/index.html (accessed on 14 June 2019).
66. Højsgaard, S. Graphical Independence Networks with the gRain Package for R. *J. Stat. Softw.* **2012**, *1*. [CrossRef]
67. Zhao, Q.; Zhang, Y.; Xu, S.; Ji, X.; Wang, S.; Ding, S. Relationships between Riparian Vegetation Pattern and the Hydraulic Characteristics of Upslope Runoff. *Sustainability* **2019**, *11*. [CrossRef]
68. Kelly, J.M.; Kovar, J.L.; Sokolowsky, R.; Moorman, T.B. Phosphorus uptake during four years by different vegetative cover types in a riparian buffer. *Nutr. Cycl. Agroecosyst.* **2007**, *78*, 239–251. [CrossRef]
69. Missaoui, A.M.; Boerma, H.R.; Bouton, J.H. Genetic variation and heritability of phosphorus uptake in Alamo switchgrass grown in high phosphorus soils. *Field Crops Res.* **2005**, *93*, 186–198. [CrossRef]
70. Oniśko, A.; Druzdzel, M.J.; Wasyluk, H. Learning Bayesian network parameters from small data sets: Application of Noisy-OR gates. *Int. J. Approx. Reason.* **2001**, *27*, 165–182. [CrossRef]
71. Pitchforth, J.; Mengersen, K. A proposed validation framework for expert elicited Bayesian Networks. *Expert Syst. Appl.* **2013**, *40*, 162–167. [CrossRef]
72. Cooper, J.R.; Gilliam, J.W. Phosphorus redistribution from cultivated fields into riparian areas. *Soil Sci. Soc. Am. J.* **1987**, *51*, 1600–1604. [CrossRef]
73. Stutter, M.I.; Langan, S.J.; Lumsdon, D.G. Vegetated buffer strips can lead to increased release of phosphorus to waters: A biogeochemical assessment of the mechanisms. *Environ. Sci. Technol.* **2009**, *43*, 1858–1863. [CrossRef]
74. Uusi-Kämppä, J.; Jauhiainen, L. Long-term monitoring of buffer zone efficiency under different cultivation techniques in boreal conditions. *Agric. Ecosyst. Environ.* **2010**, *137*, 75–85. [CrossRef]
75. Uusi-Kämppä, J. Phosphorus purification in buffer zones in cold climates. *Ecol. Eng.* **2005**, *24*, 491–502. [CrossRef]
76. Shipitalo, M.J.; Bonta, J.V.; Dayton, E.A.; Owens, L.B. Impact of Grassed Waterways and Compost Filter Socks on the Quality of Surface Runoff from Corn Fields. *J. Environ. Qual.* **2010**, *39*, 1009–1018. [CrossRef]

77. Newbold, J.D.; Herbert, S.; Sweeney, B.W.; Kiry, P.; Alberts, S.J. Water Quality Functions of a 15-Year-Old Riparian Forest Buffer System. *JAWRA* **2010**, *46*, 299–310. [CrossRef]
78. Dillaha, T.A.; Sherrard, J.H.; Lee, D.; Mostaghimi, S.; Shanholtz, V.O. Evaluation of vegetative filter strips as a best management practice for feed lots. *J. Water Pollut. Control Fed.* **1988**, *60*, 1231–1238.
79. Kassam, A.; Friedrich, T.; Derpsch, R. Global spread of Conservation Agriculture. *Int. J. Environ. Stud.* **2019**, *76*, 29–51. [CrossRef]
80. Prestele, R.; Hirsch, A.L.; Davin, E.L.; Seneviratne, S.I.; Verburg, P.H. A spatially explicit representation of conservation agriculture for application in global change studies. *Glob. Chang. Biol.* **2018**, *24*, 4038–4053. [CrossRef] [PubMed]
81. Jarvie, H.P.; Johnson, L.T.; Sharpley, A.N.; Smith, D.R.; Baker, D.B.; Bruulsema, T.W.; Confesor, R. Increased Soluble Phosphorus Loads to Lake Erie: Unintended Consequences of Conservation Practices? *J. Environ. Qual.* **2017**, *46*, 123–132. [CrossRef] [PubMed]
82. Daryanto, S.; Wang, L.; Jacinthe, P.-A. Impacts of no-tillage management on nitrate loss from corn, soybean and wheat cultivation: A meta-analysis. *Sci. Rep.* **2017**, *7*, 12117. [CrossRef]
83. Eghball, B.; Gilley, J.E.; Kramer, L.A.; Moorman, T.B. Narrow grass hedge effects on phosphorus and nitrogen in runoff following manure and fertilizer application. *J. Soil Water Conserv.* **2000**, *55*, 172–176.
84. Raffaelle, J.B., Jr.; McGregor, K.C.; Foster, G.R.; Cullum, R.F. Effect of narrow grass strips on conservation reserve land converted to cropland. *Trans. ASAE* **1997**, *40*, 1581–1587. [CrossRef]
85. Dillaha, T.A.; Sherrard, J.H.; Lee, D.; Shanholtz, V.O.; Mostaghimi, S.; Magette, W.L. *Use of Vegetative Filter Strips to Minimize Sediment and Phosphorus Losses from Feedlots*; Phase 1, Experimental Plot Studies; Virginia Polytechnic Institute and State University Blacksburg: Blacksburg, VA, USA, 1986; p. 68.
86. Blanco-Canqui, H.; Gantzer, C.J.; Anderson, S.H. Performance of grass barriers and filter strips under interrill and concentrated flow. *J. Environ. Qual.* **2006**, *35*, 1969–1974. [CrossRef]
87. White, M.J.; Arnold, J.G. Development of a simplistic vegetative filter strip model for sediment and nutrient retention at the field scale. *Hydrol. Process.* **2009**, *23*, 1602–1616. [CrossRef]
88. Vidon, P.G.F.; Hill, A.R. Landscape controls on nitrate removal in stream riparian zones. *WRR* **2004**, *40*. [CrossRef]
89. Habibiandehkordi, R.; Lobb, D.A.; Sheppard, S.C.; Flaten, D.N.; Owens, P.N. Uncertainties in vegetated buffer strip function in controlling phosphorus export from agricultural land in the Canadian prairies. *Environ. Sci. Pollut. Res.* **2017**, *24*, 18372–18382. [CrossRef] [PubMed]
90. Verstraeten, G.; Poesen, J.; Gillijns, K.; Govers, G. The use of riparian vegetated filter strips to reduce river sediment loads: An overestimated control measure? *Hydrol. Process.* **2006**, *20*, 4259–4267. [CrossRef]
91. Helmers, M.J.; Eisenhauer, D.E.; Dosskey, M.G.; Franti, T.G.; Brothers, J.M.; McCullough, M.C. Flow pathways and sediment trapping in a field-scale vegetative filter. *Trans. ASAE* **2005**, *48*, 955–968. [CrossRef]
92. Dosskey, M.G.; Helmers, M.J.; Eisenhauer, D.E.; Franti, T.G.; Hoagland, K.D. Assessment of concentrated flow through riparian buffers. *J. Soil Water Conserv.* **2002**, *57*, 336–343.
93. Daniels, R.B.; Gilliam, J.W. Sediment and chemical load reduction by grass and riparian filters. *Soil Sci. Soc. Am. J.* **1996**, *60*, 246–251. [CrossRef]
94. Yadav, B.; Hatfield, K. Stream network conflation with topographic DEMs. *Environ. Model. Softw.* **2018**, *102*, 241–249. [CrossRef]
95. Hösl, R.; Strauss, P.; Glade, T. Man-made linear flow paths at catchment scale: Identification, factors and consequences for the efficiency of vegetated filter strips. *Landsc. Urban Plan.* **2012**, *104*, 245–252. [CrossRef]
96. Beven, K.J.; Kirkby, M.J. A physically based, variable contributing area model of basin hydrology. *Hydrol. Sci. B* **1979**, *24*, 43–69. [CrossRef]
97. Beven, K. TOPMODEL: A critique. *Hydrol. Process.* **1997**, *11*, 1069–1085. [CrossRef]
98. De Vente, J.; Poesen, J.; Arabkhedri, M.; Verstraeten, G. The sediment delivery problem revisited. *Prog. Phys. Geogr.* **2007**, *31*, 155–178. [CrossRef]

Article

Modification of the MONERIS Nutrient Emission Model for a Lowland Country (Hungary) to Support River Basin Management Planning in the Danube River Basin

Zolt Jolánkai *, Máté Krisztián Kardos and Adrienne Clement

Department of Sanitary and Environmental Engineering, Budapest University of Technology and Economics, 1111 Budapest, Hungary; kardos.mate@epito.bme.hu (M.K.K.); clement.adrienne@epito.bme.hu (A.C.)

* Correspondence: jolankai.zsolt@epito.bme.hu; Tel.: +36-1-463-2955

Received: 31 January 2020; Accepted: 10 March 2020; Published: 19 March 2020

Abstract: The contamination of waters with nutrients, especially nitrogen and phosphorus originating from various diffuse and point sources, has become a worldwide issue in recent decades. Due to the complexity of the processes involved, watershed models are gaining an increasing role in their analysis. The goal set by the EU Water Framework Directive (to reach "good status" of all water bodies) requires spatially detailed information on the fate of contaminants. In this study, the watershed nutrient model MONERIS was applied to the Hungarian part of the Danube River Basin. The spatial resolution was 1078 water bodies (mean area of 86 km^2); two subsequent 4 year periods (2009–2012 and 2013–2016) were modeled. Various elements/parameters of the model were adjusted and tested against surface and subsurface water quality measurements conducted all over the country, namely (i) the water balance equations (surface and subsurface runoff), (ii) the nitrogen retention parameters of the subsurface pathways (excluding tile drainage), (iii) the shallow groundwater phosphorus concentrations, and (iv) the surface water retention parameters. The study revealed that (i) digital-filter-based separation of surface and subsurface runoff yielded different values of these components, but this change did not influence nutrient loads significantly; (ii) shallow groundwater phosphorus concentrations in the sandy soils of Hungary differ from those of the MONERIS default values; (iii) a significant change of the phosphorus in-stream retention parameters was needed to approach measured in-stream phosphorus load values. Local emissions and pathways were analyzed and compared with previous model results.

Keywords: MONERIS; nitrogen; phosphorus; diffuse nutrient emission; empirical modeling; river basin management plan of Hungary

1. Introduction

1.1. The Nutrient Problem

Anthropogenic activities such as intensive agriculture and communal and industrial wastewater discharges lead to nutrient over-enrichment in waters. Nitrogen and phosphorus play primary roles in the deterioration of the water quality of rivers and lakes. Many countries have implemented strict measures to protect the status of waters. Two examples are the Clean Water Act in the US and the Water Framework Directive in Europe. According to the latter, EU (European Union) member states are obliged to reach a "good" status of each water body by 2027 at the latest. River basin management plans (RBMPs) prepared in a six year cycle describe both the current status and the measures that must be taken to reach this goal.

In Hungary, the ecological status of surface waters is generally worse than the EU average. In the first RBMP, approximately 6% of the surface water bodies of Hungary were classified as poor, and 35% were classified as moderate based on physicochemical conditions [1,2]. In these water bodies, the inadequate status was predominantly due to the nitrogen and phosphorus concentrations, with a stronger influence of the latter.

Due to the various sources and pathways from/through which nitrogen and phosphorus can reach surface and subsurface waters, it is difficult to choose the optimal mitigation option. Watershed models provide a useful tool with which to tackle this problem [3].

There are several available watershed models with the capacity to model nutrient emissions to surface waters. The scale of models spreads from event-based dynamic plot scale models to static, annual time-scale, basin-scale models [4]. Model review studies have been carried out by multiple authors to compare and assess the accuracy of these models [5–7]. The most widespread models in this field include AGNPS, AnnAGNPS, ANSWERS, CASC2D, DWSM, HSPF, KINEROS, MIKE SHE, PRMS, SWAT, and MONERIS. The review of phosphorus emission models led the authors of the paper to the conclusion that it would be favorable to use models that use parsimonious approaches, where yearly time steps are used with a relatively low number of input parameters combined with a stochastic framework [7]. The EUROHARP project aimed specifically to give a better insight into the similarities and differences among watershed-scale nutrient emission models and stream water quality models (including NL-CAT, REALTA, and NOPOLU, among others) [8]. River retention estimates of the separate approaches were compared, and it was found that the variability of the estimates was large, and therefore it increased the uncertainty of the model predictions. A most recent study compared three frequently used nutrient emission models of the Danube basin for 18 ICPDR (International Commission for the Protection of the Danube River) regions [9]. The study aimed to compare and conclude the results of the models and not to assess the models themselves. The study revealed that all three models (SWAT, MONERIS, GREEN) were capable of estimating yearly nutrient loading; they showed coherent results with each other, but the GREEN model consistently overestimated TP (Total Phosphorus) loads.

The model system MONERIS was developed in the Leibnitz Institute of Freshwater Ecology and Inland Fisheries in Berlin [10]. It is a nutrient emission model developed to achieve three goals: (i) to identify the sources and pathways of nutrient emissions of AUs (spatial units for which input data are aggregated), (ii) to analyze nutrient transport and retention in river systems, and (iii) to create a framework to assess management alternatives for river systems. It has been used for German river catchments [11], for other catchments in Europe (Vistula, Po, Odra) [12] and outside Europe [13], and for the whole Danube catchment [14,15].

1.2. Aims of the Study

Even though a country-scale assessment of phosphorus emissions has already been carried out in Hungary [16], the same is not true for nitrogen. Nitrogen loads have not been estimated with pathway-specific modeling. Neither has an integrated point-/diffuse-source emission model been applied for either of the substances.

The MONERIS model was chosen for our study as it has been previously used and validated in the Danube Basin, including Hungary, and because it seemed to have a balance between model complexity and data demand, as was also concluded by other authors [17]. Even though the MONERIS model has been used in many river basins, it has to be stressed that this is a semi-empirical model, with tens of parameters set by data collected in Germany and wider Europe. They should be applied with caution in regions where the hydrological, hydrogeological, soil conditions, land use structures, and infrastructures characterizing of the calibration area are significantly different [17].

This study aimed to

- determine the degree to which the model system MONERIS is capable of estimating nutrient fluxes in a lowland country such as Hungary;

- identify the model components (equations, parameters) that had to be adjusted to better describe processes in the study area;
- give country-wide, waterbody scale estimates for nitrogen and phosphorus loads of various pathways.

2. Materials and Methods

2.1. Study Area

The analysis was carried out across the whole area of Hungary, which is located within the Carpathian Basin and has a total area of 93,000 km^2. The country is dominated by two flat, lowland areas with an average altitude below 200 m a.s.l. (the Alföld in the southeast, and the Kis-Alföld in the northeast). There are also several hilly and mountainous regions (altitudes still below 1000 m a.s.l.) in the central western, southern, and the northeastern parts of the country. The country lies entirely within the Danube Basin, and it can be divided into four subcatchments: the Danube, Tisza, and Drava River catchments, and Lake Balaton subcatchment (Figure 1). The Tisza River subcatchment covers the largest area of the country, while Lake Balaton is the most important standing water in the country. The average runoff rate is small (10–50 mm in the lowlands, and 50–200 mm in the hilly/mountainous parts [18], which is due to the relatively low precipitation, high evapotranspiration, and small reliefs, while the inflowing river flows are high (~1200 mm) [19]. Besides the natural river bodies, it is relevant to mention that the lowlands of Hungary are drained by an extensive network of artificial channels (total length is approx. 42,000 km) to protect the farmlands from excess water.

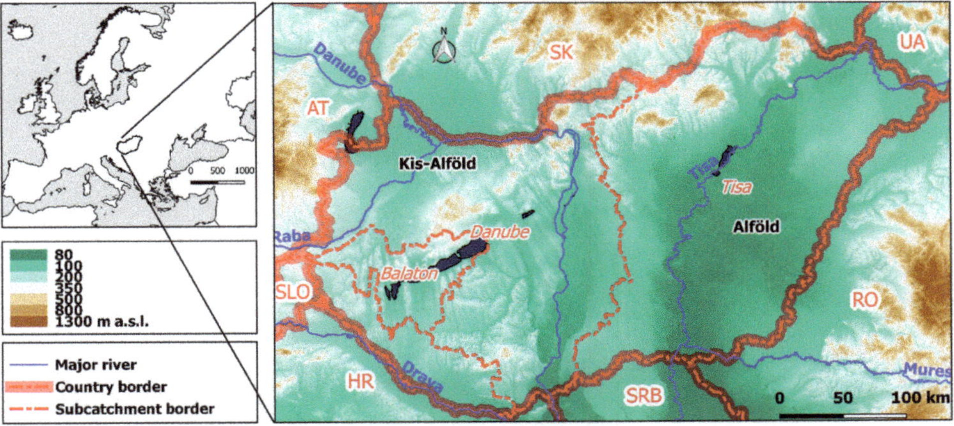

Figure 1. Overview of Hungary and its main rivers and subcatchments.

Hungary lies in the temperate climatic zone and its climate is very erratic, as three different climates have a strong effect on it: the oceanic climate; the dry continental climate, with extreme temperatures; and the Mediterranean climate, with dry summers and wet winters. Any of these can rule the climate of the Carpathian Basin for shorter or longer periods within the year [20]. The long-term average precipitation in Hungary is between 500 and 750 mm with a strong spatial variability, with the wettest region in the southwest and the driest in the central eastern part of the lowland regions. The most precipitation occurs in the mountainous regions throughout the country. Within this spatial distribution, the Drava and Danube catchments receive significantly more precipitation than the Tisza catchment in the eastern part of Hungary [20]. The wettest season is the summer (June being the wettest with around 70 mm) and the driest season is the winter (below 30 mm in February). The yearly average annual precipitation is tending towards a slight decrease, with a 10% decrease in the last century. Extremes both in precipitation and temperature are frequent phenomena, often occurring within a single year in the same region (e.g., the year 2000).

In the upmost 10 m, the country is predominantly (ca. 85%) covered by loose sediments, mostly loess and sand, while clay also makes up a significant proportion. Solid rocks are only present in small mountainous areas, in the northeastern part of the country and near the Balaton, while porous limestone regions are slightly more frequent in the mountainous regions. The majority of the country (especially the lowlands) is covered by good, fertile soils suitable for agricultural production [21].

2.2. Model Description

MONERIS is an empirical, spatially semi-distributed, time-aggregated catchment model. The smallest spatial unit of calculation is an *administrative unit* (AU), which comprises one or more hydrologically interrelated river stretches and their direct catchment area. Since only a tiny part of the AU is monitored (either for water quantity or for quality), the term *catchment* is used throughout this paper for the catchments of monitoring points comprising one or more conterminous, hydrologically interrelated AUs.

In this section, only those parts of the model that were investigated or adjusted for the Hungarian conditions are described. For further information on the model's structure and equations, the reader is referred to Reference [22], which is also the source for the relationships described below.

2.2.1. Flow Components/Water Balance

The MONERIS model assumes seven main water pathways, according to Equation (1).

$$Q = Q_{atm} + Q_{surf} + Q_{sm} + Q_{tile} + Q_{gw} + Q_{urb} + Q_{ww} \quad (1)$$

where Q. is the total runoff in the AU (model input), Q_{atm} is runoff from precipitation falling directly on open water surfaces, Q_{sm} is runoff from snowmelt, and Q_{gw} is the natural groundwater flow. Q_{urb} is runoff (originating from precipitation + wastewater) from unsewered urban areas and Q_{ww} is wastewater discharge (model input). Q_{surf} is the surface runoff and Q_{tile} is tile drainage flow. Flow is expressed in m^3/s.

$$Q_{surf} = k_{w1} \cdot Q^{k_{w2}} \quad (2)$$

$$Q_{tile} = A_{tile}(k_{w3} \cdot P_s + k_{w4} \cdot P_w) \quad (3)$$

where A_{tile} is the area of tile-drained agricultural fields; P_s and P_w are summer and winter precipitation (mm) values, respectively; and k_{w1}, k_{w2}, k_{w3}, and k_{w4} are model constants (-). Q_{gw} is calculated as the difference between Q and the six other terms.

$$Q_{gw} = Q - \left(Q_{atm} + Q_{surf} + Q_{sm} + Q_{tile} + Q_{urb} + Q_{ww}\right) \quad (4)$$

2.2.2. Subsurface Pathway of Nitrogen and Phosphorus

Concerning the subsurface fate of nitrogen, the MONERIS approach separates four hydrogeological types: unconsolidated porous media with deep or shallow groundwater, and consolidated rock with good or poor permeability. The equation for the calculation of groundwater nitrate levels is as follows.

$$C_{GW} = \left(\sum_{i=1}^{4} \frac{1}{1 + k_{n1i} \cdot R^{kn2i}} \cdot \frac{A_{HG,i}}{A_{AU}} \right) \cdot (C_{LWPOT})^{kn3} \quad (5)$$

where

C_{GW} is the nitrate–nitrogen concentration of groundwater (mg N L^{-1}), $A_{HG,i}$ is the area of the catchment of hydrogeological class i (km^2), R is the long-term average recharge of the catchment (mm y^{-1}), A_{AU} is the size of the AU (km^2), C_{LWPOT} is the potential nitrate–nitrogen concentration of the leachate (mg L^{-1}), $k_{n1i,}$ and k_{n2i} are model constants varying by hydrogeological type, and k_{n3} is the model coefficient for denitrification.

Regarding subsurface phosphorus, the following five soil types are distinguished by the model: sandy soils, clayey soils, fen soils, bog soils, woodland, and open areas.

$$C_{GWAG}^{SRP} = \frac{\sum_{i=1}^{4} C_{GWi}^{SRP} \cdot A_i}{\sum_{i=1}^{4} A_i} \quad (6)$$

where C_{GWAG}^{SRP} is the groundwater soluble reactive phosphorus (SRP) concentration for agricultural land (mg P L^{-1}); C_{GWi}^{SRP} is the groundwater SRP concentration for sandy, loamy, fen, and bog soil types defined as model constants; and A_i is the area of sandy, loamy, fen, and bog soils (km^2). In a second step, the average SRP concentrations in groundwater are calculated as an area-weighted average of agricultural and non-agricultural areas.

$$C_{GW}^{SRP} = \frac{C_{GWAG}^{SRP} \cdot A_{AG} + C_{GW\ WOOP}^{SRP} \cdot A_{WOOP}}{A_{AG} + A_{WOOP}} \quad (7)$$

where C_{GW}^{SRP} and A_{WOOP} are the groundwater SRP concentration and area of woodlands and open areas, respectively, and C_{GW}^{SRP} is the groundwater SRP concentration (mg P L^{-1}).

2.2.3. Nutrient Retention in Tributaries

In the MONERIS model, in-stream concentration reducing processes such as settling, denitrification, etc. are considered to be aggregated under the term "retention". Retention coefficients are calculated as a function of the hydraulic loads, mean water temperature, and area-specific discharge (Equations (8)–(10)).

$$R_{TN} = \frac{1}{1 + k_{r1} \cdot e^{\mathrm{kr2 \cdot T}} \cdot HL^{kr3}} \quad (8)$$

$$R_{DIN} = \frac{1}{1 + k_{r4} \cdot e^{\mathrm{kr5 \cdot T}} \cdot HL^{kr6}} \quad (9)$$

$$R_{TP} = \frac{1}{1 + \mathrm{k}_{r7} \cdot q^{kr8}} + \frac{1}{1 + \mathrm{k}_{r9} \cdot HL^{kr10}} \quad (10)$$

where R_{TN} is the retention coefficient for TN, k_{r1} to k_{r3} are the retention parameters for TN, k_{r4} to k_{r6} are the retention parameters for DIN (dissolved inorganic nitrogen), HL is the hydraulic load for the water body (river flow/water surface, m y^{-1}), T is the yearly average water temperature (°C), R_{TP} is the retention coefficient for TP, k_{r7} to k_{r10} are the retention parameters for TP, and q is the area-specific runoff (l s^{-1} km^{-2}).

Validation of the model is generally achieved using surface water quality monitoring points, where sufficient data are available both for flow and water quality parameters. In the current case, total nitrogen and total phosphorus measurements were the critical parameters for validation of the nutrient emission estimation, as their measurements are generally scarce both spatially and temporally.

2.2.4. Modeling Environment

The calculation of local emissions according to the equations described in the model version 2.14.1 [22] was carried out in a Microsoft Excel environment [17]. For calibration purposes, the cumulative river load calculations along the river hierarchy and the river retention were calculated in the MATLAB environment. MATLAB and R [23] were used to process water quality data. The ArcMap 10.1 model version was used for the preparation of spatial data. This software, along with ArcMap and QGIS, was used to visualize the results.

2.3. *Preparation of Model Input Data*

In the present study, the MONERIS model was applied to the Hungarian part of the Danube River Basin for two subsequent 4 year periods, 2009–2012 and 2013–2016. However, the input data—depending on availability—comprised shorter or longer periods. Only some of the input data were updated between the two periods.

2.3.1. Delineation of AUs and Calculation of Their Basic Characteristics

Extensive use of spatial data was necessary to initialize the model. Model AUs were water bodies determined by the RBMP of Hungary. This meant a total of 1078 catchments, 189 of which belonged to lake water bodies, and the rest of which belonged to river water bodies. The average size of the catchments was 83 km^2 (median was 51.7 km^2) and the maximum size was 1166 km^2 (Lake Balaton), the smallest being only 0.3 km^2.

Concerning hydrometeorology, the National Meteorological Service provided long-term mean precipitation values for the years 1981 to 2010 in a grid-based format. Monthly actual evapotranspiration for the period 2000 to 2009 was also involved [24,25]. Daily precipitation and mean temperature data for 245 stations were provided by the General Directorate of Water Management, Hungary, for the period 1991–2016. These data were interpolated using inverse distance-weighted interpolation and kriging.

Topographical properties such as elevation and slope were calculated by processing digital elevation data [26]. Water network characteristics were calculated based on the River Water Body and River Water Segment spatial databases [27]. Land use data were included based on the Corine Land Cover Database [28]. Soil hydrogeological and topsoil physical properties were processed using the AGROTOPO [29] and DOSOREMI [30] databases.

2.3.2. Runoff

As Hungary lacks a country-wide hydrological model, runoff values were estimated using the following five-tiered approach. For gauged headwater AUs, time series' mean values were used. For ungauged AUs of gauged catchments, long-term mean flow at the catchment outflow point was spatially weighted by the difference between long-term mean precipitation (P) and evapotranspiration (ET). On some AUs (especially the ones along the Tisza River), long-term values of evapotranspiration exceeded those of precipitation; for such catchments, long-term mean flow was spatially weighted by long-term mean runoff values [31]. On ungauged catchments, where both sandy and limestone areas occupied less than 50% of the catchment area, the runoff was estimated by establishing a linear regression between model period mean flow of gauged catchments and their channel length values. For AUs where neither of the above methods were applicable, long-term mean runoff values [31] were corrected by the proportion of the investigated periods and the long-term rainfall.

2.3.3. Soil Loss and Nutrient Surplus

The universal soil loss equation [32] was used for the mean annual soil loss estimation in each of the AUs. C and R factors were determined based on the European scale JRC maps [33] and K factor maps were prepared by the Research Institute of Soil Sciences and Agricultural Chemistry [34]. L and S factors were determined from a 50 m resolution hydrologically corrected digital elevation model [35].

Yearly nutrient balance estimations were calculated for the period 1961–2016, according to the OECD (Organisation for Economic Co-Operation and Development) method used to calculate gross nutrient balance based on county-scale agro-statistical data on fertilizer inputs, harvested yields, and animal husbandry [36,37]. Soil nutrient conditions for each county were allocated to the catchments directly (Figure 2).

For a detailed description of the data sources and their processing, see Tables A1 and A2 in the Appendix A.

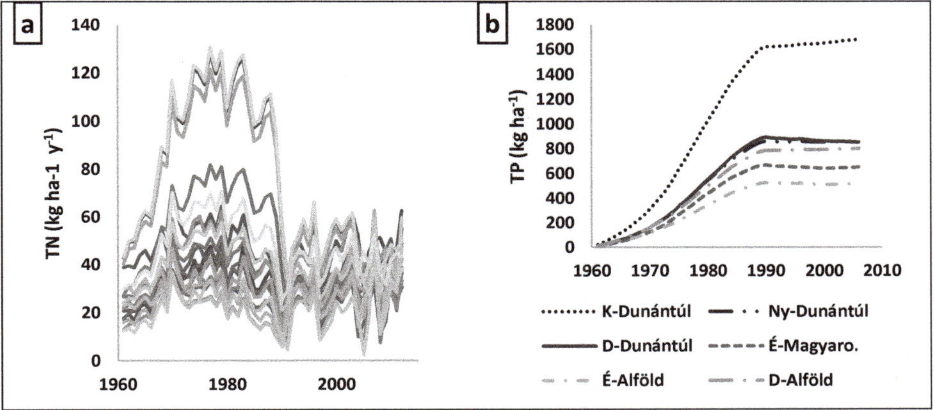

Figure 2. Total nitrogen (TN) surplus by the Hungarian counties (each county is represented by one line) (kg ha^{-1} y^{-1}) (**a**) and accumulated total phosphorus (TP) by the Hungarian regions (kg ha^{-1} É,D,K,NY stand for north, south, east and west) (**b**).

2.3.4. Nutrient Emission Pathways

Concerning atmospheric deposition, the EMEP (European Monitoring and Evaluation Programme) model results were used for oxidized and reduced nitrogen compounds [38]. Values increased from southeast to northwest from 360 to 570 mg N/m^2 for NO_x, and from 460 to 850 mg/m^2 from east to west for reduced nitrogen forms. Values for the 2013–2016 period were around 3% higher, presumably due to the higher precipitation values.

Concerning point-source loads, the national wastewater information system [39] data were processed to ascertain wastewater discharge and emission loads. Each operating wastewater treatment plant's (WWTP) emissions were linked to the water body it discharges into. Discharge, TN, and TP load values were used. There was a reduction in country-wide emission values between the two modeled periods: 9000 t y^{-1} to 7384 t y^{-1} and 1028 t y^{-1} to 852 t y^{-1} for total nitrogen and total phosphorus loads, respectively.

Urban diffuse emissions estimations require extensive information about the separate and combined sewer network, the connected population, etc. Detailed data were not available for the whole country; only county-scale rough estimates were available from the National Water Directorate and the Hungarian Central Statistical Office (KSH). Information about inhabitants connected to sewer systems and WWTPs was available from the national sewage information system [39].

2.4. *Water Quality Data Used for Calibration and Validation Purposes*

2.4.1. Groundwater Well Data

To support the recalibration of nitrogen retention parameters (Section 2.2.2) and to review the average phosphorus levels in groundwater, a monitoring database of the shallow groundwater was processed. Water quality data were gathered for the period 2004–2018 for the whole country. Wells within 500 m of artificial areas and within 1500 m of larger rivers were excluded from the analysis (Figure 3). The rest of the wells were overlapped with the Corine Land Cover Map [28], the map of topsoil USDA classes [30], the hydrogeological type [29], and the groundwater depth map [40]. Using the spatial information of the maps, wells were classified into four nitrogen and four phosphorus categories (Tables 1 and 2).

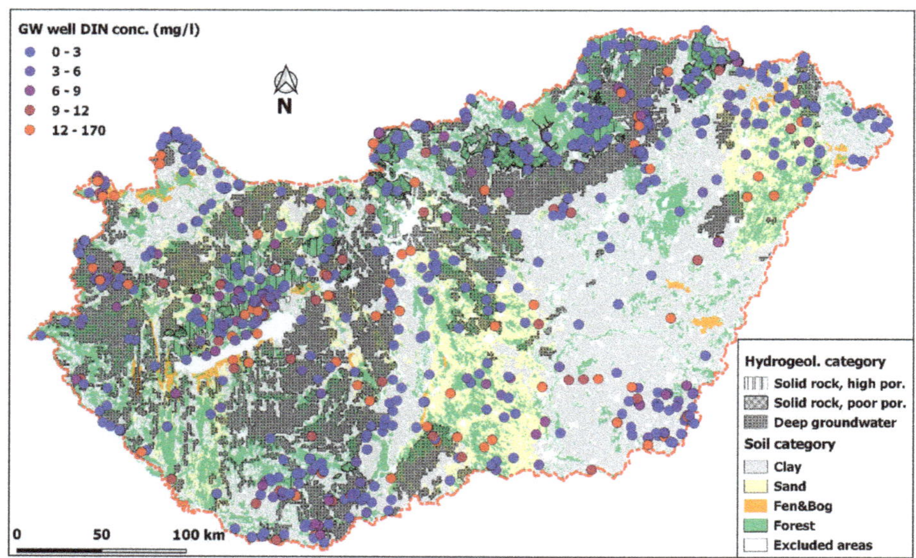

Figure 3. Shallow groundwater wells used for model calibration and geological and soil types according to MONERIS classification (GW – Groundwater, DIN – dissolved inorganic nitrogen).

Table 1. Well categories for nitrogen.

Well Category for Nitrogen	Groundwater Depth	Soil Geology	Number of Wells
Unconsolidated rock, shallow groundwater	<3 m *	any other	1022
Unconsolidated rock, deep groundwater	>3 m *	any other	55
Solid rock, high porosity	any	limestone	69
Solid rock, impermeable	any	granite, andesite	36

* the horizon for the division between the two groundwater depth classes was not defined in the literature. For this calculation, the threshold was set as 3 m.

Table 2. Well categories for phosphorus. SL = sandy loam; LS = loamy sand.

Well Category for Phosphorus	Land Cover	Soil Texture	Soil Physical Category	Number of Wells
Sandy soils	agricultural	SL, LS, or sand	not peat or mull	81
Clayey soils	agricultural	any other	not peat or mull	482
Fen & bog soils	any	any	peat or mull	5
Woodland & open land	forest & seminatural	any	not peat or mull	250

Time series mean DIN and PO4-P concentrations were calculated for each well. Category mean and median values were plotted on the histogram, and category median values were used for further processing.

2.4.2. Surface Water Quality

The monitoring concept in Hungary, but also in other countries in Europe, uses different ways to monitor general water quality status (averages), to identify concentration maxima at point-source discharges or to collect information about the state of smaller river sections. For the latter, monthly measurements are organized for a sequenced (1 year) period, and then repeated in the next period of revision of the water quality of the rivers. Besides this, there are other shifts of the monitoring points according to the practice in Hungary. The frequency of monitoring changes along time at other points, which means a change in the reliability of load estimation.

For calibration and validation purposes, hydrological data were obtained from the National Hydrological Database (MAHAB). Water quality data from the surface water module of the National Environmental Information System (OKIR–FEVISS) were also processed and utilized. Monitoring points for the current analysis is shown on Figure 4.

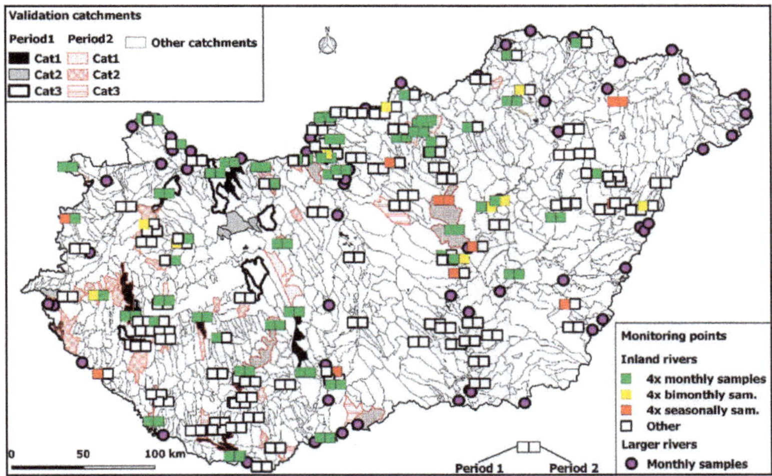

Figure 4. Surface water flow and quality monitoring stations.

2.5. Adjustments of the Model Parameters and Structure

2.5.1. Runoff Separation with Digital Filter

As surface runoff and groundwater flow are characterized by different concentrations of nitrogen and phosphorus, it is important to estimate their proportions as accurately as possible. In the MONERIS model, the groundwater flow was estimated using Equation (4). This equation is the result of subtracting all other flow components from the total AU runoff (model input), many of which are calculated with empirical formulas (see Equations (2)–(3)). Due to errors in the input data and the estimation formulas, the estimated groundwater flow can be unrealistically small.

In this study, groundwater flow proportion values calculated by the MONERIS equations were compared to those calculated by the digital-filter-based method [41] developed by Arnold [42]. For this purpose, catchments with a flow gauge (daily discharge data) described by monthly TN/TP measurements and consisting of fewer than three AUs were selected. For the sake of comparison, atmospheric runoff and urban runoff were considered to be part of the surface runoff component, while tile drainage and point-source discharge were considered to be part of the baseflow. The aim of this comparison was to get a view of the model's sensitivity of the nutrient loads to the division of surface and subsurface waters.

2.5.2. Groundwater Nitrogen Concentration

Parameter calibration was carried out against median DIN (Dissolved Inorganic Nitrogen) concentrations (mg N L^{-1}) measured in the wells. As the accuracy and especially the spatial resolution of nitrogen surplus of agricultural sites are very poor, calibration simply aimed for the improvement of catchment groundwater nitrate levels against groundwater well averages. In the calibration process, the AUs were used as the domain for input data aggregation. This means that the calculation of groundwater recharge, surplus, and leachate values are carried out for the AUs. The AUs with groundwater well data were selected and included in the calibration dataset. Over 300 AUs were included and they covered all of the geological categories, but most AUs had varying geology in their area; therefore all of the geological categories were calibrated simultaneously. The parameter fitting was carried out with the generalized reduced gradient (GRG2) non-linear optimization method [43]. The objective function for parameter fitting was the sum of square errors.

2.5.3. River Retention

Model improvement was carried out through the implementation of parameter fitting of the river nutrient retention parameters (kr_1–kr_{10}, Equations (8)–(10)) for all nutrient components described in the model. Optimization of the parameters was done with the Matlab software "fmincon" function, using the interior point algorithm [44]. For DIN and TP, the objective function was the sum of mean square error, while for TN the sum of relative errors was used. Calibration points were classified into three categories and all categories were taken into account with equal weights.

2.6. Model Validation

As Hungary is a "downstream country", many of the available monitoring stations are placed on large rivers with transboundary catchments (the Danube, the Tisza, and the Drava, to name just the biggest ones) that are of no use if the aim is to calibrate or validate processes happening inside the country. Only monitoring stations for which the entire catchment falls within the country borders can be considered for validation. The available monitoring data were divided into three classes based on the reliability of yearly average loads calculated from the time series.

The most reliable class contained the data from stations with monthly sampling regimes (at least 40 measurements throughout each of the 4 year periods). The second group was formed from the data with at least bi-monthly sampling frequencies, whereas the third group contained the data from stations with seasonal measurements. Considering the fact that the number of stations represented only 5.75% of the total (modeled) catchments, the number of stations with adequate data was very limited (Table 3).

Table 3. The number of monitoring stations per uncertainty category and model period. Per. 1: 2009–2012; Per. 2: 2013–2016.

Category	Yearly Sample Number	Per. 1	Per. 2	Total (w/o Overlap)
Cat. 1	10+	8	16	20
Cat. 2	6–9	16	18	32
Cat. 3	4–5	9	9	18
Cat. 1–3		33	43	62

Weekly sampling is only carried out at one of the studied stations, at the lower section of the Zala River, which is the primary tributary to Kis-Balaton wetland and Lake Balaton. For this reason, the Zala River catchment was the most important validation catchment in this study, while other mid-sized catchments with lower sampling size followed, such as Kapos, Zagyva, Babócsai-Rinya, Fekete-Víz, Marcal, and Lónyai-csatorna catchments. These catchments also collect daily flow monitoring data that were helpful to increase the accuracy of the yearly average estimations.

In model validation, the annual average loads (t/y) were compared to the loads calculated by the model. The scale of the catchments and therefore the loads also varied with several orders of magnitude. This gave a good indication of the model's validity.

3. Results

3.1. Flow Components/Water Balance

Most of the AUs fell in measured catchments and thus were assigned to Tier 1 (43 AUs) or Tier 2 (537 AUs). In Tier 3, 208 AUs were identified. The regression equation (Tier 4, 276 AUs) was mainly used along the larger rivers: Danube, Drava, Upper Tisza, Rába, and Nádor Channel (Figure 5a). The relationship (Equation (11)), Figure 5b) proved to be indicative ($R^2 = 0.93$, $p < 0.0001$).

$$Q = 0.003 \cdot L - 0.030 \tag{11}$$

where Q is the flow ($m^3\ s^{-1}$) and L the channel length (km). Tier 5 was used only with a 14 AUs. The average difference between the actual period and long term mean runoff was −2.8 and 27.6 mm^{-s} for Periods 1 and 2, respectively.

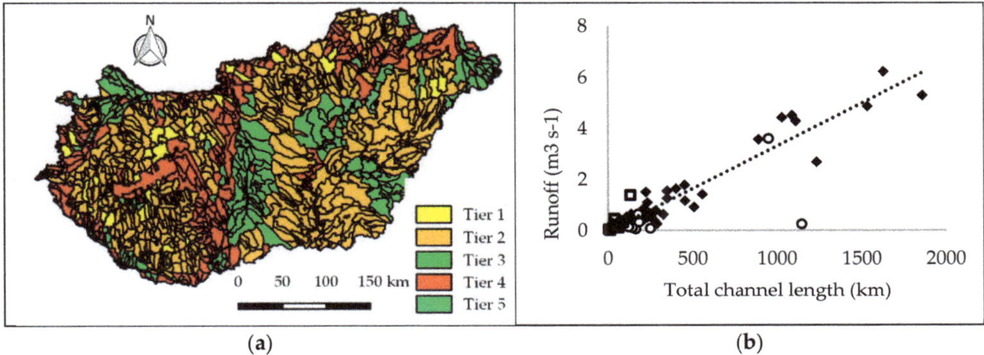

Figure 5. (**a**) Methods used for runoff estimation throughout the country. (**b**) measured average annual runoff vs. total channel length on the catchment. Hollow rounds: outliers from sandy catchments; squares: karstic areas.

The groundwater flow proportions calculated using different approaches showed strong similarities (Table 4). According to the MONERIS method, mean groundwater flow (total subsurface flow) share values were 0.73, 0.72, and 0.78 for the slope categories 0%–1%, 1%–5%, and >5%, respectively. According to the digital filter method, mean baseflow index values were 0.79, 0.68, and 0.64 for the same slope categories, respectively. Because surface runoff is higher on areas with higher slope (e.g., see Reference [45]), the digital-filter-based groundwater flow separation provided a more realistic result (where measured data were available), and it was selected for use in the model calculations.

Table 4. Comparison of surface runoff and baseflow index values calculated for gauged catchments as determined by the MONERIS (MON) and the digital filter (DF) methods.

	Catchment		Total	Surface Runoff		GW Flow Share	
Catchment Name	**Area**	**Slope**	**Runoff**	**MON**	**DF**	**MON**	**DF**
	(km^2)	**(%)**	**($m^3\ s^{-1}$)**	**($m^3\ s^{-1}$)**	**($m^3\ s^{-1}$)**	**(-)**	**(-)**
Arany Creek	36	1.9	0.33	0.022	0.043	0.85	0.70
Kenyérmezei Creek	125	11.3	0.15	0.000	0.000	0.79	0.76
Kígyós Channel	594	0.8	0.25	0.010	0.002	0.73	0.78
Tapolca Creek	51	4.1	0.28	0.061	0.000	0.84	0.85
Tetves Creek	88	8.2	0.18	0.022	0.034	0.84	0.69
Torna Creek	176	7.2	0.56	0.099	0.035	0.84	0.82
Únyi Creek	172	9.2	0.29	0.019	0.077	0.73	0.62
Villány–Pogányi c.	202	5.6	0.26	0.054	0.077	0.80	0.62
Zagyva Creek (upper)	168	14.1	0.94	0.112	0.439	0.84	0.48
Average	179	6.9	0.36	0.04	0.08	0.81	0.70

3.2. Groundwater Pathway of Nutrients

Groundwater nitrogen and phosphorus concentrations showed a large variation between the wells. The distribution of the concentrations showed that the vast majority of the wells were within a narrow concentration range of relatively low concentration values (except soils with a small number of wells, Figure 6). It was assumed that higher concentrations represented local extremities due to contamination from unregistered sources. Due to their small number, their impact on surface water loads might be negligible. For this reason, median values (as opposed to the mean values) were used in application of the model.

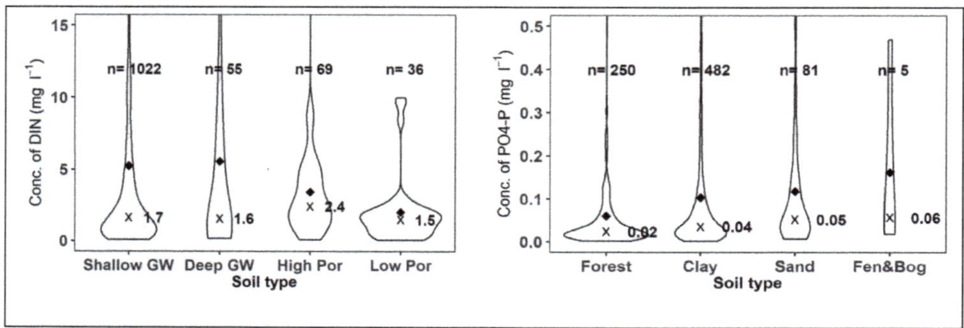

Figure 6. Kernel density plot showing the distribution of the measurements of dissolved inorganic nitrogen (**left**) and orthophosphate (**right**) concentrations measured in groundwater wells. Diamond: mean value, x: median value.

The calibration of groundwater nitrogen retention parameters resulted in new values (Table 5) that balanced out subsurface pathways (Figure 7) in Hungarian regions from hilly to mountainous parts of the country. The calibration resulted in a more robust estimation throughout the load scales in the river monitoring points, especially in the higher load range (Figure 7, Table 6). In the lower load range, slight overestimation was observed, while at the higher load values, some underestimation was still present.

Values for forests and open spaces were as advised by the original model dataset based on German monitoring results. For sandy agricultural soils, the values found in Hungarian wells were lower compared to the original model data (0.05 versus the original value of 0.1 mg L^{-1}). For clayey agricultural soils, the data were also identical to the original data. Several studies have stated that bog soils have higher nutrient values due to their high organic matter content [10]. The low number of wells situated in fen/bog soils of Hungary did support this statement. Due to a lack of representativeness, original P concentrations were not changed.

The adjustment of the groundwater nitrogen retention parameters improved the model's accuracy in terms of the coefficient of determination and in terms of the relative errors (Figure 7a,b, Table 6).

Table 5. Model constants before and after adjustment. The letters in the brackets correspond to those of chapter 2.1. Original: according to [21]; Adjusted: as considered in the present study. UC = unconsolidated; GW = groundwater.

Process/Constant Name	Soil Category	Units	Original	Adjusted
Subsurface Nitrogen				
Nitrogen constant 1 (k_{n1})	UC rock, shallow GW	-	2752	84.24
	UC rock, deep GW	-	68,560	7917
	Solid rock, high porosity	-	60.23	67.33
	Solid rock, poor porosity	-	78.54	99,787
Nitrogen constant 2 (k_{n2})	UC rock, shallow GW	-	−1.540	−1.216
	UC rock, deep GW	-	−1.959	−3.750
	Solid rock, high porosity	-	−0.903	−1.124
	Solid rock, poor porosity	-	−0.662	−2.747
Denitrification in topsoil (k_{n3})	any	-	0.6368	0.4340
Subsurface Phosphorus				
P conc. in groundwater C_{GW}^{SRP}	Sandy agricultural soils	mg L^{-1}	0.10	0.05
	Clayey agricultural soils	mg L^{-1}	0.03	0.03
	Fen agricultural soils	mg L^{-1}	0.10	0.10
	Bog agricultural soils	mg L^{-1}	0.50	0.50
	Woodland and open areas	mg L^{-1}	0.02	0.02
River Retention				
TN constant 1 (k_{r1})		-	4.74	78.8
TN constant 2 (k_{r2})		-	0.067	−0.31
TN constant 3 (k_{r3})		-	−1	−0.53
DIN constant 1 (k_{r4})		-	8.58	14.39
DIN constant 2 (k_{r5})		-	0.067	−0.06
DIN constant 3 (k_{r6})		-	−1	−0.51
TP constant 1 (k_{r7})		-	5.07	200
TP constant 2 (k_{r8})		-	−1	−9.69
TP constant 3 (k_{r9})		-	25.74	4.87
TP constant 4 (k_{r10})		-	−1	−0.89

Table 6. Calibration results for groundwater nitrogen retention parameters.

	R^2		Absolute Relative Error	
	Original	**Adjusted**	**Original**	**Adjusted**
Cat. 1	0.78	0.95	52%	21%
Cat. 2	0.74	0.92	23%	31%
Cat. 3	0.86	0.98	82%	17%
Cat. 1–3	0.48	0.95	46%	25%

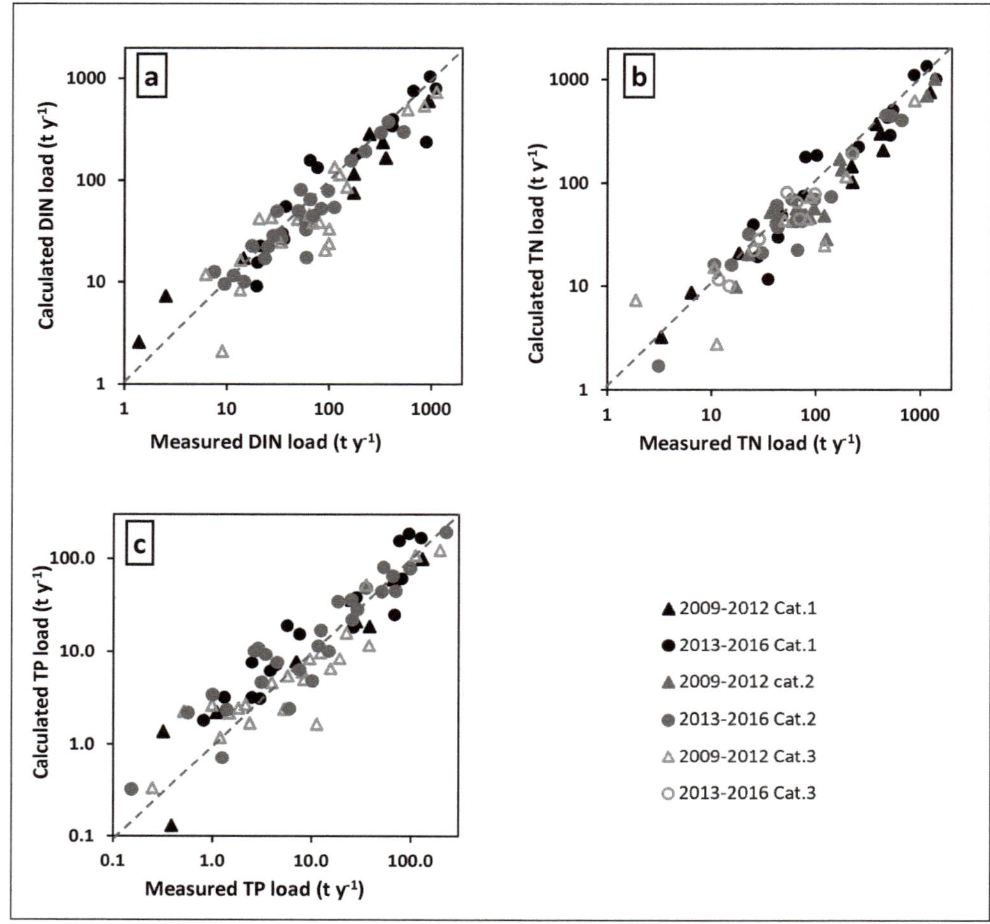

Figure 7. Measured and calculated loads at all monitoring stations. (**a**,**b**,**c**) shows the validation results of DIN/TN and TP loads respectively.

3.3. River Nutrient Retention

The calibration of TN retention parameters resulted in a 1%–3% improvement of the sum of relative errors (Figure 7). Since these were rather small numbers, no strict recommendation can be made to use the adjusted retention parameters. In the case of DIN, the change in the retention was larger, but there was also very little improvement in the model accuracy (R^2 improved from 0.81 to 0.82 for the whole dataset). For TP, calibration of the retention parameters resulted in a significant improvement of model efficiency (Figure 7). Root mean square error changed from 1.12 to 0.67, from 0.88 to 0.48, and from 0.42 to 0.57 for load classes <5, 5–20, and >20 t y^{-1}, respectively. As a mean of all load categories, root mean square error changed from 0.81 to 0.57.

3.4. Model Validation

The validation of the model showed a relatively good agreement for all compounds (Figure 7, Table 7). The Category 3 validation dataset showed a larger relative error for all compounds except in the second period. For TN and DIN, the highest load range showed a slight underestimation. For TP, the lower load range showed a stronger overestimation of loads, and the overall goodness of the estimation was slightly worse than for TN and DIN.

Table 7. Validation results of all compounds for the two periods.

	Per. 1			Per. 2			Per. 1 + Per. 2		
Category	**DIN**	**TN**	**TP**	**DIN**	**TN**	**TP**	**DIN**	**TN**	**TP**
Square of Pearson correlation coefficient (R^2)									
Cat. 1	0.93	0.94	0.95	0.77	0.89	0.78	0.79	0.85	0.74
Cat. 2	0.99	0.98	0.93	0.89	0.94	0.85	0.96	0.97	0.89
Cat. 3	0.98	0.99	0.85	0.94	0.94	0.66	0.97	0.98	0.72
Cat. 1–3	*0.96*	*0.97*	*0.91*	*0.86*	*0.92*	*0.76*	*0.9*	*0.93*	*0.78*
Mean relative error									
Cat. 1	0.76	0.35	1.18	0.49	0.48	1.04	0.6	0.44	1.09
Cat. 2	0.48	0.39	1.01	0.41	0.35	1.41	0.44	0.37	1.25
Cat. 3	3.13	1.07	4.36	0.37	0.26	1.14	2.23	0.78	3.19
Cat. 1–3	*1.45*	*0.6*	*2.18*	*0.43*	*0.36*	*1.2*	*1.09*	*0.53*	*1.84*

4. Discussion

4.1. Comparison of Nutrient Load Estimation Results on Gauged Catchments with Different Baseflow Separation Methods, Using the MONERIS Model

Baseflow indices (BFI, the ratio of the average baseflow to average total runoff) were close to each other in seven of the sub-catchments (<5% diff.), while in seven cases, the MONERIS method overestimated BFI by over 10%, and in two catchments it underestimated BFI by over 10%. The total average BFI was calculated to be 0.78 by MONERIS, while it was estimated to be 0.71 by the digital filter method.

Even though the baseflow indices calculated by the two methods differed significantly, the loads in the river at the monitoring points differed by less than 4% for all of the substances. For nitrogen, this can be explained by the fact that concentrations in the surface and subsurface pathways did not differ significantly. In the case of total phosphorus, this can be explained by the small share of these two pathways combined in total diffuse emissions (less than 15% combined).

Differences were more substantial with regard to the distribution of the emissions via the surface and subsurface pathways (Table 8, Figure 7).

Table 8. Total nitrogen and total phosphorus loads in unique pathways before and after the correction of the groundwater flow values.

	Surface Runoff		Groundwater		Total Diffuse
	($t\ y^{-1}$)	**Ratio to Total (%)**	**($t\ y^{-1}$)**	**Ratio to Total (%)**	**($t\ y^{-1}$)**
Total nitrogen					
Original	16	*3.4*	242	*52*	468
Adjusted	52	*12*	185	*41*	447
Difference (%)	*69*		*31*		*4.7*
Total phosphorus					
Original	1.6	*2.6*	5.7	*9.0*	63
Adjusted	5.4	*8.3*	4.1	*6.3*	65
Difference (%)	*70*		*36*		*3.5*

For nitrogen, the share of surface runoff in total diffuse nitrogen emissions increased from 3.4% to 11.7%, while groundwater share dropped around 10%. This finding raised some concern regarding the surface runoff pathway as it leveled up with urban diffuse emissions (56.2 and 46.9 $t\ y^{-1}$ respectively), and combined with agricultural erosion, it almost leveled up with groundwater.

For total phosphorus, the share of surface runoff increased from 2.6% to 8.3%, while the share of groundwater decreased from 9% to 6.3%. This was not a dramatic change, especially given that agricultural erosion dominates the pathways followed by urban runoff emission. According to these results, however, dissolved phosphorus in agricultural runoff was three times more important than the original model results indicated. Agricultural runoff would have been even more significant if agricultural soils were closer to phosphorus saturation, because dissolved P levels in surface runoff increase nonlinearly with P saturation of soils [15,46].

It is evident therefore that the right baseflow index or the right portion of surface and subsurface runoff is important for the determination of the share of surface and subsurface emission pathways. A more accurate estimate for this as made here by using digital filters directly where possible, e.g., where measured runoff was available at the outlet of an analytical unit. This would improve the accuracy of the load estimates and possibly highlight the need for measures to be taken to control surface runoff and erosion on agricultural areas.

4.2. Subsurface Processes

According to the results, nitrogen retention in subsurface pathway is higher in high porosity consolidated rocks and lower in unconsolidated rocks with shallow groundwater than in previous studies (Figure 8). The latter can be caused due to the very low runoff from lowland parts of the country. In the Alföld area (Figure 1), there is a concentration increase due to two separate phenomena: (1) the concentration increase due to the re-evaporation of groundwater and (2) the higher background concentrations in the groundwater due to geochemical reasons.

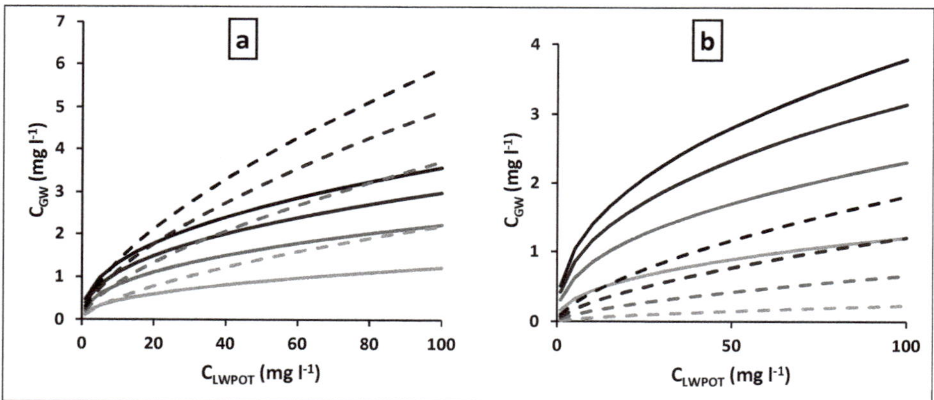

Figure 8. Groundwater nitrogen concentration vs. potential leachate concentration and their dependency of annual recharge using old and new retention parameters. (**a**) Consolidated rocks with high porosity; (**b**) unconsolidated rock with shallow groundwater. Dashed line: original parameters; continuous: adjusted parameters. Recharge values 10, 20, 30, 40 mm y^{-1} from lighter to darker colors.

4.3. River Retention

Total phosphorus load values were systematically underestimated at all scales with the original parameters. Possible causes might have been the underestimation of the emissions (possibly higher loads from the agricultural areas) or disregarding of the internal loads from river sediments. The latter is well-known in shallow lakes [47] and can be observed after external load reduction [48]. Generally, phosphorus retention is regulated by a variety of physical, chemical, and biological factors, by the physical processes in streams, such as flow velocity, discharge, and water depth, are dominated. Abiotic sorption reactions controlling P retention in streams are the same as in wetlands; however, long-term storage of P in stream sediments is inhibited by the rapid mobilization and transport that occurs during storm events [49]. TP retention is more sensitive to the specific total runoff than to the hydraulic loads (Figure 9).

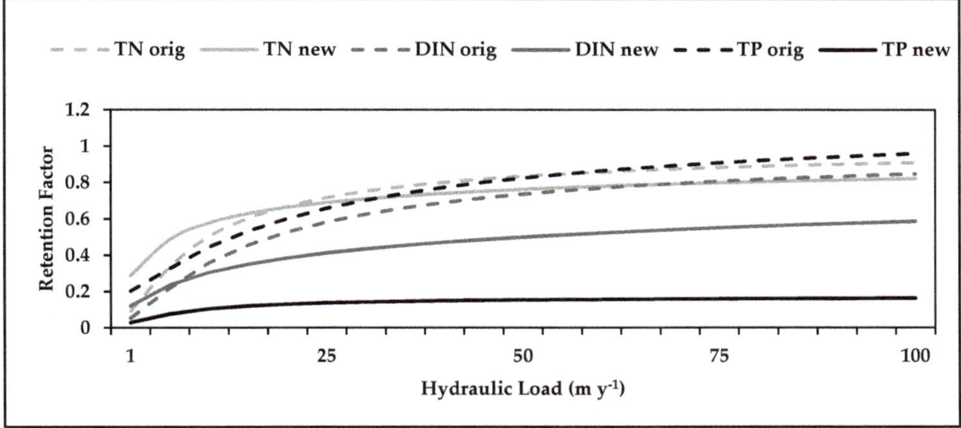

Figure 9. Effect of original and adjusted retention parameters on the retention factor as a function of the hydraulic load (specific total runoff for TP retention was 1 l s^{-1} km^{-2}).

It was a strong result, however, that the change of tributary retention of the catchments resulted in a much better fit at almost every monitoring point. However, we noted that this was not true for the lower load ranges, where decreased retention resulted in an overestimation of the loads. The other possible sources of TP that the model missed with the original retention values might have been higher diffuse-source loads and point-source loads, but these would differ among the study catchments, and it is therefore unlikely that this was the real reason. Another possible cause for the systematic underestimation of river TP loads may have been the underestimation of the sediment delivery ratio, as sediment transports the largest fraction of TP to the river outlets.

This latter, however, is in conflict with previous findings on this topic, and the overestimation of the total phosphorus loads in the lower load ranges suggested that the retention equation adjustment might compensate other loads not taken into account in the model. From the current analysis, it was difficult to make a conclusion on this issue; it should be subjected to further investigation on possible internal loads from river sediments or other pathways, including sediment delivery. The former was anticipated as there was a significant drop in point-source (PS) loads due to the increase in the number of wastewater treatment plants in the country, and the efficiency of sewage treatment. The drop in PS emissions is generally followed by the release of phosphorus from the sediments, as was found for lakes and wetlands [50], but this might also be the case for river sediments.

4.4. Model Validation in Context of the Literature

It is well known that the number of measurement data needed to produce reliable yearly averages differ for different compounds and different sampling locations [51]. Total nitrogen and total phosphorus load estimations depend on flow and sediment conditions and are thus more event-based, while dissolved nutrient loads are more stable across the seasons. The sampling frequency recommended to give accurate load estimations for dissolved compounds (error < 5%) is weekly–biweekly sampling. A monthly sampling frequency produces a 20% error for yearly sediment load estimations using sediment-flow rating curves [52]. Other studies have shown that in order to produce average annual load estimates with less than 10% error, a 15 day sampling interval might be necessary for nitrate, 10 days for soluble phosphorus and total phosphorus, and 5 days in the case of particulate phosphorus [53]. In light of this information, the sampling frequency of even the best group of catchments available in

Hungary produced an error much higher than 10%. Other studies have suggested that up to 20% to 50% error can be expected [54], even more for suspended sediments [55]. Another problem is that the smaller the catchment size, the larger the estimation error might become.

The model performance seemed to be acceptable when compared to other studies [12,17,56]. Even though several low-accuracy monitoring stations were involved in the model validation, most of the stations were within a relatively uniform range of error along with the load scales. It seems that there was no clear relationship between the accuracy of the load measurements and the accuracy of the model estimations (Table 7). This might have been due to the fact that to improve the accuracy of measurements, an even higher number of measurements would be necessary, or that there were several sources of uncertainty in the model.

4.5. Local Emissions

Total emissions at small catchments had a recognizable spatial pattern for both nutrients (Figures 10 and 11). For total phosphorus, the pattern was easier to understand, as agricultural areas that are prone to erosion dominated the emissions. On the flatlands, however, where erosion is close to zero, phosphorus emission was also low. In the case of total nitrogen, emissions are strongly linked to nutrient surplus and groundwater emissions, while in larger cities, urban pathways also play an important role. Use of adjusted groundwater parameters resulted in a more even distribution of emissions between hilly and mountainous areas, while flatlands had smaller emissions generally due to smaller runoff and smaller surpluses.

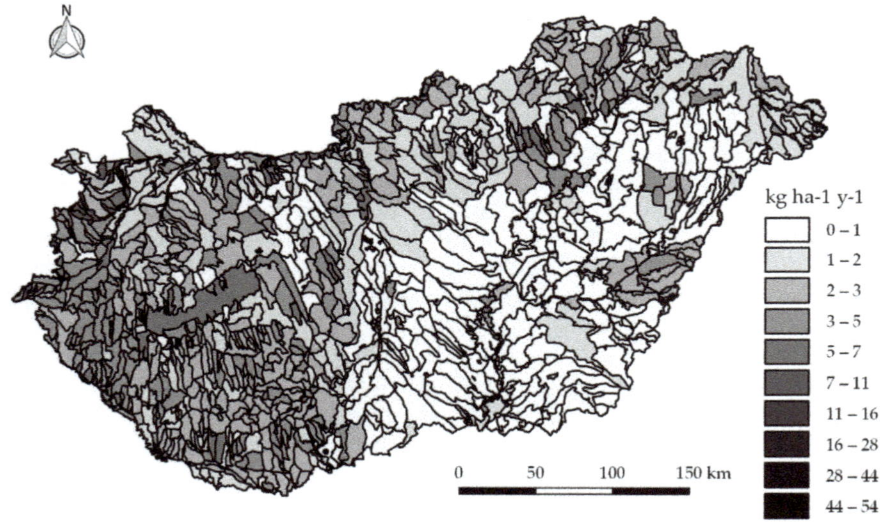

Figure 10. Diffuse-source area-specific TN emissions (totals of all pathways).

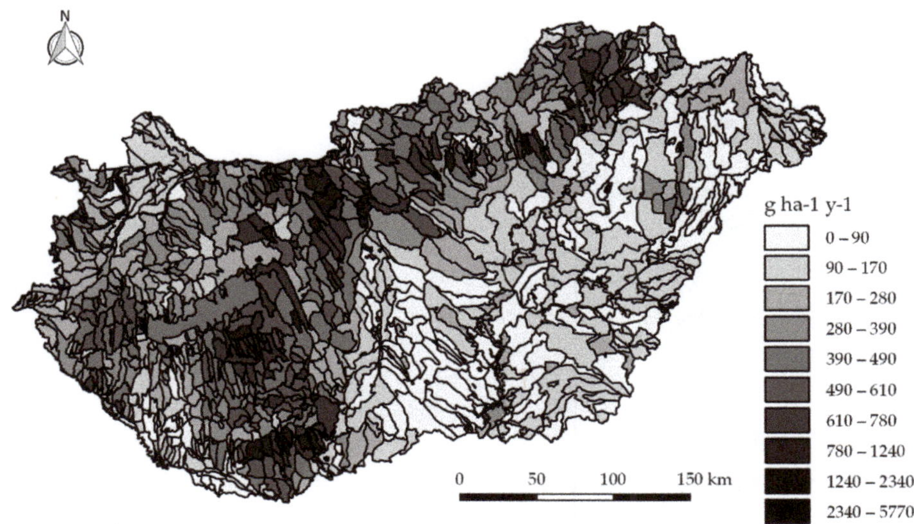

Figure 11. Diffuse-source area-specific TP emissions (sums of all pathways).

In terms of the distribution of nitrogen emissions between pathways (Figure 12), there was a very strong dominance of groundwater-related emissions (57%–70%) in all parts of the country. This agreed with previous studies from Germany [56] and the Danube River Basin [15]. Atmospheric deposition differs between slope classes due to the difference in surface water areas. Large lakes like Lake Balaton have huge emissions due to atmospheric depositions (660 t y^{-1} from atmospheric deposition), as they are considered as part of the catchment. It should be noted, however, that the simple empirical relationship used in the current model might not be precise enough to describe the retention processes of large lakes. Agricultural erosion also differs among slope classes. This pathway was not considered significant in previous studies. However, in the current study, the higher values of organic matter and organic-matter-bound nitrogen caused higher values of erosion-bound losses in the medium slope category. The share of agricultural area decreased with higher slope, and agricultural erosion was thus less significant in the steepest slope category.

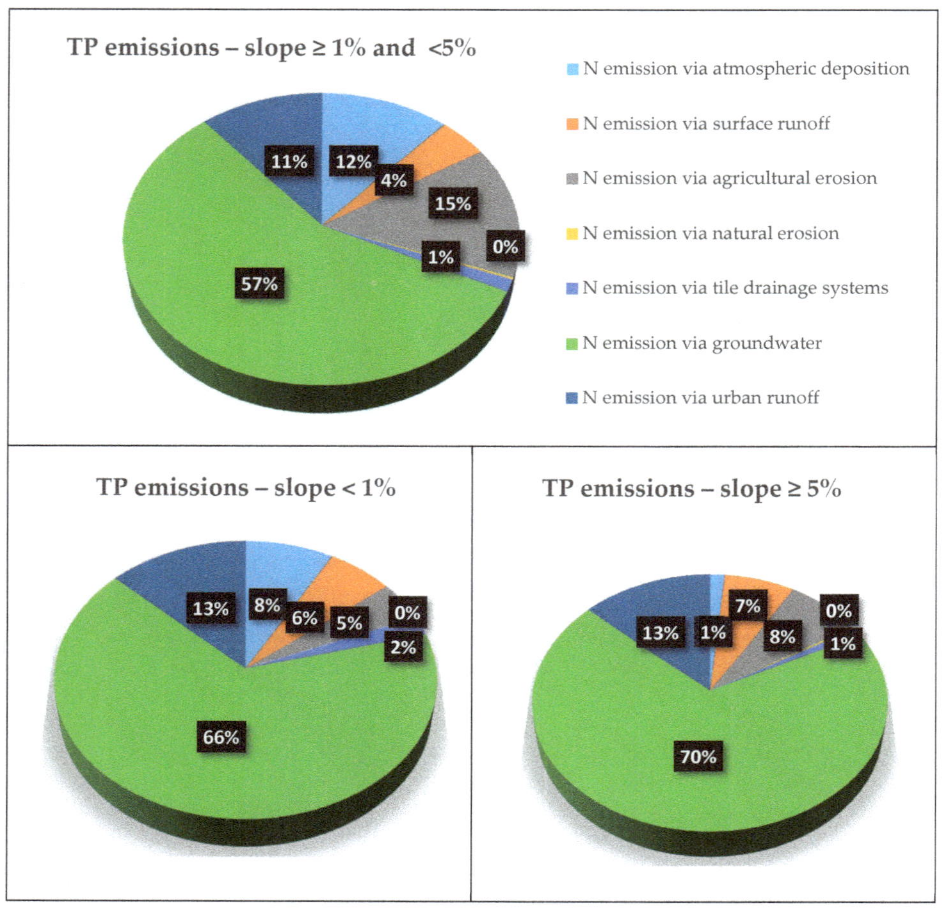

Figure 12. Division of total diffuse-source TN emission per pathway and slope class.

In agreement with previous model applications, the share of different pathways for total phosphorus differed largely from that for nitrogen. Agricultural erosion had a strong dominancy in all slope classes, but the magnitude of this dominancy differed between the slope classes (Figure 13). It is worth highlighting the significant portion of urban runoff in all slope classes. The third substantial pathway was groundwater, which was more critical in sandy catchments and wetlands than in clayey soils.

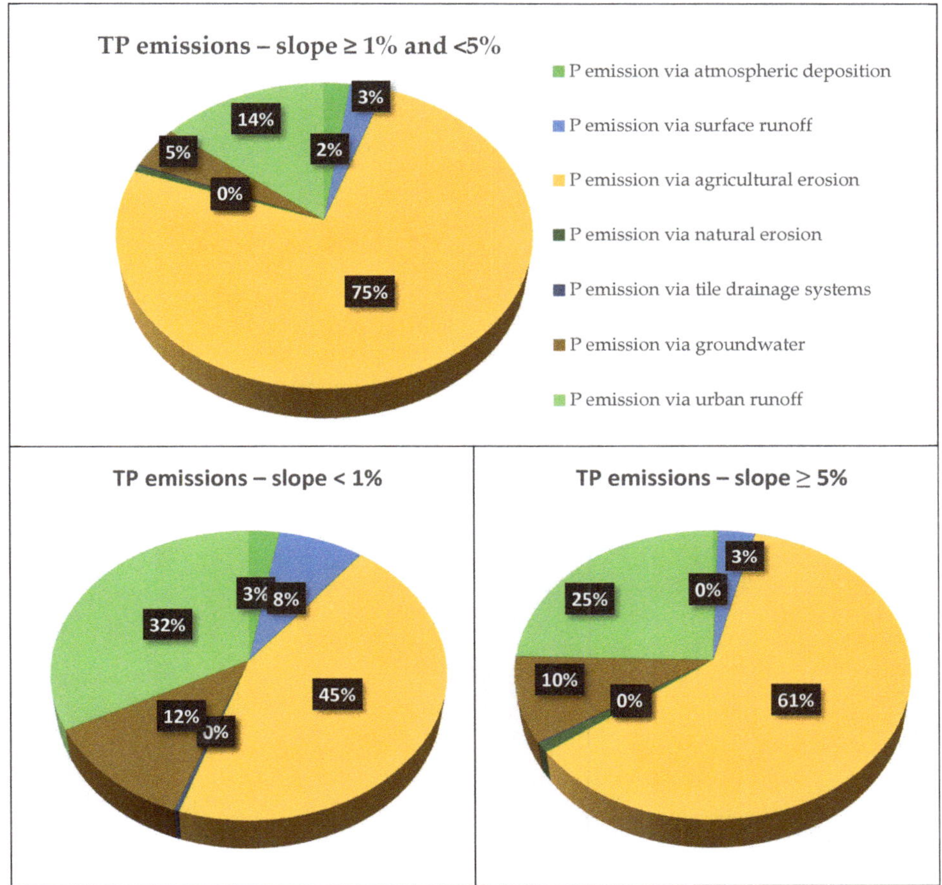

Figure 13. Division of total diffuse-source TP emissions per pathway and slope class.

4.6. Comparing Nutrient Emission Results with Results from Previous Studies

In this section, modeled emission values were compared to those of a previous model application. The previous application was delivered for the whole Danube Basin [15]. The average AU size in the previous study was 1660 km^2, while it was 86 km^2 in the present study.

A general remark regarding the large-scale application is that runoff patterns across Hungary seem not to follow the topography of the country. This can be explained by the relatively low specific runoff of this area compared to the Alps and the Carpathian Mountains, thus having a lower weight in the objective function during calibration. It was therefore proposed that further subregional flow monitoring data should be included in the calibration process of the next large-scale application of the model. In addition, the following differences between the results of both models should be mentioned.

Concerning nitrogen, surface runoff loads were 3.8 times higher than in the present study (Table 9). The larger-scale application calculated smaller erosion-bound emissions by much higher surface runoff. Without the precise knowledge of intermediate data for both of the models, it is difficult to see the exact reasons behind this anomaly. The large-scale application also estimated higher groundwater loads (Table 9). This was more than likely caused by differences in groundwater nitrogen retention parameters that caused higher concentrations in mountainous areas with higher specific runoff values.

Table 9. Total nitrogen and total phosphorus loads per pathway for the two modeled periods and the ICPDR application [15]. AD = Atmospheric deposition; SR = Surface runoff; AE = agricultural erosion; NE = natural erosion; TD = tile drainage; GW = groundwater; UR = urban runoff. ICPDR = the ICPDR application; HUN = the present study. Per. 1 = years 2009–2012; Per. 2 = years 2013–2016.

	AD	SR	AE	NE	TD	GW	UR	Total Diffuse	Point Source	Total
Total nitrogen loads (1000 t y^{-1})										
ICPDR Per. 1	1605	3403	1006		483	16,142	2527	25,168	7852	33
HUN Per. 1	1646	894	3535	12	555	13,332	1876	21,849	10,314	32.2
HUN Per. 2	1701	980	4597	18	314	11,053	2147	20,809	8629	29.5
Differences (%)										
HUN–ICPDR	3	*−74*	*253*		*15*	*−17*	*−26*	*−13*	*31*	*−2*
Per. 2–Per. 1	3	10	*30*	*50*	*−43*	*−17*	*14*	*−5*	*−16*	*−8*
Total phosphorus loads (t y−^{1})										
ICPDR Per. 1	0	19.1	773		3.5	585	541	1922	1062	2986
HUN Per. 1	60	127	1083	7	4	270	460	2010	1253	3264
HUN Per. 2	60	123	1279	11	5	282	534	2294	1066	3360
Differences (%)										
HUN–ICPDR		565	41		14	−54	−15	5	18	9
Per. 2–Per. 1	0	−3	18	57	25	4	16	14	−15	3

Concerning phosphorus, surface runoff loads were only 15% of those in the present model (Table 9). Agricultural erosion was also lower, like in the case of nitrogen. Groundwater loads, in contrast, were 2.2 times higher.

Comparing the results of the two modeling periods, we concluded that changes in total emissions were not significant, but there were notable differences in the contribution of different pathways. Emission from point sources decreased by about 15% for both TN and TP. There was also a slight decrease in total diffuse nitrogen emission, while the total phosphorus emission was increased. This highlights the evidence of the nonlinear effects of the hydrological factors (e.g., a significant increase of erosion), which could not be clearly separated from the possible impact of the mitigation efforts (e.g., reducing soil nutrient balances.

5. Conclusions

The MONERIS model concept was applied to the surface water catchments of Hungary at the spatial scale set by the national river basin management plan. As it is important to have a model estimate on diffuse nutrient emissions that is as accurate as possible, some of the equations/parameters of the original model were reviewed and adjusted. These were

- the water balance equation with regard to the ratio of surface and subsurface runoff,
- the nitrogen retention parameters of the subsurface pathways (excluding the tile drainage),
- phosphorus concentrations in shallow groundwater,
- retention parameters for the retention in surface waters (rivers and lakes).

Even though some improvement in all the examined parts was achieved, not all of them proved to be significant.

Concerning water balance, the ratio of the surface and subsurface waters was improved when the digital-filter based separation was applied directly to catchments with monitoring points. After refinement of the water balance equation, load estimates were recalculated. It was found that the overall accuracy of model prediction did not change significantly, but the ratio between the pathways did change considerably. Surface runoff became more important with a larger share in the total emissions. This has consequences for mitigation actions, as the return period of subsurface waters was larger than that of surface pathways.

Subsurface nitrogen retention seemed to have a significant effect on model accuracy. In the current study, it was found that there were no big differences in subsurface nitrogen retentions between regions with higher recharge rates (northern and western mountains). Calibration of the retention parameters caused the drop of groundwater concentrations in the mountainous regions, while it slightly increased the concentrations in the lowlands of Hungary. Altogether, these changes improved the nitrogen load estimates across the whole calibration dataset, including catchments from all over the country.

Subsurface phosphorus concentrations were reviewed using all the available groundwater well data for shallow wells. Mean and median concentrations had large differences for clayey and sandy soils and forest, with a large portion of the values moving around the medians. For this reason, the concentration medians are proposed for use as representative values for soil or land use. The median values of Hungarian groundwater wells did not differ significantly from the original values except for sandy soils. Due to the higher infiltration rates and lower sorption capacity of sandy soils, the subsurface phosphorus concentrations were more sensitive to the surplus history of the region under study. Therefore it is recommended that groundwater well data be checked regionally.

Surface water nutrient retention turned out to be the most important part of the model. While in the current application, the nitrogen retention parameters did not improve the model accuracy significantly, adjustment of phosphorus retention parameters improved the overall model performance by 20%–30%.

The comparison of the present calculations with the larger-scale application of the same model led to the conclusion that the accuracy of total load estimates differed in the distribution of the loads among pathways. The difference of total diffuse loads was primarily caused by the insufficient spatial representation of runoff in the larger scale application. Therefore it is recommended that in the future, a more rigorously reviewed network of monitoring stations be used for flow calibration.

Author Contributions: Conceptualization, Z.J.; writing—original draft preparation, Z.J.; writing—review and editing, M.K.K. & A.C; visualization, M.K.K. & Z.J.; supervision, A.C.; funding acquisition, A.C. All authors have read and agreed to the published version of the manuscript.

Funding: The research reported in this paper was supported by the Higher Education Excellence Program of the Ministry of Human Capacities in the frame of the Water sciences & Disaster Prevention research area of BME (BME FIKP-VÍZ). Financial support from the General Directorate of Water Management Hungary is greatly acknowledged.

Acknowledgments: The authors thank Márta Bagi, István Bíró, Szabina Pelyhe, and György István Tóth for the preparation of the monitoring databases of wastewater, surface, and subsurface water quality.

Conflicts of Interest: The authors declare no conflict of interest.

Appendix A Input Data Tables

Table A1. Spatial data used for the MONERIS model input.

Data Class	Detail	Data Source and Method	Source/Comment
Catchment area		GIS data for RBMP 2015	[27]
Land use data	Urban area, arable land areas in slope classes, grassland, woodland, shrubland, water surface area, mines, open areas, wetlands	Spatial statistics based on GIS analysis of CORINE LAND COVER grid	[28]
Soil classification	Sandy agricultural soils Loamy agricultural soils Silty agricultural soils Clayey agricultural soils	The classification made based on national soil texture database	[33]
	Fen agricultural soils Bog agricultural soils	Agrotopo fen type Agrotopo bog type	[27]
Underlying geology	Unconsolidated rock areas near groundwater Unconsolidated rock areas far groundwater	Statistics were based on raster data, combining Agrotopo database and groundwater table depth from RBMP 1 in 2009	[29]
	Solid rock areas with good porosity	Agrotopo based statistics	
	Solid rock areas with poor porosity	Agrotopo based statistics	[29]
Average elevation		Based on a 50 m resolution hydrodem raster	[26]
Average slope		Based on a 50 m resolution hydrodem raster	[26]

Table A2. Temporal data used for the MONERIS model.

Data Class	Detail	Input Data and Method	Comment
Net total runoff	Average riverbed runoff in the modeled period	GDWM long term average runoff data	Corrected by rainfall ratios and measured discharge values, where available
Average temperature	Average water temperature for the modeled period	FEVISZ database [57]	
Yearly precipitation current	Average yearly precipitation in the model period	Precipitation monitoring network data operated by GDWM	Precipitation distribution by Thiessen polygon method.
Summer half-yearly precipitation current	Average summer half-year precipitation in the modeled period	Precipitation monitoring network data operated by GDWM	
Summer half-yearly precipitation long term	Average summer half-year precipitation in between 1981–2010	Precipitation monitoring network data operated by GDWM	
Evapotranspiration	Mean annual evapotranspiration of the catchments	Country scale evapotranspiration distribution map 2001–2009	Data source: [22]
Measured yearly TN load	Product of yearly average water quality data and yearly average of measured discharge data	FEVI database [57]	
Measured yearly DIN load			
Measured yearly TP load			
NH4-N, NO3-N deposition rate-current	Yearly average deposition rates of nutrients in different forms. (mg m^{-2} y^{-1})	EMEP European air quality data maps for each component	[38]
NH4-N, NO3-N deposition rate-long term			
TP deposition rate	Default value by MONERIS		
Soil loss from agricultural areas, grasslands and natural covered land	Average annual soil loss per land use (t ha^{-1} y^{-1})	JRC maps & USLE method [30]	K factor [58] K factor (2013–2016) [32] R Factor [59]
N-content of topsoil	Total N content of the topsoil (mg kg^{-1})	AGROTOPO [29] & DOSOREMI [60] databases	Calculated based on soil organic carbon content
N-surplus-current	Nitrogen surplus in the topsoil for the 2009–2012 and 2013–2016 periods (kg ha^{-1} y^{-1})	County scale statistics of N balance (average of 2009–2012 and 2013–2016 periods) [36]	

Table A2. *Cont.*

Data Class	Detail	Input Data and Method	Comment
N-surplus-residence time	Nitrogen surplus in the topsoil for the average groundwater residence time of the catchments (kg ha^{-1} y^{-1})	County scale statistics of N balance for the 1961 to 2016 period [36]	
Accumulated P-surplus	Phosphorus surplus in the topsoil from 1961 to 2016 (kg ha^{-1} y^{-1})	Large regional, later county scale statistics of P balance for the entire country	Catchment average is calculated based on the country scale raster
External discharge	Yearly average river Q data at the upper boundary of a river transboundary sub-catchment Period: 2009–2012 and 2013–2016	Data derived from the discharge data of the national river monitoring network	Some large rivers lack satisfactory Q and/or WQ data (Vág, Garam, Ipoly tributaries etc.) TNMN is used [61]
External TN load	Yearly average TN load at the upper boundary of a transboundary sub-catchment	FEVI database [57]	Some large rivers lack satisfactory Q and/or WQ data (Vág, Garam, Ipoly tributaries etc.), TNMN is used [61]
External DON load	Yearly average DON load at the upper boundary of a transboundary sub-catchment	FEVI database [57]	Some large rivers lack satisfactory Q and/or WQ data (Vág, Garam, Ipoly tributaries etc.) [61]
External TP load	Yearly average TN load at the upper boundary of a transboundary sub-catchment	FEVI database [57]	Some large rivers lack satisfactory Q and/or WQ data (Vág, Garam, Ipoly tributaries etc.) [61]

References

1. GDWM Hungarian Part of the Danube River Basin—River Basin Management Plan. *Hungarian: A Duna-vízgyűjtő magyarországi része;* Vízgyűjtőgazdálkodási Terv; General Directorate of Water Management: Budapest, Hungary, 2015; pp. 1–666.
2. Kardos, M.K.; Clement, A. Prediciting small water courses ' physico-chemical status with two multivariate statistical methods. *Open Geosci.* **2020**, *12*, 71–84. [CrossRef]
3. Kardos, M.K.; Koncsos, L. A stochastic approach for regional-scale surface water quality modeling. *Pollack Period.* **2017**, *12*, 17–27. [CrossRef]
4. Jolánkai, G.; Bíró, I. *Basis of Water Quality Protection with Special Regard of Systematic Ecohidrological Approaches (In Hungarian: Vízminőségvédelem alapjai különös tekintettel a rendszerszemléletű ökohidrológiai módszerekre)*; University Press, Eotvos Lorant Sci. Uni.: Budapest, Hungary, 1999.
5. Borah, D.K.; Bera, M. Watershed-scale hydrologic and nonpoint-source pollution models: Review of mathematical bases. *Trans. Am. Soc. Agric. Eng.* **2003**, *46*, 1553–1566. [CrossRef]
6. Novotny, V. *Water Quality: Diffuse Pollution and Watershed Management*, 2nd ed.; Novotny, V., Ed.; Wiley: Hoboken, NJ, USA, 2002; ISBN 978-0-471-39633-8.
7. Radcliffe, D.E.; Freer, J.; Schoumans, O. Diffuse phosphorus models in the United States and Europe: Their usages, scales, and uncertainties. *J. Environ. Qual.* **2009**, *38*, 1956–1967. [CrossRef]
8. Silgram, M.; Anthony, S.G.; Collins, A.L.; Stromqvist, J.; Bouraoui, F.; Schoumans, O.; Lo Porto, A.; Groenendijk, P.; Arheimer, B.; Mimikou, M.; et al. Evaluation of diffuse pollution model applications in EUROHARP catchments with limited data. *J. Environ. Monit.* **2009**, *11*, 554–571. [CrossRef]
9. Malagó, A.; Venohr, M.; Vigiak, O.; Bouraoui, F.; Kovacs, A. *Modelling Nutrient Pollution in the Danube River Basin : A Comparative Study of SWAT, MONERIS and GREEN Models*; JRC Technical Reports EUR 27676 EN; JRC: Ispra, Italy, 2015; ISBN 9789279542398.
10. Venohr, M.; Hirt, U.; Hofmann, J.; Opitz, D.; Gericke, A.; Wetzig, A.; Natho, S.; Neumann, F.; Hürdler, J.; Matranga, M.; et al. Modelling of nutrient emissions in river systems—MONERIS—Methods and background. *Int. Rev. Hydrobiol.* **2011**, *96*, 435–483. [CrossRef]
11. Behrendt, H.; Kornmilch, M.; Opitz, D.; Schmoll, O.; Scholz, G. Estimation of the nutrient inputs into river systems—Experiences from German rivers. *J. Mater. Cycles Waste Manag.* **2002**, *3*, 107–117. [CrossRef]

12. Behrendt, H.; Dannowski, R.; Deumlich, D.; Dolezal, F.; Kajewski, I.; Kornmilch, M.; Korol, R.; Mioduszewski, W.; Opitz, D.; Steidl, J.; et al. *Point and Diffuse Emissions of Pollutants, Their Retention in the River System of the Odra and Scenario Calculations on Possible Changes*; UBA Research report; Weißensee Verlag: Berlin, Germany, 2001; Project 298-28-299.
13. De Lima Barros, A.M.; Do Carmo Sobral, M.; Gunkel, G. Modelling of point and diffuse pollution: Application of the Moneris model in the Ipojuca river basin, Pernambuco State, Brazil. *Water Sci. Technol.* **2013**, *68*, 357–365. [CrossRef]
14. Schreiber, H.; Behrendt, H.; Constantinescu, L.T.; Cvitanic, I.; Drumea, D.; Jabucar, D.; Juran, S.; Pataki, B.; Snishko, S.; Zessner, M. Nutrient emissions from diffuse and point sources into the River Danube and its main tributaries for the period of 1998–2000—Results and problems. *Water Sci. Technol.* **2005**, *51*, 283–290. [CrossRef]
15. Venohr, M.; Gericke, A. *Further Development of the MONERIS Model with Particular Focus on the Application in the Danube Basin*; Technical Report, JOINTISZA Project, Project Code: DTP1-152-2.1; International Commission for the Protection of the Danube River: Vienna, Austria, 2015.
16. Kovács, Á.; Honti, M. Estimation of diffuse phosphorus emissions at small catchment scale by GIS-based pollution potential analysis. *Desalination* **2008**, *226*, 72–80. [CrossRef]
17. Zessner, M.; Kovács, Á.; Schilling, C.; Hochedlinger, G.; Gabriel, O.; Natho, S.; Thaler, S.; Windhofer, G. Enhancement of the MONERIS Model for Application in Alpine Catchments in Austria. *Int. Rev. Hydrobiol.* **2011**, *96*, 541–560. [CrossRef]
18. Stelczer, K. *Hydrological Basics of Water Resource Management (In Hungarian: A vízkészletgazdálkodás hidrológiai alapjai)*; ELTE Eötvös Kiadó: Budapest, Hungary, 2000; ISBN 9634632491.
19. Clement, A.; Somlyódy, L. Vízminőség-szabályozás. In *Magyarország vízgazdálkodása: Helyzetkép és stratégiai feladatok*; Somlyódy, L., Ed.; MTA: Budapest, Hungary, 2011; pp. 169–206. ISBN 978-963-508-608-5.
20. Bihari, Z.; Babolcsai, G.; Bartholy, J.; Ferenczi, Z. Climate. In *The National Atlas of Hungary—Natural Environment*; Kocsis, K., Ed.; MTA CSFK Geographical Institute: Budapest, Hungary, 2018; ISBN 978-963-9545-56-4.
21. Mozsgai, K.; Podmaniczky, L.; Skutai, J.; Deák, J.; Clement, A.; Simonffy, Z. *Assessing Achievements and Impacts on Water Quality of Hungarian Rural Development Program*; Mechanical Engineering Letters, Szent István University: Gödöllö, Hungary, 2019.
22. Venohr, M.; Hirt, U.; Opitz, D.; Gericke, A.; Wetzig, A.; Ortelbach, K.; Natho, S.; Neumann, F. *The Model System MONERIS—Version 2.14.1vba—Manual*; Leibniz-Institute of Freshwater Ecology and Inland Fisheries: Berlin, Germany, 2009; pp. 1–118.
23. R Core Team. R: A Language and Environment for Statistical Computing. Available online: https://www.r-project.org/ (accessed on 1 January 2019).
24. Szilagyi, J.; Kovacs, A. A calibration-free evapotranspiration mapping technique for spatially-distributed regional-scale hydrologic modeling. *J. Hydrol. Hydromech.* **2011**, *59*, 118–130. [CrossRef]
25. Szilágyi, J.; Kovács, Á. Complementary-relationship-based evapotranspiration mapping (cremap) technique for Hungary. *Period. Polytech. Civ. Eng.* **2010**, *54*, 95–100. [CrossRef]
26. GDWM. *HydroDEM—Hydrologically Corrected Digital Elevation Model of Hungary*; Digital Data; General Directorate of Water Management: Budapest, Hungary, 2014.
27. GDWM. *Spatial Database of River Water Bodies, Lake Water Bodies and Their Watershed Areas (In Hungarian: Víztestek és víztest-vízgyűjtők téradat állománya. Az OVGT digitális melléklete)*; General Directorate of Water Management: Budapest, Hungary, 2015.
28. Copernicus Land Monitoring Service Corine Land Cover (CLC) 2012, Version 18. Available online: https://land.copernicus.eu/pan-european/corine-land-cover/clc-2012 (accessed on 1 January 2018).
29. Research Institute for Soil and Agricultural Chemistry(RIISAC). *Agrotopo Database (In Hungarian: Agrotopográfiai Adatbázis)*; RIISAC: Budapest, Hungary, 1991.
30. Pásztor, L.; Laborczi, A.; Takács, K.; Szatmári, G.; Dobos, E.; Illés, G.; Bakacsi, Z.; Szabó, J. Compilation of novel and renewed, goal oriented digital soil maps using geostatistical and data mining tools. *Hung. Geogr. Bull.* **2015**, *64*, 49–64. [CrossRef]
31. Szalay, M. *Quantitative Characterization of Surface Waters—Small Water Courses. RBMP Background Document No 2.3 (In Hungarian: A felszíni vizek mennyiségi jellemzése—Kisvízi készlet—OVGT 2.3 háttéranyag)*; National Water Directorate (OVF): Budapest, Hungary, 2009.

32. Wischmeier, W.H.; Smith, D.D. Predicting rainfall erosion losses—A guide to conservation planning. *USDA Agric. Handb.* **1978**, *537*, 1.
33. Panagos, P.; Borrelli, P.; Poesen, J.; Ballabio, C.; Lugato, E.; Meusburger, K.; Montanarella, L.; Alewell, C. The new assessment of soil loss by water erosion in Europe. *Environ. Sci. Policy* **2015**, *54*, 438–447. [CrossRef]
34. Pásztor, L.; Waltner, I.; Centeri, C.; Belényesi, M.; Takács, K. Soil erosion of Hungary assessed by spatially explicit modelling. *J. Maps* **2016**, *12*, 407–414. [CrossRef]
35. Jolánkai, Z.; Muzelák, B.; Kardos, M.K. *RBMP Background Document No 3-1: Modeling of Nutrient Loads in Surface Water Bodies—Application of the MONERIS Model to Estimate Diffuse Nutrient Emissions to Water Bodies in Hungary (In Hungarian: Felszíni víztestek tápanyagterhelésének modellezése)*; General Directorate of Water Management: Budapest, Hungary, 2015.
36. Deák, J. Assessment of the RDP Impact Indicators on National Level in the Field of Water Quality. In *Ex-Post Revision of the Rural Development Programme*; Final Report; Consortium of AAM-AKI-Collectivo: Budapest, Hungary, 2016. (In Hungarian)
37. Kremer, A.M. *Methodology and Handbook Eurostat/OECD Nutrient Budgets*; Version 1.02; European Comission Eurostat: Luxemburg, 2013; pp. 1–112.
38. EMEP The European Environmental Monitoring and Evaluation Programme. Available online: https://www.emep.int/mscw/index.html (accessed on 15 March 2015).
39. GDWM Wastewater Load Data. Supplement No. 3-1 to the Hungarian River Basin Management Plan. (In Hungarian: 3.1 melléklet az Országos Vízgyűjtőgazdálkodási Tervek 2015. évi felülvizsgálatához: Szennyvízterhelés jellemzői: Kommunális és ipari szennyvízkibocsátás. Available online: https://www.vizugy.hu/vizstrategia/documents/10B9EE2E-D889-4C94-815D-5CB2D53C846A/3_1_melleklet_szennyvizterheles.xls (accessed on 1 June 2018).
40. Ács, T.; Simonffy, Z. A new deterministic method for groundwater mapping using a digital elevation model. *Water Sci. Technol. Water Supply* **2013**, *13*, 1146–1153. [CrossRef]
41. Jolánkai, Z.; Koncsos, L. Base flow index estimation on gauged and ungauged catchments in Hungary using digital filter, multiple linear regression and artificial neural networks. *Period. Polytech. Civ. Eng.* **2018**, *62*, 363–372. [CrossRef]
42. Arnold, J.G.; Allen, P.M. Automated methods for estimating baseflow and ground water recharge from streamflow records. *J. Am. Water Resour. Assoc.* **1999**, *35*, 13353–13366. [CrossRef]
43. Ladson, L.S.; Fox, R.L.; Ratner, M.W. Nonlinear optimization using the generalized reduced gradient method. *Rev. française d'automatique, informatique, Rech. opérationnelle. Rech. opérationnelle* **1974**, *8*, 73–104.
44. Byrd, R.H.; Gilbert, J.C.; Nocedal, J. A trust region method based on interior point techniques for nonlinear programming. *Math. Program. Ser. B* **2000**, *89*, 149–185. [CrossRef]
45. Chow, V.T.; Maidment, D.R.; Mays, L.W. *Applied Hydrology*; McGraw-Hill: New York, USA, 1988.
46. Vadas, P.A.; Kleinman, P.J.A.; Sharpley, A.N.; Turner, B.L. Relating soil phosphorus to dissolved phosphorus in runoff: A single extraction coefficient for water quality modeling. *J. Environ. Qual.* **2005**, *34*, 572–580. [CrossRef] [PubMed]
47. Søndergaard, M.; Jensen, J.P.; Jeppesen, E. Role of sediment and internal loading of phosphorus in shallow lakes. *Hydrobiologia* **2003**, *506–509*, 135–145. [CrossRef]
48. Clement, A. Modeling the trophic response of a shallow lake following external load reduction: A case study. In *Proceedings of the International Association of Theoretical and Applied Limnology*; Williams, W., Ed.; E Schweizerbart'sche Verlagsbuchhandlung: Stuttgart, Germany, 2001; Volume 27, pp. 819–822.
49. Reddy, K.R.; Kadlec, R.H.; Flaig, E.; Gale, P.M. Phosphorus retention in streams and wetlands: A review. *Crit. Rev. Environ. Sci. Technol.* **1999**, *29*, 83–146. [CrossRef]
50. Clement, A.; Somlyódy, L.; Koncsos, L. Modeling the phosphorus retention of the Kis-Balaton upper reservoir. *Water Sci. Technol.* **1998**, *37*, 113–120. [CrossRef]
51. Rode, M.; Suhr, U. Uncertainties in selected surface water quality data. *Hydrol. Earth Syst. Sci. Discuss.* **2006**, *3*, 2991–3021. [CrossRef]
52. Horowitz, A.J. An evaluation of sediment rating curves for estimating suspended sediment concentrations for subsequent flux calculations. *Hydrol. Process.* **2003**, *17*, 3387–3409. [CrossRef]
53. Moatar, F.; Meybeck, M. Compared performances of different algorithms for estimating annual nutrient loads discharged by the eutrophic River Loire. *Hydrol. Process.* **2005**, *19*, 429–444. [CrossRef]

54. Webb, B.W.; Phillips, J.M.; Walling, D.E.; Littlewood, I.G.; Watts, C.D.; Leeks, G.J.L. Load estimation methodologies for British rivers and their relevance to the LOIS RACS(R) programme. *Sci. Total Environ.* **1997**, *194–195*, 379–389. [CrossRef]
55. Walling, D.E.; Webb, B.W.; Woodward, J.C. Some sampling considerations in the design of effective strategies for monitoring sediment-associated transport. *Eros. Sediment Monit. Program. River Basins. Proc. Int. Symp. Oslo* **1992**, *1992*, 279–288.
56. Behrendt, H.; Dannowski, R. *Nutrients and Heavy Metals in the Odra River System*; Behrendt, H., Dannowski, R., Eds.; Weissensee Verlag Ökologie: Stuttgart, Germany, 2005.
57. Agricultural Ministry of Hungary. *Surface Water Quality Database of Hungary (In Hungarian: OKIR-FEVISz: Országos Környezetvédelmi Információs Rendszer, Felszíni Vízminőség Szakmodul)*; Agricultural Ministry of Hungary: Budapest, Hungary, 2015.
58. Panagos, P.; Meusburger, K.; Alewell, C.; Montanarella, L. Soil erodibility estimation using LUCAS point survey data of Europe. *Environ. Model. Softw.* **2012**, *30*, 143–145. [CrossRef]
59. Panagos, P.; Ballabio, C.; Borrelli, P.; Meusburger, K.; Klik, A.; Rousseva, S.; Tadić, M.P.; Michaelides, S.; Hrabalíková, M.; Olsen, P.; et al. Rainfall erosivity in Europe. *Sci. Total Environ.* **2015**, *511*, 801–814. [CrossRef] [PubMed]
60. Pásztor, L.; Laborczi, A.; Takács, K.; Szatmári, G.; Fodor, N.; Illés, G.; Farkas-Iványi, K.; Bakacsi, Z.; Szabó, J. *Compilation of Functional Soil Maps for the Support of Spatial Planning and Land Management in Hungary*; Elsevier Inc.: Amsterdam, The Netherlands, 2017; ISBN 9780128052013.
61. ICPDR Transnational Monitoring Network Data. Available online: https://www.icpdr.org/main/activities-projects/tnmn-transnational-monitoring-network (accessed on 15 March 2015).

Article

Lessons Learnt from the Long-Term Management of a Large (Re)constructed Wetland, the Kis-Balaton Protection System (Hungary)

Mark Honti [1,*], Chunni Gao [2], Vera Istvánovics [1] and Adrienne Clement [3]

1 MTA-BME Water Research Group, H-1111 Budapest, Hungary; istvanovics.vera@gmail.com
2 School of Soil and Water Conservation, Beijing Forestry University, Beijing 100083, China; chunni@bjfu.edu.cn
3 Department of Sanitary and Environmental Engineering, Budapest University of Technology and Economics, H-1111 Budapest, Hungary; clement@vkkt.bme.hu
* Correspondence: mark.honti@gmail.com; Tel.: +36-1-463-1894

Received: 13 January 2020; Accepted: 26 February 2020; Published: 29 February 2020

Abstract: Environmental management decisions should be made based on solid scientific evidence that relies on monitoring and modeling. In practice, changing economic, societal, and political boundary conditions often interfere with management during large, long, and complex projects. The result may be a sub-optimal development path that may finally diverge from the original intentions and be economically or technically ineffective. Nevertheless, unforeseen benefits may be created in the end. The Kis-Balaton wetland system is a typical illustration of such a case. Despite tremendous investments and huge efforts put in monitoring and modeling, the sequence of decisions during implementation can hardly be considered optimal. We use a catchment model and a basic water quality model to coherently review the impacts of management decisions during the 30-year history. Due to the complexity of the system, science mostly excelled in finding explanations for observed changes after the event instead of predicting the impacts of management measures a priori. In parallel, the political setting and sectoral authorities experienced rearrangements during system implementation. Despite being expensive as a water quality management investment originally targeting nutrient removal, the Kis-Balaton wetland system created a huge ecological asset, and thereby became worth the price.

Keywords: nutrient retention; constructed wetland; water resources management; eutrophication

1. Introduction

Water resources management—just like any other type of decision-making—should principally rely on rationality and scientific evidence [1,2]. In rational decision theories, scientific knowledge influences both the objectives of management and the concepts of how these could be achieved [3]. Scientific influence is more explicit on the latter through monitoring and mathematical modelling. According to the best available management practice, decisions should be based on careful analysis of system behavior and on model-based forecasts of system alterations driven by external impacts [1]. However, decisive boundary conditions for management include the political context, and long-term shifts in societal attitudes that arise from a multitude of often contradictory interests of various stakeholder groups [4]. These conditions determine priorities, which may or may not comply with the natural operation mode of the managed system. In any case, changing priorities gradually alter the objectives of management. Large-scale management structures, which need large investment costs and have long lifetimes are prone to the consequences of changing priorities. This first of all applies to structures designed for relatively soft management purposes, such as water quality protection. In this paper we demonstrate how the interplay of rationality and changing environmental and sociopolitical boundary conditions may distort the outcomes of management using the case of the Kis-Balaton Protection System.

1.1. The Doom and Adventurous Rebirth of the Kis-Balaton Wetlands

Balaton is a large (surface area is 596 km^2), shallow (mean depth is 3.2 m) lake, and the second tourist and recreational attraction in Hungary. Given its vital ecological and socioeconomic importance, it has been the most prominent subject of water resources and water quality management since decades [5]. Lake Balaton is a semiastatic lake, which has had climate-driven water level fluctuations of up to 4 m during the past three centuries [6]. This, together with the elongated shape and the asymmetrical catchment (Figure 1a) supported a cascade of adjacent aquatic habitats ranging from wet meadows to periodically inundated swamps and to shallow but permanent open waters (Figure 1b). The regulation of Lake Balaton from the early 1800s resulted in a 3 m drop of the mean water level [7]. Wet habitats dried up on higher altitudes. The shallowest westernmost basin fed by the largest inflow of the lake, the Zala River, was disconnected from the present Lake Balaton. In a century, it developed into a reed dominated swamp. This area and its surrounding wetlands are called the Kis-Balaton wetlands. Regulation of the lower Zala River and extensive drainage works from the early 1900s further deteriorated the state of the wetlands. Reflecting a partial shift in the political attitude, 3 ha of open water and 1400 ha of reed become protected in 1951. A few years later, the fruitless efforts to gain good quality agricultural land by drainage of wetlands had been abandoned, and a new function was assigned to the Kis-Balaton wetlands. The new function was the retention of suspended solids to ease increasing siltation in Lake Balaton, which was a consequence of increasing soil erosion in the catchment.

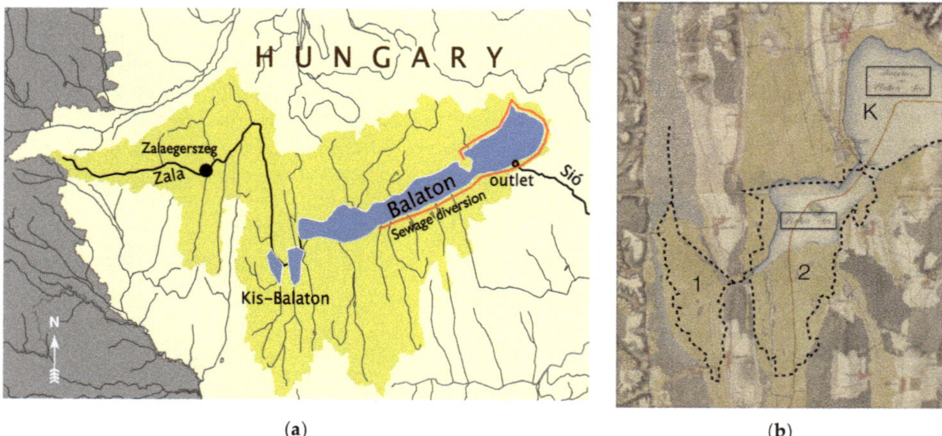

Figure 1. (**a**) Position of the Kis-Balaton wetlands within the Lake Balaton catchment. Zala is the largest tributary; Zalaegerszeg is the largest town upstream of Kis-Balaton. The red line indicates the sewage collector belt around Lake Balaton; (**b**) the western parts of Lake Balaton in the First Military Survey of Hungary (1782–1785, map slides with permission of Österreichischen Staatsarchiv) and the present outlines of the Kis-Balaton system. K: Keszthely Basin, presently the westernmost basin of Lake Balaton; 1 and 2: phases of the present Kis-Balaton system. "Lake Balaton" (in German, "*Platten See*" is written in the thin frames).

While designing the "anti-siltation" Kis-Balaton project, Lake Balaton has experienced a well-documented and serious eutrophication from the late 1960s due to the boom of tourism, the lagging development of sewerage construction and wastewater treatment behind public water supply, and the intensive agricultural production in the catchment. By the mid 1970s, the western area of the lake was hypertrophic with regular summer blooms of N_2-fixing cyanobacteria and occasional fish kills [8]. The government did its best to keep eutrophication in secret because West German tourists meeting their East German relatives at Lake Balaton were a vitally important source of hard currency income of Hungary. A high-level comrade demonstrated good water quality by publicly drinking a glass of water directly drawn from the lake and containing over 100 mg m^{-3} of chlorophyll *a* (Chl) and over 90% filamentous cyanobacteria! Finally, the unprecedented bloom of *Cylindrospermopsis raciborskii* in 1982 forced the government to take eutrophication seriously [9]. The action plan for management dates back to the decision of the Council of Ministers taken in 1983. In spite of the socialism-specific birth of the decision, its scientific background was particularly solid thanks to the project coordinated by the International Institute for Applied Systems Analysis, in which Lake Balaton was the case study for modelling and managing shallow lake eutrophication. To reduce point loads, 40 farms producing liquid manure were immediately closed up along the Zala River, treated sewage was diverted from the lake in about two thirds of the shoreline settlements by 1986 (Figure 1a), and the largest municipal wastewater treatment plants (WWTPs) of the catchment were to be upgraded with phosphorus removal [10]. As agriculture was a prominent, untouchable sector of the socialist economy, the management of diffuse pollution could not rely on reduction of fertilizer application. Instead, an end-of-pipe solution was conceived [11]. The Kis-Balaton wetlands project was redesigned to retain both nutrients and suspended solids.

The overall area of the Kis-Balaton protection system was planned to reach 147 km^2. The motivation for designing such a large area was twofold: (i) to increase hydraulic residence time thereby maximizing the efficiency of retention processes and (ii) to prevent rapid loss of the volume due to siltation. The magnitude of works made it necessary to split the project into consecutive phases.

The ancient, naturally dried-up wetlands situated furthest upstream from Lake Balaton were inundated in phase 1 in 1985 (Lake Hídvégi, H; Figure 2). Influential ecology experts suggested that a macrophyte-dominated lake would retain nutrients and suspended solids with the highest efficiency [12]. To the surprise of these experts and managers, Lake H (surface area 18 km^2, mean depth ~1 m, mean water residence time 30 days) has quickly developed into a hypertrophic pond. From 1988, annual mean concentration of Chl fluctuated around 150 mg m^{-3} at the outflow. The "surprise" could easily be explained. First, Lake H received nearly two times higher area-specific nutrient load than the westernmost basin of Lake Balaton, in which this lower load maintained hypertrophic conditions. The scientifically sound management plan correctly determined the sequence of various measures and emphasized the importance of inundating the Kis-Balaton wetlands only after upgrading the WWTP in the largest town of the catchment (Zalaegerszeg, ca. 60 thousand inhabitants). However, insufficient regulation and the intricacies of institutional and political interests overrode rationality. Second, ecology experts failed to provide sufficient design criteria to facilitate the spread of aquatic macrophytes. Lake H is situated in a wind channel open to the prevailing NW winds and its main axis is nearly parallel with the channel. Therefore, establishment of aquatic macrophytes has efficiently been prevented by the long fetch and frequent breaking waves.

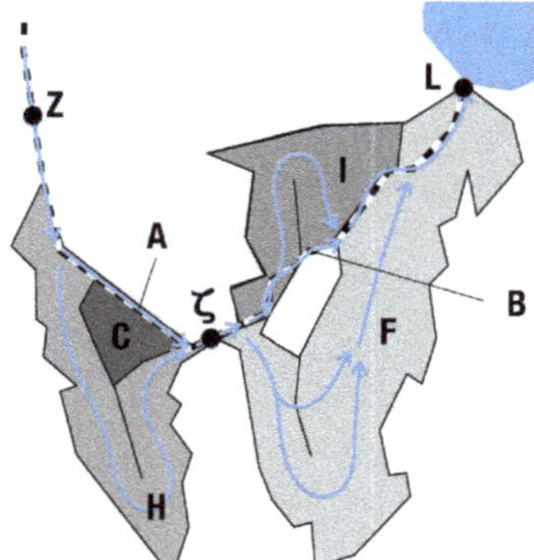

Figure 2. The present Kis-Balaton Water Protection System. Z: Zala River; H: Lake Hídvégi inundated in phase 1; C: emergency reservoir to enclose and treat pollution (e.g., oil spill) from Z; A: bypass of H; ζ: connector between phases 1 and 2 (a section of the Zala River); I: Ingói Reeds; F: Lake Fenéki inundated in phase 2, B: bypass of F; L: inflow into Lake Balaton. Black dots indicate major monitoring sites of inputs and outputs, blue arrows indicate alternative, adjustable pathways of flow.

In harmony with global trends, the National Agency for the Protection of Nature and Environment declared the entire territory of the Kis-Balaton project protected in 1986. Protection did not extend to the adjacent wetlands to the North and South of the project area, despite their ecological values that slowly deteriorated in the lack of a sufficient water supply.

In 1991, the WWTP in Zalaegerszeg was finally upgraded and the external phosphate load of Lake H suddenly dropped to about half of its former value. As a result of the enhanced internal P load, the efficiency of P retention decreased dramatically, and hypertrophic conditions persisted.

Following the inundation of Lake H, implementation of the Kis-Balaton project slowed down due to the economic crisis that later led to the political conversion of the country. After the political system collapsed in 1989–1990, the new legislation urged the completion of phase 2. According to the original plan, an area of 51 km^2 should have been inundated. Due to the lack of funds, only the 16 km^2 northern part, the Ingói Reeds, were inundated in 1993 (Figures 2 and 3). The result was catastrophic. Masses of algae leaving the hypertrophic Lake H died in a few days during their passage through the closed reeds of the Ingói, where no sufficient light was available for aquatic photosynthesis. Phosphate that liberated during fast and temperature-dependent mineralization of algal detritus flowed out from the system in the absence of sediments that might have adsorbed the nutrient [13]. Total P retention was slightly positive on an annual basis due to rapid sedimentation of suspended solids and inorganic particulate P. Decomposition of the detritus, together with the lack of aquatic photosynthesis turned the water anoxic. End products of anaerobic metabolism and increased water depth caused a large-scale die-back of reed in the upstream areas of the Ingói Reeds [14]. To address these problems, the original design of the Kis-Balaton was revised in 1996. The revision concluded that the Ingói Reeds have to be disconnected from the main flow path and dedicated to nature protection. Simultaneously, the not yet inundated lower area (Lake Fenéki, F; Figure 2) should be implemented so that the algal-rich outflow of Lake H should not reach Lake Balaton. No significant P retention was anticipated in the freshly inundated Lake F before a steady state develops in this area. The steady state was perceived as a heterogeneous wetland habitat and mosaics of shallow open water, where long-term P retention is due to P adsorption by the carbonate-rich lacustrine sediments [15,16]. The final revision of the Kis-Balaton Water Protection System took place between 2005 and 2013. It kept the main findings of the previous revision regarding the Ingói Reeds, but suggested to maintain a lower and fluctuating water level in the future Lake F to increase habitat diversity and preserve macrophyte coverage [17]. This shift in the suggested design and operation of Lake F returned to the original concept of the

Kis-Balaton system, according to which macrophytes would retain nutrients with higher efficiency than an open-water-dominated lake [17].

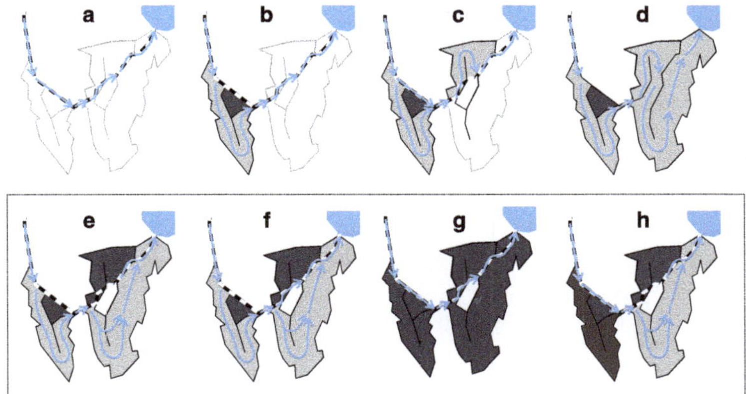

Figure 3. Historical (**a–c**), originally planned (**d**), and present (**e–h**) operational setups of the Kis-Balaton. (**a**): before 1985; (**b**): 1985–1992; (**c**): 1993–2013; (**d**): original plan of the complete system, (**e–h**): main flow routing options since 2014; (**e**) shows the normal way of operation. Dark shading indicates stagnant/bypassed water bodies, blue arrows show flow paths. Flow sequences using the codes from Figure 2 are as follows: in (**a**) ZAζBL, in (**b**) ZHζBL, in (**c**) ZHζIL, in (**d**) ZHζIFL, in (**e**) ZHζFL, in (**f**) ZHζ{2/B}L, in (**g**) ZAζBL, in (**h**) ZAζFL. The original plan of the Kis-Balaton was ZHζIFL. At present the following flow patterns are also possible: ZHζ{I/F}L, ZAζIL, ZAζ{I/2}L, ZAC, where curly brackets indicate parallel alternative pathways.

The redesigned Lake F was completed in late 2014 thanks to EU funding, yet water had occasionally been diverted to it already from January 2013. The project had the same triad of aims as the original initiative: (i) protecting the water quality of Lake Balaton by retaining nutrients, (ii) protecting and enhancing the ecological value of the Kis-Balaton area, and (iii) reducing flood risks. However, the emphasis moved from objective (i) to objective (ii) and the project was actually listed as a Nature Protection investment. In accordance with the suggested revisions, the Ingói Reeds were isolated from the main flow to protect the ecosystem and to prevent phosphate emission. The Balaton-Felvidéki National Park has been appointed as the manager of the Ingói Reeds.

To prepare for the variability of water quality and nutrient retention, the completed Kis-Balaton system provides several options of operation (Figure 3). Lake F became the U-shaped main flow path, yet its depth was not sufficient to create a fully aquatic habitat, it became a swamp. Both Lakes H and F can be bypassed if, for example, there is a net nutrient emission from the system.

1.2. Management Context and Study Objectives

In summary, the Kis-Balaton Protection System has evolved during a rather long and complex process including several shifts in priorities and objectives. Thus, despite the now "complete" system, it is unclear if the present state and operation of the wetlands provide an optimal solution to the water quality problems of Lake Balaton and nature protection. At the moment, management of the wetlands lacks answers to substantial questions: Has nutrient retention increased along the various setups of the wetlands? Could the present system be controlled better with regard to the objectives? Is another rebuild necessary?

The objective of this study is to provide answers to some of these questions by evaluating the nutrient retention of the wetlands during the various stages of its development (Table 1) and schemes of management. Retention will be assessed retrospectively using a model-based standardized approach including a catchment and a basic water quality model. The ultimate aim is to clarify the contribution of monitoring and modeling to the management of a highly complex system in a changing sociopolitical environment.

Table 1. Timeline of development.

Date	Event/Action
1970s	Eutrophication of Lake Balaton
1982	Unprecedented large bloom of *Cylindrospermopsis raciborskii* in Lake Balaton
1983	Decree of the Council of Ministers
1985	Completion of Lake H (phase 1)
1986	The whole Kis-Balaton project area gets nature protection status
1990	Collapse of agriculture in relation to the change of the political system
1991	Upgrade of WWTP in the town of Zalaegerszeg
1993	Inundation of the Ingói Reeds (start of phase 2)
1996	Revision of plans of phase 2
2013	Start of flow diversion from Ingói Reeds to Lake F
2014	Official completion of Lake F
2019	Unprecedented large bloom of *Aphanizomenon flos-aquae* and *Ceratium furcoides* in Lake Balaton

2. Materials and Methods

The assessment consisted of the following steps:

1. Compilation of water, suspended solids (SS), and total phosphorus (TP) balances for the entire system and phases 1 and 2 separately. This included estimating non-monitored components (hydraulic, sediment, and nutrient loads from smaller tributaries) by catchment modelling.
2. Calculation of retentions efficiencies of SS and TP for the entire system and for phases 1 and 2 separately. Retention in Lake H was also simulated by a dynamic model to characterize conditions when a net release of TP was likely.
3. Analysis of retention and the efficiency of management with respect to the declared objectives of the Kis-Balaton system.

2.1. Balances of Water, SS, and TP

The Kis-Balaton system is one of the best-monitored wetland systems in the world. Extensive monitoring and research have started before the inundation of Lake H. Daily discharge data were available in the Zala River and the six largest secondary tributaries of the Kis-Balaton from the 1950s. Of these six tributaries, three enter Lake H, the largest one enters the Ingói Reeds, and the remaining two discharge into Lake F. At the key sections of the Zala River, Z and L, daily water quality measurements have been done from 1977 and 1975, respectively. At ζ, daily load measurements started immediately upon inundation of Lake H in 1985. Similarly, daily water chemistry was measured at the outflow of the Ingói Reeds from 1993 (inundation) to 2013 (exclusion of the Ingói Reeds from the main flow path). Appropriate flow data were missing from the outflow since it was a temporary spillway. Biweekly water quality data has been collected in the largest secondary inflow and two to six samples were taken yearly in 10 small tributaries from 1989. The present water quality monitoring scheme was introduced in 2013. At Z, the frequency of sampling was decreased from seven to three per week. Simultaneously, the frequency was increased in the secondary inflows. The largest secondary inflow is sampled weekly; the largest inflow of Lake H and the two largest inflows of Lake F are sampled biweekly, whereas monthly samples are taken from eight and three small inflows of Lake H and Lake F, respectively. Additional water samples are taken after large rains.

In spite of the large-scale monitoring efforts, material balances are not closed either for the whole system or for its phases 1 and 2. Evaporation, groundwater exchange, and inflows from numerous smaller tributaries as well as from the direct shoreline catchment are not monitored. Discharge from some tributaries (other than Z and ζ) are monitored too sparsely compared to their dynamics.

Concentration measurements on small tributaries are less frequent than discharge measurements, thereby rendering direct inputs of TP and SS uncertain. Additionally, changes in the frequency of water quality monitoring during the operation period introduce heterogeneity into the time series.

To bridge these gaps, we compiled a database of hydraulic, sediment, and nutrient loads to the system using the same calculation and quality control methods for the entire existence of the system. First, we estimated the contributions from non-monitored sources by modeling. We used the PhosFate catchment model [18,19] for this purpose. PhosFate is a static, GIS-based model system calculating seasonal mean flows and P loads. Inputs are maps of elevation, land use, physical soil type, meteorology, and nutrient budgets of agricultural soils. The model does not require calibration, except for the region-specific overland and in-stream retention coefficients. As PhosFate had already been calibrated for the Zala catchment upstream of the Kis-Balaton [18], it did not require further adjustments. Contributions of the direct catchments were estimated for both a wet and a dry year (2001 and 2012, respectively) for flow, and as a long-term average for SS and TP.

The estimated relative contribution of the direct catchments and meteorological fluxes were added to the measured inputs at Z and ζ to complete the water balance. Before comparing the in- and outflowing fluxes of SS and TP, an additional correction step was necessary. Despite the good accuracy (error < 5%) of discharge measurements at Z, ζ, and L, the net changes of the completed water balance of Lake H considerably exceeded possible fluctuations in storage. Thus, flow was corrected based on setting the long-term cumulated water balance (e.g., the change in storage) to zero. This approach assumed that the change in storage on a multi-annual basis was negligible relative to the magnitude of water balance errors. Water balance of Lake F was corrected similarly.

2.2. Annual and Seasonal Retention of TP and SS

Since sediment and nutrient retention are the raisons d'être of the Kis-Balaton system, management measures taken during the 35 years of operation were evaluated on the basis of their impacts on the retention. Interventions that had no or detrimental impacts on retention of either TP or SS were considered as unsuccessful.

Annual retention efficiency (R [–]) of TP and SS was calculated from the corrected water and material balances for the whole system and its two phases:

$$R = \frac{\sum F_{in} - \sum F_{out}}{\sum F_{in}} \tag{1}$$

where F_{in} and F_{out} are the in- and outflowing fluxes at the boundaries of the system, respectively.

To explore boundary conditions that are likely to result in negative retentions, that is, net release from the system, we assessed seasonal dynamics of retention using a simple, dynamic mass balance model. The model was applied to Lake H using the corrected daily boundary fluxes as inputs. The model described the change in the amount of total phosphorus in the water:

$$\frac{dM_{\mathrm{TP}}}{dt} = L_{\mathrm{TP}} - M_{\mathrm{TP}}\frac{Q_{\zeta}}{V_{\mathrm{H}}} - M_{\mathrm{TP}}\frac{v_s}{V_{\mathrm{H}}}A_{\mathrm{H}} + k_{\mathrm{r}}M_{\mathrm{SP}} \tag{2}$$

where M_{TP} is the amount of total phosphorus in Lake H [kg], L_{TP} is the daily TP load at Z [kg d^{-1}], V_{H} is the volume of Lake H [m^3], A_{H} is the surface area of Lake H [m^2], Q_{ζ} is the discharge at ζ [m^3 d^{-1}], M_{SP} is the active P content in the surface layer of the sediments [kg], vs. is the apparent settling velocity of TP [m d^{-1}], and k_{r} is the resuspension rate of the surface sediments [d^{-1}]. The dynamics of the P stock in the sediments is:

$$\frac{dM_{\mathrm{SP}}}{dt} = M_{\mathrm{TP}}\frac{v_s}{V_{\mathrm{H}}}A_{\mathrm{H}} - k_{\mathrm{r}}M_{\mathrm{SP}} - k_{\mathrm{b}}M_{\mathrm{SP}} \tag{3}$$

where k_{b} is the burial rate of the active P content in the sediments [d^{-1}].

If the model with a single set of the three free parameters (v_s, k_r, and k_b) reasonably describes the multiyear behavior of Lake H, it could directly be used to evaluate the impact of internal load on retention processes.

3. Results and Discussion

3.1. Balances of Water, SS, and TP

The PhosFate model estimated that direct inflows other than the Zala River contributed 13% of the total flow into Lake H (Table 2). Contrary to Lake H, the secondary tributaries of Lake F have a relatively large catchment area (755 km^2, 30% of total) and the proportion of secondary inflow is larger, 29% and 40% in wet and dry years, respectively. These results highlight that the larger Zala catchment is water deficient compared to the smaller secondary tributaries. Due to sporadic measurements on many secondary tributaries, these estimates cannot systematically be compared with data. Yet, the mean ratio of flows from simultaneous observations in secondary tributaries and in the Zala River broadly supports model results; measured secondary contributions to Lake H made up 9.2% of the outflow in the period covered with data, modelled contribution was 7.1%. For Lake F, the respective values were 31.9% and 29.1%.

Table 2. Contribution of secondary tributaries.

	Type of Year	Flow	SS	TP
Lake H	Average	9%	12%	14%
	Dry	13%		
	Wet	9%		
Lake F	Average	33%	13%	20%
	Dry	40%		
	Wet	29%		

Model results showed that in Lake H secondary tributaries contributed 12% and 14% of SS and TP loads, respectively. The respective values were 13% and 20% in Lake F. While in Lake H the contribution of small tributaries was roughly similar in terms of both discharge and loads, in Lake F secondary loads of SS and TP were disproportionately low relative to flow (Table 2).

The long-term cumulated water budget of Lake H showed several longer and shorter periods of systematically increasing or decreasing trends, even after correcting for the inputs of secondary tributaries and meteorological water fluxes (Figure 4). We calculated how much mean water depth would have changed if these trends have been related to a change in storage. Evidently, the systematic error in the cumulated water budget greatly exceeded the storage capacity of Lake H (Figure 4). Accordingly, periodically constant correction terms had to be applied to the water balance to keep the closing error in the range, which could realistically be accounted for by the dynamics of storage volume (Figure 4). Although ζ is situated downstream of the large storage volume of Lake H, daily flows occasionally showed rapid oscillations when the flow at Z changed smoothly. This indicated that flow measurements might be less reliable at ζ than at Z and therefore, only the ζ-discharge was corrected. Noticeably, the necessary mean correction was only 3.2% of the mean flow of Z, a value definitely within the accuracy of best-conducted discharge measurements.

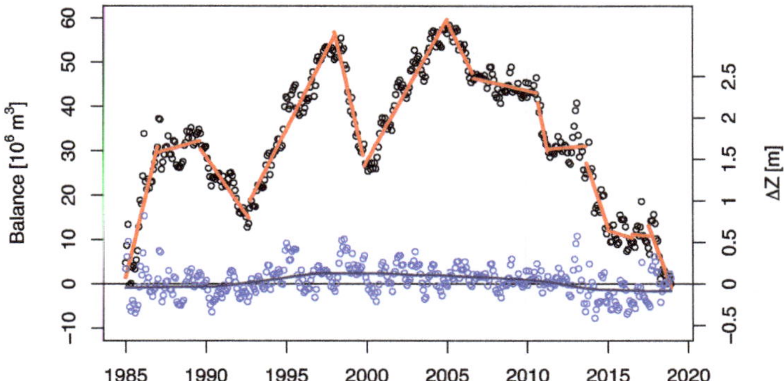

Figure 4. Cumulated water balance of Lake H based on raw measurements and a simple correction for long-term trend (**black circles**) and corrected discharges (**blue circles**). Red straight lines indicate manually delineated periods of constant systematic errors, blue line is a local polynomial regression (LOESS) on the corrected discharge. Right axis shows approximate change in storage level.

The water budget of phase 2, that is the joint hydrological balance of Lake F and the Ingói Reeds was corrected using the same approach, but with considering the corrected flow at ζ. The long-term mean correction was nearly 0, but, in certain periods, a correction as high as 1 m^3 s^{-1} (21% of mean flow) was necessary. This could be due to the complicated hydrology of this patchy area and the relatively large contribution of weakly monitored tributaries to the total inflow (Table 2). As there was no single suspect to assign the corrections, they were halved between the outflow (L) and the direct tributaries.

3.2. Retention and Evaluation of the Operational Success

The SS load of Lake H showed large fluctuations ever since the inundation (Figure 5). Outflowing fluxes of SS were always lower than the inflowing ones; efficiency of annual retention varied between 50% and 75%. This indicated that Lake H, and especially its northern area near the inflow, functioned as a settling tank [20]. In contrast to the fluctuating SS load, TP load decreased systematically in the period 1987 to 1993 from about 100 t yr^{-1} to below 50 t yr^{-1} (Figure 5). Upgrading of the WWTP in Zalaegerszeg (nominal capacity is 15,000 m^3 d^{-1}) in 1991 (Table 1) significantly contributed to this trend. In about two to three years after inundation, a steady state had established with respect to TP retention, which stabilized at around 50%. Similar to other lakes, the external loads of which were reduced significantly [21], retention efficiency of Lake H decreased to 30% on average upon the drop of its external TP load during the 1990s. Negative retention efficiencies (net release) were observed during the drought in 2000–2003, when the TP flux at Z decreased below 10 t yr^{-1} due to the diminutive surface runoff and diffuse TP emission. Decreases of TP retention could be related to enhanced internal P load, which is a result of the imbalance between the P content of the freshly forming sediments and that of the surface sediments accumulated during the period of high external load [22]. Obviously, the absolute fluxes during the periods of low and negative retention were much smaller than in wet years when TP retention was stable and high.

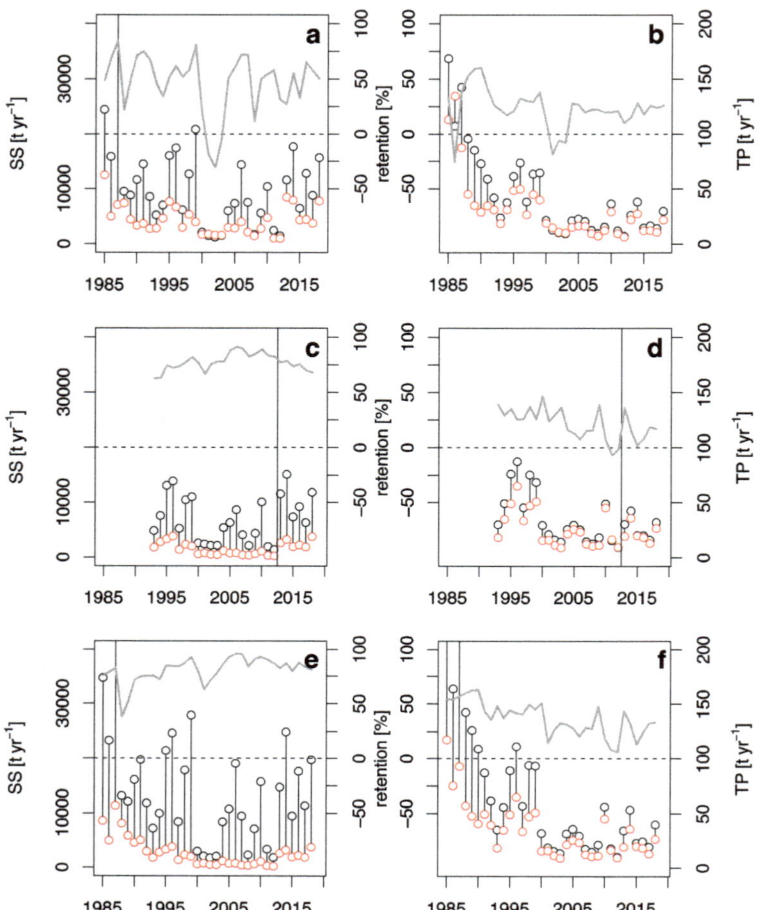

Figure 5. Fluxes (black circles: incoming, red circles: outgoing, connected for clarity) and retentions (gray line) of suspended solids (SS) and total phosphorus (TP) in Lake H (**a**,**b**), in the area of phase 2 (**c**,**d**), and in the entire system (**e**,**f**). In c and d, the vertical line indicates diversion of flow from the Ingói Reeds to Lake F in January 2013.

Similar to Lake H, the annual SS load of phase 2 of the Kis-Balaton also fluctuated widely. Until 2013, the Ingói Reeds intercepted this load with efficiencies over 80%. This high retention was due to the lack of resuspension in the dense reed stands, which fully sheltered the water from the wind [23]—similarly to other systems [24]. A slight reduction could be observed in SS retention, when the flow was turned from the Ingói Reeds to Lake F. Since the latter is also covered by dense macrophyte vegetation, it is not clear, whether this slight deterioration is a transient phenomenon. TP retention zigzagged between 0 and 30% both before and after 2013 and did not show any correlation with meteorological boundary conditions. Thus, phase 2 of the Kis-Balaton played a decisive role in the retention of SS, governed solely by physical processes, and a minor role in TP retention that depends on intricate interactions of physical, chemical, and biotic processes.

The behavior of the entire system was more stable than that of its components. SS retention was stabilized between 75% and 90% by the reliable performance of phase 2. Overall annual retention efficiency of TP has been usually low from 2000, but always in the positive domain (10–60%).

The simple dynamic mass balance model reasonably reproduced the observed daily TP flux at the outflow of Lake H (Figure 6). Optimal parameter values were $v_s = 0.002$ m d^{-1}, $k_r = 7.3 \cdot 10^{-5}$ d^{-1}, and $k_b = 1.2 \cdot 10^{-8}$ d^{-1}. These suggest that assuming a mean depth of 1 m, the half-life of TP would be 346 days in the water column. Moreover, the sediment stock practically never gets exhausted, and it can re-enter the water whenever conditions favor internal P load. These imply that the load reduction during the recent decade might increase the relative weight of internal load in the outflowing P flux. According to the model, negative retention (net release) of TP is likely to occur in months when the incoming P load is less than 2 t month^{-1}.

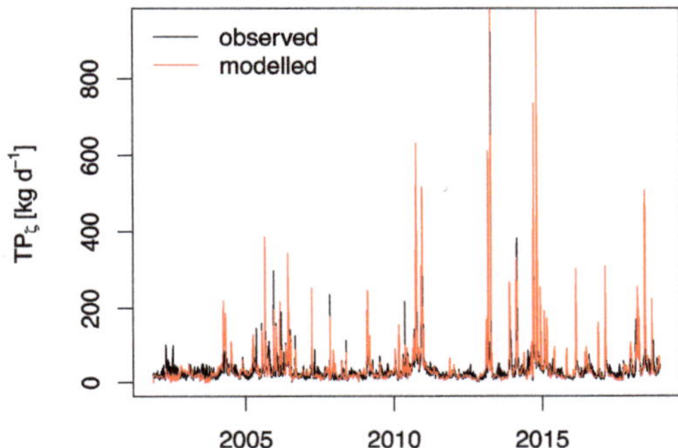

Figure 6. Observed and modelled TP fluxes at site ζ.

3.3. Futile Management Actions

Several interventions aimed at optimizing the operational objectives of retention and nature protection. One such intervention was the gradual drawdown of Lake H. The water level of this lake dropped by 30–50 cm corresponding to about a 30–50% decrease in storage volume during the summers of 2003, 2012, 2013, and 2014 because inflow could not compensate evaporation and outflow. From 2015, the newly introduced operational guidelines directed a planned reduction of the summer water level by 50 cm in 10 cm yr^{-1} increments. The reasoning was that Lake H typically does not retain nutrients in the dry and hot summer months, so ecological objectives could be pursued without compromising water quality targets. It was envisioned that a lower water level would favor the spread of macrophytes and increase habitat diversity. Noticeably, neither local experience nor monitoring data gave any support to this vision. Macrophytes failed to invade a larger area in Lake H during previous years when the water level was low. Moreover, monitoring data indicated that monthly mean TP concentrations at the outflow tended to increase with decreasing monthly mean water levels. The increase was roughly threefold (from 20 to 60 mg P m^{-3}) in the range of 20 cm to 50 cm drop in water level, greatly decreasing the efficiency of P retention. The unsuccessful and unreasonable intervention was terminated after two years.

Another futile intervention was the construction of the bypass of Lake H. The present design of the Kis-Balaton provides a high operational flexibility (see Figure 3 panels e–h). Low or negative TP retention during months of low flow conceived the idea of activating bypass mode in Lake H when the TP flux at the outflow (ζ) exceeded that at the inflow (Z) by more than 30% for more than a week [22]. During the summer of 2015, bypass operation was tested in Lake H for six weeks. During this period, the mean concentration of nitrate-N and chlorophyll *a* was 2 g m^{-3} and 6 mg m^{-3}, respectively in the Zala River at Z. At the inflow section of the bypass channel A, the respective values were 0.16 g m^{-3} and 112 mg m^{-3}. Thus, water quality measurements indicated that the bypass weir failed to lead the water of the Zala River into channel A; it rather drained waters from the northern area of Lake H. This might happen because Lake H was inundated by cutting a sequence of openings into the right-hand bank of the Zala River, whereas the bypass weir is situated a few kilometers downstream from the openings. The insufficient design has most likely been chosen to make the investment cheaper, yet the result was costly: the bypass of Lake H does not work and needs to be redesigned.

Yet another ineffective intervention was the bypass of Lake F. As mentioned above, revision of phase 2 was motivated by the catastrophic effects of inundating the Ingói Reeds with the algal-rich outflow of Lake H that included summer release of phosphate and die-back of the reed. After the redesign of phase 2, ecological objectives have become dominant in the operation of the entire Kis-Balaton. The objective of managing the Ingói Reeds were (i) to satisfy the ecological water demand, and (ii) to mimic the "natural" fluctuations of water level. Criteria for ecological water requirements have not been defined to govern flow routing. This has led to a paradoxical practice. In line with the two objectives, 13% to 21% of the annual flow at ζ was diverted to the Ingói with the exception of 2015, when the diversion was negligible (<2%). However, for unknown reasons the bypass weir was kept open in 2017 to 2019. As a consequence, the supplied water did not spread out in the Ingói Reeds but flowed through the old Zala River channel to bypass channel B. In contrast to management objective (ii), variance of daily water level decreased by a factor of 3.6 in the period 2014–2018 compared to 2005–2013. As a side-effect of the apparent water supply to the Ingói Reeds, elevated SS concentrations were observed at L, the vivid fluctuations of which were synchronous with the fluctuations at ζ (Figure 7). Although Lake F is a densely vegetated wetland system, the overall retention of SS decreased after its inundation because of the direct passage of a significant volume of water along bypass B.

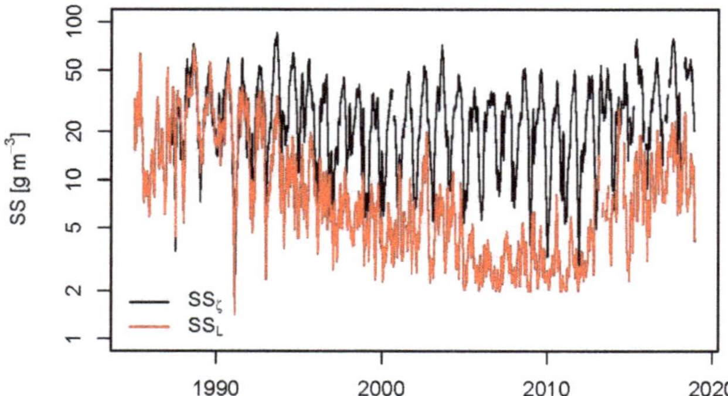

Figure 7. Observed SS concentrations at sites ζ and L (1 month moving averages). The Ingói reeds operated between the two dashed lines.

Finally, the design of Lake F does not seem to ensure spreading out the water in this large area. Lake F is twice as large as the Ingói Reeds and the planned mean water residence time was correspondingly higher. Therefore, according to the plans, the opening of Lake F should have increased the retention of TP, since (a) TP retention in the Ingói Reeds was low (Figure 5) due to the conversion of algal-bound P to phosphate and (b), the relatively large direct tributaries of Lake F formerly passed the wetlands and joined the Zala River without stretching out. In contrast to the plans, TP retention of the entire system did not increase in comparison to the period before opening Lake F (Figure 5). One of the reasons was the new operational scheme of the Ingói Reeds. Another reason was the sub-optimal hydraulic profile of Lake F. The trapezoidal channels of the direct tributaries of Lake F were kept nearly intact for flood protection purposes. These longitudinal openings have much smaller hydraulic resistance than the densely vegetated shallows of the wetland, and therefore most of the water (80–90%) flowed along the channels towards L. For the time being, the real water residence time in Lake F is unknown but certainly much less than the theoretical residence time estimated from lake and inflow volumes.

3.4. Climate as the Chief Manager

The timelines of SS and TP retentions and their dynamics indicate that climatic factors were the ultimate drivers of retention processes in the Kis-Balaton system (Figure 8). In dry years, loads and retentions plummeted, in wet years both increased. The primary role of climate-related factors behind these dynamics such as discharge, residence time, erosion, and diffuse P loads suggests that management actions were of secondary importance and could not significantly influence the efficiency of retention. The only exception was the reduction of external TP load between 1987 and 1993 related to years long cessation of fertilizer application due to the collapse of agriculture after the change in the political system in 1989–1990 and to upgrading the WWTP in Zalaegerszeg (Table 1). The initial TP retention of the entire system was 50–60% on average, and it decreased to 30–40% after this load reduction (Figure 9). No interventions could restore the retention to its value before the load reduction. Conclusively, this huge and complex quasi-natural system follows its own way of operation autonomously; management actions targeting the operation of the system resulted in only marginal improvements (if any) in retention.

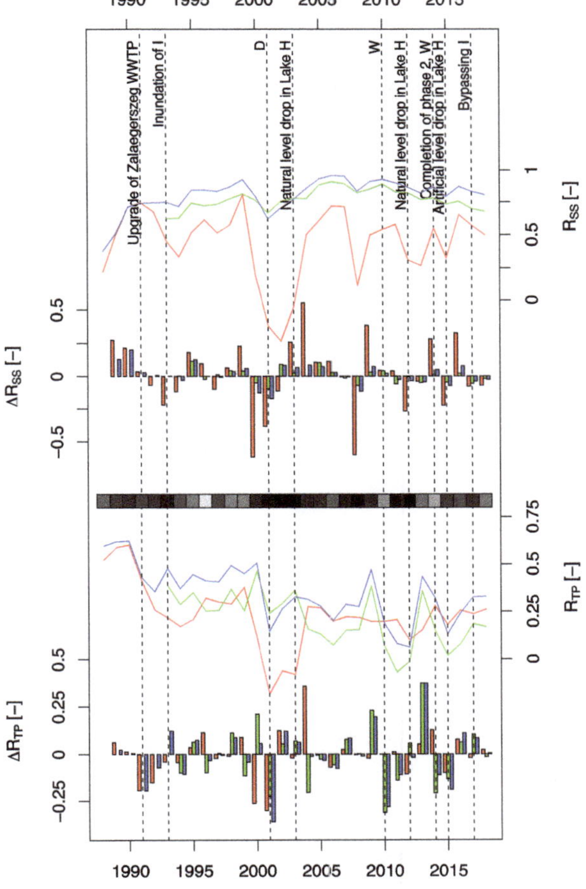

Figure 8. Retentions of SS (R_{SS}) and TP (R_{TP}) and their annual changes (ΔR_{SS} and ΔR_{TP}, respectively). Red: Lake H, green: phase 2, blue: entire system. The grayscale ribbon indicates the relative magnitude of flow at Z, dark: minimal, light: maximal. "W" and "D" indicate extremely wet and dry years, respectively.

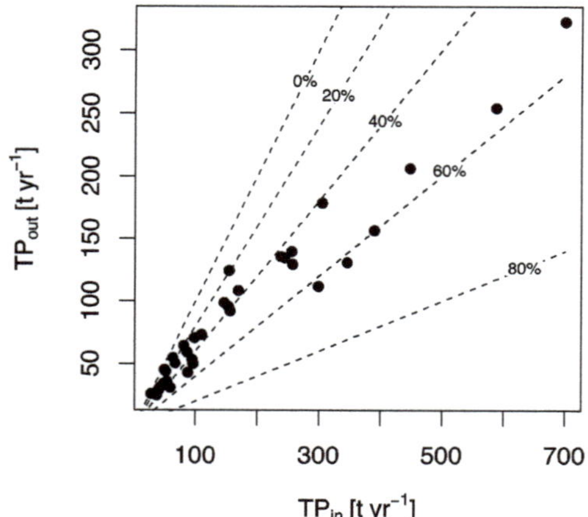

Figure 9. Outflowing flux of TP (TP_{out}) as a function of inflowing TP flux (TP_{in}) for the entire system. Isolines show constant retention rates.

3.5. Does Science-Based Rational Environmental Management Exist in Practice?

Despite the immense monitoring and scientific efforts that supported management, numerous scientifically justified and apparently rational management actions proved futile. It was only the initial concept of creating a prestorage that proved to be robust in the long-term. The predictive power of science was generally weak, only retrospective analysis could explain:

- the absence of macrovegetation cover in Lake H after inundation;
- the negligible TP retention efficiency of dense reedstands when load is mostly algal P;
- TP retention efficiency of the complex habitat in the hydraulic boundary conditions of Lake F;
- the efficiency of bypass without a hydraulically well-defined upstream section.

Monitoring was exceptionally intensive, yet insufficient to

- close the material balances and provide a solid basis for retention assessments;
- follow the behavior of the complex lake F.

During the design, implementation, and operation of the Kis-Balaton project, the political weights of the interested sectors changed too. Before 1990, the National Water Authority was a highly autonomous powerful state organization with a mixed mandate that covered both management and legal control. After 1990, most of its former power suddenly vanished; ecology and nature protection became more prominent. Between the inundation of the Ingói Reeds (1993) and the final revision of the plans of Lake F (2005–2013), in parallel with the recognition that the (re)constructed wetland system created a huge ecological asset, nature protection perspectives dominated management and planning. The realization of Lake F was even funded with the primary objective of increasing the ecological potential by saving the vegetation of the Ingói Reeds and creating a diverse habitat in Lake F. Nutrient retention and the management of the nutrient load of Lake Balaton (itself a Ramsar site during the winter months) has always been of secondary importance for the National Park. In recent years, due to the boom of infrastructural investments and cuts on the state financing of nature protection, the roles seem to change again soon. The likelihood of this outcome increased substantially in 2019, when an unprecedentedly large late summer bloom of phytoplankton developed in the southwestern areas of Lake Balaton (Honti and Istvánovics, unpublished data, peak concentration of chlorophyll was around 300 mg m^{-3}; Table 1) after 25 years of acceptable water quality, during which time water quality management was considered to successfully arrest eutrophication [25]. Conclusively, the political background was a far more important driver of management actions than science-based data analysis and prediction.

3.6. Final Balance of the Kis-Balaton Project

Since the inundation of Lake H, the Kis-Balaton system retained approximately 140,000 tons of SS and 366 tons of TP. No estimates are available about the total costs of the Kis-Balaton project, but the low retention efficiency resulted in a very high specific cost in the order of 1000 € (kg P)$^{-1}$. The low direct cost efficiency was counterbalanced by a series of socioeconomic and environmental benefits from development of recreational tourism at Lake Balaton including its southwestern areas to boost of ecotourism in the Kis-Balaton region and to the renewal of a large, undisturbed wetland habitat.

4. Conclusions

The management history of Kis-Balaton illustrates the difficulty of operating a complex system along different and often conflicting objectives. Despite efforts to manage the system on the basis of scientific evidence and forecasts, the coherent, model-based review of material balances showed that most management actions were futile and did not result in a better fulfilment of the principal objectives. Therefore, the question remains open: what kind of institutional setup could ensure that operational experience and simple, target-oriented analysis of existing monitoring data a priori excluded fruitless management efforts?

The high autonomy of the system during its history suggests that a less proactive management could operate the system in a more natural (and less managed) way with roughly the same efficiency but with significantly less efforts and running costs. Nevertheless, considering the lessons learnt in this large and complex wetland system (namely: the importance of implementation sequence, vegetation dynamics, weather-induced retention variability, and steady political boundary conditions) can facilitate the design and operation of other large constructed wetlands dedicated to improve water quality.

Author Contributions: Conceptualization was done by V.I. and M.H.; the methodology was designed by V.I., A.C. and M.H.; all authors contributed to the formal analysis; data curation was done by A.C., V.I. and C.G.; for writing, all authors contributed to both original draft preparation, review and editing; visualization was done by M.H. All authors have read and agreed to the published version of the manuscript.

Funding: This research was funded by National Research, Development and Innovation Office, grant number NNE 129990 and FIKP-VÍZ, and General Water Directorate of Hungary under contract "Analysis of the background load of the Kis-Balaton".

Acknowledgments: Data and advice from the West Transdanubian Water Directorate is gratefully acknowledged.

Conflicts of Interest: The authors declare no conflict of interest. The funders had no role in the design of the study; in the analyses, or interpretation of data; in the writing of the manuscript, or in the decision to publish the results.

References

1. McIntosh, B.S.; Ascough, J.C.; Twery, M.; Chew, J.; Elmahdi, A.; Haase, D.; Harou, J.J.; Hepting, D.; Cuddy, S.; Jakeman, A.J.; et al. Environmental decision support systems (EDSS) development—Challenges and best practices. *Environ. Model. Softw.* **2011**, *26*, 1389–1402. [CrossRef]
2. Eisenführ, F.; Weber, M.; Langer, T. *Rational Decision Making*; Springer: Berlin, Germany, 2010; ISBN 978-3642-028-489.
3. von Winterfeldt, D. Bridging the gap between science and decision making. *Proc. Natl. Acad. Sci. USA* **2013**, *110* (Suppl. S3), 14055–14061. [CrossRef] [PubMed]
4. Reichert, P.; Langhans, S.D.; Lienert, J.; Schuwirth, N. The conceptual foundation of environmental decision support. *J. Environ. Manag.* **2015**, *154*, 316–332. [CrossRef] [PubMed]
5. Somlyódy, L.; van Straten, G. *Modeling and Managing Shallow Lake Eutrophication: With Application to Lake Balaton*, 1st ed.; Springer: Berlin, Germany, 1986; ISBN 978-3-642-82709-9. [CrossRef]

6. Zlinszky, A.; Timár, G. Historic maps as a data source for socio-hydrology: A case study of the Lake Balaton wetland system, Hungary. *Hydrol. Earth Syst. Sci.* **2013**, *17*, 4589–4606. [CrossRef]
7. Virág, Á. *Past and Present of Lake Balaton [A Balaton Múltja És Jelene]*; Egri Press Ltd.: Eger, Hungary, 1998; (In Hungarian). ISBN 9639060216. Available online: https://library.hungaricana.hu/en/view/VizugyiKonyvek_244a/ (accessed on 28 February 2020).
8. Herodek, S. Phytoplankton Changes during Eutrophication and P and N Metabolism. In *Modeling and Managing Shallow Lake Eutrophication: With application to Lake Balaton*; Somlyódy, L., van Straten, G., Eds.; Springer: Berlin, Germany, 1986; ISBN 978-3-642-82709-9. [CrossRef]
9. Láng, I. Impact on Policymaking: Background to a Government Dcision. In *Modeling and Managing Shallow Lake Eutrophication: With Application to Lake Balaton*; Somlyódy, L., van Straten, G., Eds.; Springer: Berlin, Germany, 1986; ISBN 978-3-642-82709-9. [CrossRef]
10. Istvánovics, V.; Somlyódy, L. Factors influencing lake recovery from eutrophication—The case of Basin 1 of Lake Balaton. *Water Res.* **2001**, *35*, 729–735. [CrossRef]
11. Somlyódy, L. Eutrophication Management Models. In *Modeling and Managing Shallow Lake Eutrophication: With Application to Lake Balaton*; Somlyódy, L., van Straten, G., Eds.; Springer: Berlin, Germany, 1986; ISBN 978-3-642-82709-9. [CrossRef]
12. Kárpáti, I. *Biological Foundations for the Operational. Technological and Organisational Concepts of Kis-Balaton Protection System ["A Kisbalaton Védőrendszer Üzemeltetésének. Technológiai És Üzemszervezési Koncepcióinak Biológiai Megalapozása"]*; Agricultural University of Keszthely: Keszthely, Hungary, 1980. (In Hungarian)
13. Istvánovics, V.; Kovács, A.; Vörös, L.; Herodek, S.; Pomogyi, P. Phosphorus cycling in a large, reconstructed wetland, the lower Kis-Balaton Reservoir (Hungary). *SIL Proc. 1922–2010* **1997**, *26*, 323–329. [CrossRef]
14. Čížková, H.; Pechar, L.; Husák, Š.; Květ, J.; Bauer, V.; Radová, J.; Edwards, K. Chemical characteristics of soils and pore waters of three wetland sites dominated by Phragmites australis: Relation to vegetation composition and reed performance. *Aquat. Bot.* **2001**, *69*, 235–249. [CrossRef]
15. Somlyódy, L.; Herodek, S. *Revision of the Lower Reservoir of Kis-Balaton—Synthesis Report ["A Kis-Balaton Alsó-Tározó Felülvizsgálata. Szintézis Jelentés"]*; Budapest University of Technology and Economics, Department of Sanitary and Environmental Engineering: Budapest, Hungary, 1997. (In Hungarian)
16. Somlyódy, L. Eutrophication Modeling, Management and decision making: The Kis-Balaton Case. *Water Sci. Technol.* **1998**, *37*, 165–175. [CrossRef]
17. BioAquaPro. *Proposal on the Operation Regime of the Kis Balaton Water Protection System. Biomonitoring System for the Implementation of the 2nd Phase of Kis Balaton" Project. [A Kis-Balaton Vízvédelmi Rendszer Üzemeltetéssel Kapcsolatos Előzetes Javaslatok. «Kis-Balaton Vízvédelmi Rendszer II. Ütem Megvalósítása» Projekthez Kapcsolódó Biomonitoring Rendszer Kialakítása"]*; KEOP-2.2.1/2F/09-2009-0001 project report; BioAquaPro Ltd.: Debrecen, Hungary, 2018. (In Hungarian)
18. Kovács, Á.; Honti, M.; Clement, A. Design of best management practice applications for diffuse phosphorus pollution using interactive GIS. *Water Sci. Technol.* **2008**, *57*, 1727–1733. [CrossRef] [PubMed]
19. Kovács, Á.; Honti, M. Estimation of diffuse phosphorus emissions at small catchment scale by GIS-based pollution potential analysis. *Desalination* **2008**, *226*, 72–80. [CrossRef]
20. Clement, A.; Somlyódy, L.; Koncsos, L. Modeling the phosphorus retention of the Kis-Balaton upper reservoir. *Water Sci. Technol.* **1998**, *37*, 113–120. [CrossRef]
21. Sas, H. *Lake Restoration and Reduction of Nutrient Loading: Expectations Experiences and Extrapolations*; Academia Verlag Richarz: St. Augustin, Germany, 1989; ISBN 978-3883453798.
22. Istvánovics, V.; Somlyódy, L. The role of sediments in P retention of the Kis-Balaton reservoir. *Int. Rev. Hydrobiol.* **1998**, *83*, 225–234.
23. Gaál, R.; Bencsics, A.; Puskás, Z. *Provisional Operational Procedures for the Kis-Balaton Water Protection System ["KBVR Ideiglenes Üzemeltetési Szabályzat"]*; West Transdanubian Water Directorate, West Transdanubian Environmental and Nature Protection Authority and Balaton-felvidéki National Park: Szombathely, Hungary, 2014; p. 40. (In Hungarian)

24. Rodrigo, M.A.; Valentín, A.; Claros, J.; Moreno, L.; Segura, M.; Lassalle, M.; Vera, P. Assessing the effect of emergent vegetation in a surface-flow constructed wetland on eutrophication reversion and biodiversity enhancement. *Ecol. Eng.* **2018**, *113*, 74–87. [CrossRef]
25. Istvánovics, V.; Clement, A.; Somlyódy, L.; Specziár, A.; G.-Tóth, L.; Padisák, J. Updating water quality targets for shallow Lake Balaton (Hungary), recovering from eutrophication. *Hydrobiologia* **2007**, *581*, 305–318. [CrossRef]

MDPI
St. Alban-Anlage 66
4052 Basel
Switzerland
Tel. +41 61 683 77 34
Fax +41 61 302 89 18
www.mdpi.com

Water Editorial Office
E-mail: water@mdpi.com
www.mdpi.com/journal/water

www.ingramcontent.com/pod-product-compliance
Lightning Source LLC
LaVergne TN
LVHW071034170726
843515LV00006B/1278

* 9 7 8 3 0 3 6 5 1 5 4 8 9 *